Ken Albrecht

D1651491

TALL FESCUE

AGRONOMY

A Series of Monographs

The American Society of Agronomy (ASA) and Academic Press published the first six books in this series. Subsequent books were published by ASA alone, but in 1978 the associated societies, ASA, Crop Science Society of America (CSSA), and Soil Science Society of America (SSSA), published Agronomy 19. The books numbered 1 to 6 on the list below are available from Academic Press, Inc., 111 Fifth Avenue, New York, NY 10003; those numbered 7 to 19 are available from ASA, 677 S. Segoe Road, Madison, WI 53711.

General Editor Monographs 1 to 6, A. G. NORMAN

1. C. EDMUND MARSHALL: The Colloid Chemistry of the Silicate Minerals, 1949
2. BYRON T. SHAW, *Editor:* Soil Physical Conditions and Plant Growth, 1952
3. K. D. JACOB: Fertilizer Technology and Resources in the United States, 1953
4. W. H. PIERRE and A. G. NORMAN, *Editor:* Soil and Fertilizer Phosphate in Crop Nutrition, 1953
5. GEORGE F. SPRAGUE, *Editor:* Corn and Corn Improvement, 1955
6. J. LEVITT: The Hardiness of Plants, 1956

7. JAMES N. LUTHIN, *Editor:* Drainage of Agricultural Lands, 1957
   General Editor, D. E. Gregg
8. FRANKLIN A. COFFMAN, *Editor:* Oats and Oat Improvement
   Managing Editor, H. L. Hamilton
9. C. A. BLACK, *Editor-in-Chief,* and D. D. EVANS, J. L. WHITE, L. E. ENSMINGER, and F. E. CLARK, *Associate Editors:* Methods of Soil Analysis, 1965.
   Part 1—Physical and Mineralogical Properties, Including Statistics of Measurement and Sampling
   Part 2—Chemical and Microbiological Properties
   Managing Editor, R. C. Dinauer
10. W. V. BARTHOLOMEW and F. E. CLARK, *Editor:* Soil Nitrogen, 1965
    Managing Editor, H. L. Hamilton
11. R. M. HAGAN, H. R. HAISE, and T. W. EDMINSTER, *Editors:* Irrigation of Agricultural Lands, 1967
    Managing Editor, R. C. Dinauer
12. R. W. PEARSON and FRED ADAMS, *Editors:* Soil Acidity and Liming, 1967
    Managing Editor, R. C. Dinauer
13. K. S. QUISENBERRY and L. P. REITZ, *Editors:* Wheat and Wheat Improvement, 1967
    Managing Editor, H. L. Hamilton
14. A. A. HANSON and F. V. JUSKA, *Editors:* Turfgrass Science, 1969
    Managing Editor, H. L. Hamilton
15. CLARENCE H. HANSON, *Editor:* Alfalfa Science and Technology, 1972
    Managing Editor, H. L. Hamilton
16. B. E. CALDWELL, *Editor:* Soybeans: Improvement, Production, and Use, 1973
    Managing Editor, H. L. Hamilton
17. JAN VAN SCHILFGAARDE, *Editor:* Drainage for Agriculture, 1974
    Managing Editor, R. C. Dinauer
18. GEORGE F. SPRAGUE, *Editor:* Corn and Corn Improvement, 1977
    Managing Editor, D. A. Fuccillo
19. JACK F. CARTER, *Editor:* Sunflower Science and Technology, 1978
    Managing Editor, D. A. Fuccillo
20. ROBERT C. BUCKNER and L. P. BUSH, *Editors:* Tall Fescue, 1979
    Managing Editor, D. A. Fuccillo

TALL FESCUE

Edited by

ROBERT C. BUCKNER
SEA, USDA
University of Kentucky
Lexington, Kentucky

and

LOWELL P. BUSH
Agronomy Department
University of Kentucky
Lexington, Kentucky

Managing Editor: D. A. FUCCILLO

Assistant Editor: KARI J. SHERMAN

Editor-in-Chief ASA Publications: MATTHIAS STELLY

**Number 20 in the series
AGRONOMY**

American Society of Agronomy,
Crop Science Society of America,
Soil Science Society of America, Inc., Publishers
Madison, Wisconsin USA
1979

Copyright © 1979 by the American Society of Agronomy, Inc., Crop Science Society of America, Inc., and Soil Science Society of America, Inc.

ALL RIGHTS RESERVED UNDER THE U.S. COPYRIGHT LAW OF 1978 (P.L. 94-553).
Any and all uses beyond the "fair use" provision of the law require written permission from the publishers and/or the authors; not applicable to contributions prepared by officers or employees of the U.S. Government as part of their official duties.

Library of Congress Cataloging in Publication Data

(Agronomy, a series of monographs; no. 20)
Includes index.
1. Tall fescue. I. Buckner, Robert Cecil. II. Bush, Lowell P., 1939–   III. Series.
SB201.T34T34   633.2   79-14392
ISBN 0-89118-057-5

The American Society of Agronomy, Inc.
677 S. Segoe Road, Madison, Wisconsin, USA 53711

Printed in the United States of America

Tall Fescue Panicle

Tall Fescue Plant

Tall Fescue Seedfield

Cattle Grazing on Tall Fescue Pasture

Photographs courtesy of the
Kentucky Agricultural Experiment Station

CONTENTS

	Page
FOREWORD	ix
GENERAL FOREWORD	xi
PREFACE	xiii
CONTRIBUTORS	xv

Chapter 1 Historical Development 1

ROBERT C. BUCKNER, JERREL B. POWELL, and ROD V. FRAKES

Origin	1
History	2
Production and Utilization in the United States	5
Production and Utilization Outside the United States	6

Chapter 2 Adaptation 9

JOSEPH C. BURNS and DOUGLAS S. CHAMBLEE

Introduction	9
Climatic and Adaptive Factors that Influence Growth and Persistence	10
Adaptation of Tall Fescue in the United States Compared with that of Other Cool-Season Grasses	16
Adaptation vs. Production	22
Adaptation of Tall Fescue in Other Regions of the World	23

Chapter 3 Taxonomy, Morphology, and Phylogeny 31

EDWARD E. TERRELL

Introduction	31
Nomenclature	32
Geographic Distribution	32
Morphology and Anatomy	33
Identification of Tall and Meadow Fescues	35
Taxonomic Relationships	36
Taxonomic and Phylogenetic Implications of Cytogenetics	36

Chapter 4 Mineral Nutrition 41

S. R. WILKINSON and D. A. MAYS

Introduction	41
Nitrogen Requirement	42

Sulfur Requirement.	54
Phosphorus Requirement.	57
Potassium Requirement.	61
Lime, Calcium, and Magnesium Requirement.	64
Micronutrient Requirements.	67
Silica Accumulation.	68

Chapter 5 Physiology of Growth and Development. ... 75

D. D. WOLF, R. H. BROWN, and R. E. BLASER

Introduction.	75
Physiology of Development.	75
Growth Principles.	81
Yield Limitations.	87

Chapter 6 Cytogenetics and Genetics. ... 93

CLYDE C. BERG, G. T. WEBSTER, and PREM P. JAUHAR

Cytology.	93
Intraspecific Hybrids.	98
Interspecific Hybrids.	100
Intergeneric Hybrids.	103
Genetics.	105

Chapter 7 Breeding and Cultivars ... 111

K. H. ASAY, ROD V. FRAKES, and ROBERT C. BUCKNER

Breeding Procedures.	111
Genetic Variation.	112
Inbreeding.	114
Germplasm Resources.	114
Forage Quality.	115
Forage Yield.	121
Disease Resistance.	122
Seed Yield.	123
Turf.	124
Cultivars.	125
Intergeneric and Interspecific Hybridization.	130

Chapter 8 Seed Production ... 141

HAROLD YOUNGBERG and HOWELL N. WHEATON

Economic Importance and Volume of Seed Control.	141
Areas of Seed Production.	142
Cultural Practices and Management.	143
Certification.	148
Harvesting.	148
Seed Storage.	151

Marketing.. 152
Economic Return from Seed Production 152

**Chapter 9　Stand Establishment and Renovation of Old Sods
　　　　　　for Forage** 155

T. H. TAYLOR, W. F. WEDIN, and W. C. TEMPLETON, JR.

Stand Establishment 156
Renovation of Old Sods 160
Summary and Conclusions................................... 167

Chapter 10　Management 171

A. G. MATCHES

Introduction... 171
Season-Long Management of Tall fescue..................... 171
Stockpiled Tall Fescue for Autumn and Winter Grazing 177
Nitrogen-Fertilized Tall Fescue and Tall Fescue-Legume
　Mixtures.. 185
Interseeding Tall Fescue into Bermudagrass 190
Tall Fescue Cultivars and Grass Species—A Comparison
　of Yield and Quality 191

**Chapter 11　Tall Fescue in Forage-Animal Production Systems
　　　　　　for Breeding and Lactating Animals** 201

R. W. VAN KEUREN and JOHN A. STUEDEMANN

Introduction... 201
Tall Fescue for Beef Cows 201
Tall Fescue for Sheep Production 216
Tall Fescue for Dairy Cattle 222
Summary 228

**Chapter 12　Tall Fescue Pasture for Growing and Finishing
　　　　　　Animals** 233

A. E. SPOONER and W. S. MC GUIRE

Backgrounding 233
Finishing... 242

Chapter 13　Animal Disorders 247

LOWELL BUSH, JAMES BOLING, and SHELLY YATES

Introduction... 247
Fescue Foot 248
Grass Tetany 264
Non-Protein Nitrogen.. 272
Fat Necrosis... 282

Chapter 14 Turf 293

J. J. MURRAY and JERREL B. POWELL

Introduction... 293
History of Use and Attributes as a Turfgrass.................. 294
Establishment 296
Maintenance 299
Pests 304

Chapter 15 Diseases and Nematodes 307

R. A. CHAPMAN

Diseases 307
Nematodes.. 313
Disease and Nematode Control................................... 314

Chapter 16 Conservation... 319

ORUS L. BENNETT

Soil Stabilization... 319
Soil Improvement 328
Special Contributions... 332

Chapter 17 The Future of Tall Fescue 341

A. A. HANSON

Index.. 345

FOREWORD

Much of our land must be dedicated to perennial ground cover in order to preserve it for use by future generations. At a time when soil erosion is considered the most serious agricultural problem in this country, tall fescue has come of age as a productive and nutritious forage crop which stabilizes our soils for agricultural and recreational uses.

The appearance of *Tall Fescue* is most timely because of the resurgence of interest in the beef cow and calf as effective users of forages. These forages are grown on land that must be grazed if it is to be utilized efficiently. Tall fescue has been identified with grazing cattle more than any other cultivated improved crop.

Most of what is known about this forage has been learned during the past 40 years—indeed, tall fescue was not classified as a species until 1950 after two improved cultivars demonstrated its merit and difference from meadow fescue. It is appropriate for eminent scientists to pause, record the progress, and stir our curiosity about this crop before continuing its development.

As pointed out by the authors, some problems still remain. If these are attacked in the same imaginative, interdisciplinary, and diverse manner characteristic of this monograph, the future is assured. The array of contributors spans the continent and the breadth of this country and the various disciplines in plant, soil, and animal sciences.

We wish to express our societies' deepest appreciation and gratitude especially to the editors, Dr. Robert C. Buckner and Dr. Lowell P. Bush, and their associate editors who orchestrated the technical aspects of the monograph. We also acknowledge and thank the authors for their cooperation and superior efforts and recognize the help of other society members who reviewed the manuscripts. We are grateful to the Headquarters editorial and production staff whose efforts made it possible to place this volume into your hands with a deep pride of accomplishment.

ROY G. CREECH
President
Crop Science Society of America

LEO M. WALSH
President
Soil Science Society of America

JOHN PESEK
President
American Society of Agronomy

Madison, Wisconsin
April 1979

GENERAL FOREWORD

Tall fescue is the predominant cool-season, perennial grass in the United States. This plant is adapted to a wide range of climatic conditions, from Canada to Florida, but grows best in the transition zone. The literature on tall fescue is widely scattered. This new monograph gathers results from extensive research on the grass, originally reported in various publications, and places the work in perspective.

The editors and authors, with the assistance of reviewers, contributed serious efforts in the writing of this book. Their efforts resulted in a text covering the history, adaptation, taxonomy, physiology, breeding, seed production, management, pasture use, and various other aspects of this widely grown grass. Each chapter was contributed by one or more specialists in a given field, thus giving this publication unique quality information.

Tall Fescue is the 20th monograph in the series prepared by the American Society of Agronomy since 1949. The first six volumes were published by Academic Press, Inc., of New York, but since 1957 the American Society of Agronomy has become the publisher and continues to be the sole publisher, including the 18th volume entitled *Corn and Corn Improvement,* 1977. A complete list of the titles in the series may be found in the beginning pages of this book. The monographs represent a significant and continuing activity of the American Society of Agronomy, its officers, and its approximately 10,600 members located in 100 countries.

The American Society of Agronomy is closely associated with the Crop Science Society of America and the Soil Science Society of America. The societies share many objectives and activities in promoting these branches of agriculture and scientific disciplines. Members of the societies contribute generously of their time and talents in producing various publications, including monographs, and in pursuing other activities in the interest of human welfare as evidence of this close cooperation. Beginning with the 19th volume, entitled *Sunflower Science and Technology,* and including this 20th volume on *Tall Fescue,* all three societies jointly served as publishers.

This volume should be of use to students as well as to researchers. As a source of important information on a valuable crop, this book will be valuable to growers, breeders, geneticists, economists, and agri-business. The presentation of up-to-date scientific and practical material undoubtedly will increase the successful production and utilization of tall fescue.

In behalf of the members of the associated societies and myself in particular, I sincerely thank Dr. Robert C. Buckner and Dr. Lowell P. Bush for their diligent work as editors, the editorial committee, the many authors, Managing Editor Domenic Fuccillo, Assistant Editor Kari Sherman, and all others who have contributed directly or indirectly to the accomplishment of this worthy project.

Madison, Wisconsin
April 1979

MATTHIAS STELLY
Editor-in-Chief
ASA Publications

PREFACE

Tall fescue (*Festuca arundinacea* Schreb.) is a valuable grass of temperate agriculture. It has become increasingly popular with farmers in Europe from where it was introduced into the United States during the late half of the 19th Century. Morphological resemblance of tall with meadow fescue, however, prevented tall fescue from attaining recognition as a taxonomic species in the United States until the middle of the 20th Century. It gained importance as an agricultural crop after the release during the 1940's of the 'Alta' and 'Kentucky 31' cultivars. The rapid acceptance by farmers of these two cultivars resulted in tall fescue being described as a distinct species by Hitchcock in 1950. Tall fescue has since become one of the most widely grown cool-season grasses in the United States and is now grown on an estimated 35 million acres.

The rapid acceptance and widespread utilization of this grass by farmers resulted in accelerated research efforts by scientists throughout its area of adaptation, particularly in Wales, United States, West Germany, Switzerland, Poland, and Holland. The published literature embodying the results of these researches includes the broad disciplines of forage production, animal nutrition and production, soil conservation, and the interdisciplinary interactions. The published results of these researches are, however, scattered through numerous journals and periodicals. While several review monographs on crops like corn, wheat, and oats, etc., have been published, no such integrated review of published information on tall fescue is available. Since tall fescue has attained a rapid and widespread usage, made a vast contribution to our economy, and is of tremendous importance to agriculture and the nation, it became essential to assemble and integrate the most pertinent available information into a single volume. This monograph covers the origin, history, morphology, taxonomy, cytology, genetics, and breeding of tall fescue along with its management and utilization for forage, turf, and conservation purposes. Thus, it provides an up-to-date, authoritative and requisite information to the scientist and specialist on technology of management and production in addition to basic information regarding the physiology, cytogenetics, and breeding of the species.

The history, utilization, and adaptation of the grass are presented in Chapters 1 and 2. Chapters 3 through 6 include detailed background information in regard to morphology and taxonomy, mineral nutrition and physiology, and cytogenetics and genetics of tall fescue. Technical and practical information regarding breeding objectives, breeding methodology, and varietal development are treated in Chapter 7. An insight into seed production, establishment, and management is provided in subsequent Chapters 8, 9, and 10. Utilization of tall fescue in various livestock management systems is given detailed attention in Chapters 11 and 12. Chapters 7 through 12 include discussions relating to varieties, establishments, fertilization, and management for pasture, hay, and seed production.

While tall fescue is a nutritious grass, disorders occur at times in livestock grazing tall fescue. In Chapter 13, the various disorders and current knowledge as to the causative agents are discussed in detail. Tall fescue is widely used for turf and conservation purposes. Management, utilization, and value of the species for these purposes are given detailed treatment in Chapters 14 and 16. Finally, an excellent insight into the value of this species for the future is discussed in Chapter 17. Thus, this monograph can be a valuable source of basic information to students interested in principles of forage production and management, as well as to researchers involved in multidisciplinary approach to tall fescue breeding and improvement.

The editors wish to express their appreciation to the authors of the different chapters. The chapters have been written by individuals who are recognized as authorities of the disciplines treated in each chapter. The authors have drawn from broad and varied experience and have provided numerous references to significant research. Appreciation is expressed to the Associate Editors—Dr. R. W. Van Keuren and Dr. R. E. Blaser—and to all those scientists who assisted in the chapter reviews.

Finally, compliments and appreciation are expressed to Dr. Matthias Stelly, Domenic Fuccillo, and other members of the American Society of Agronomy Headquarters staff for their professional and efficient services.

Lexington, Kentucky
April 1979

ROBERT C. BUCKNER
LOWELL P. BUSH
Editors

CONTRIBUTORS

K. H. ASAY, Science and Education Administration, U.S. Department of Agriculture, Utah State University, Logan, UT 84321

ORUS L. BENNETT, Science and Education Administration, U.S. Department of Agriculture, West Virginia University, Morgantown, WV 26506

CLYDE C. BERG, Science and Education Administration, U.S. Department of Agriculture, U.S. Regional Pasture Research Laboratory, University Park, PA 16802

R. E. BLASER, Agronomy Department, Virginia Polytechnic Institute and State University, Blacksburg, VA 24061

JAMES BOLING, Animal Science Department, University of Kentucky, Lexington, KY 40506

R. H. BROWN, Agronomy Department, University of Georgia, Athens, GA 30601

ROBERT C. BUCKNER, Science and Education Administration, U.S. Department of Agriculture, University of Kentucky, Lexington, KY 40506

JOSEPH C. BURNS, Science and Education Administration, U.S. Department of Agriculture, North Carolina State University, Raleigh, NC 27607

LOWELL BUSH, Agronomy Department, University of Kentucky, Lexington, KY 40506

DOUGLAS S. CHAMBLEE, Crop Science Department, North Carolina State University, Raleigh, NC 27607

R. A. CHAPMAN, Plant Pathology Department, University of Kentucky, Lexington, KY 40506

ROD V. FRAKES, Farm Crops Department, Oregon State University, Corvallis, OR 97331

A. A. HANSON, Science and Education Administration, U.S. Department of Agriculture, Beltsville Research Center, Beltsville, MD 20705

PREM P. JAUHAR, Agronomy Department, University of Kentucky, Lexington, KY 40506

A. G. MATCHES, Science and Education Administration, U.S. Department of Agriculture, University of Missouri, Columbus, MO 65201

D. A. MAYS, National Fertilizer Development Center, Tennessee Valley Authority, Muscle Shoals, AL 35660

W. S. MC GUIRE, Crop Science Department, Oregon State University, Corvallis, OR 97331

J. J. MURRAY, Science and Education Administration, U.S. Department of Agriculture, Beltsville Research Center, Beltsville, MD 20705

JERREL B. POWELL, Science and Education Administration, U.S. Department of Agriculture, Beltsville Research Center, Beltsville, MD 20705

A. E. SPOONER, Agronomy Department, University of Arkansas, Fayetteville, AR 72701

JOHN A. STUEDEMANN, Animal Science Department, Southern Piedmont Research Center, Watkinsville, GA 30677

T. H. TAYLOR, Agronomy Department, University of Kentucky, Lexington, KY 40506

W. C. TEMPLETON, JR., Science and Education Administration, U.S. Department of Agriculture, U.S. Regional Pasture Research Laboratory, University Park, PA 16802

EDWARD E. TERRELL, Science and Education Administration, U.S. Department of Agriculture, Beltsville Research Center, Beltsville, MD 20705

R. W. VAN KEUREN, Agronomy Department, Ohio Agriculture Research and Development, Wooster, OH 44691

G. T. WEBSTER, Agronomy Department, University of Kentucky, Lexington, KY 40506

W. F. WEDIN, Agronomy Department, Iowa State University, Ames, IA 50010

HOWELL N. WHEATON, Agronomy Department, University of Missouri, Columbus, MO 65201

S. R. WILKINSON, Science and Education Administration, U.S. Department of Agriculture, Soil and Water Conservation Research Center, Watkinsville, GA 30677

D. D. WOLF, Agronomy Department, Virginia Polytechnic and State University, Blacksburg, VA 24061

SHELLY YATES, Science and Education Administration, U.S. Department of Agriculture, Northern Utilization Research and Development Division, Peoria, IL 61604

HAROLD YOUNGBERG, Agronomy Department, Oregon State University, Corvallis, OR 97331

Chapter 1 Historical Development

ROBERT C. BUCKNER
SEA-USDA, and Department of Agronomy
University of Kentucky
Lexington, Kentucky

JERRELL B. POWELL
SEA-USDA
Beltsville Research Center
Beltsville, Maryland

ROD V. FRAKES
Farm Crops Department
Oregon State University
Corvallis, Oregon

ORIGIN

Tall fescue (*Festuca arundinacea* Schreb.) belongs to the Bovinae section of the genus *Festuca* and is classified as belonging to the tribe Festuceae.

Tall fescue was introduced into North and South America from Europe; Western Europe is a main center of variation of the tribe Festuceae (Borrill, 1976). Tall fescue is found growing in damp pastures and wet places throughout Europe and North Africa, extending to western Siberia. It is also distributed on mountains of East Africa and Madagascar. In Europe it is considered too coarse and unpalatable to be first-class pasture and is seldom sown (Whyte et al., 1959). Meadow fescue (*F. pratensis* Huds.) and tall fescue were originally described as separate and distinct species. Linnaeus first described meadow fescue as *F. elatior* in 1753, and in 1771 the German botanist Schreber recognized and described tall fescue as being more robust than meadow fescue and, therefore, designated it *F. arundinacea*. Tall fescue was regarded by Hitchcock (1935) as *F. elatior* var. *arundinacea* (Schreb.) Winn. The two grasses were described once again by Hitchcock (1950) as distinct species and listed according to the international rules of botanical nomenclature as *F. elatior* L. (meadow fescue) and *F. arundinacea* Schreb. (tall fescue).

Morphological similarities of the two species, difficulties encountered by taxonomists in separating the grasses, and the acceptance and utilization of meadow fescue in Europe and in certain regions of the United States as an important agricultural crop delayed the recognition of the potential value of tall fescue. The taxonomy, morphology, and phylogeny of tall fescue are discussed in detail in Chapter 3 of this monograph.

Copyright © 1979 ASA-CSSA-SSSA, 677 South Segoe Road, Madison, WI 53711 USA. *Tall Fescue.*

HISTORY

Tall fescue culture arose in the United States after the rapid rise to prominence and subsequent decline of the culture of meadow fescue. The historical development of tall fescue is intertwined with that of meadow fescue. By 1870, seed catalogues distinguished between meadow fescue, labeled *F. pratensis,* and tall fescue, *F. elatior.* The distinctive characteristics of tall fescue were described as "a robust variety of meadow fescue, succeeds admirably in moist soils where the meadows are subject to flood" (Anon., 1870).

Before 1880, nearly all seed for pastures was imported into the United States (Vinall, 1909). Kennedy (1900) believed that meadow fescue was introduced into this country from Great Britain prior to 1800. Mixtures of seed were planted for pasture use, and meadow fescue was a constituent in these mixtures; thus, it was spread throughout the United States. Commercial seed production of meadow fescue was underway at Gardner, Kan., in 1877 (Vinall, 1909) and grown in Manhattan, Kan., in 1879 (Ten Eyck, 1903). By 1896 Kansas had harvested its largest seed crop of this species, estimated at 1,524,086 kg (Ten Eyck, 1903). "Oat" rust (*Puccinia coronata* Cda.) and overproduction, however, soon reduced meadow fescue to minor importance, while special note was being taken of tall fescue.

In a report dated 1892 at Utah, 19 grasses had been tested during 3 consecutive years for yield per acre and height in inches. These included meadow fescue, tall fescue, and several fine-leaved fescues (*Festuca* spp.). Kearney (1895) reported tall fescue (*F. elatior pratensis* Hack.) growing along the roadside in Norfolk, Va., while he was on a collection and observation trip in the southern U.S.

Among the many species of grasses evaluated in a test from 1892 to 1900 at the Kentucky Agricultural Experiment Station by Garman (1900) were two fescue species referred to as *F. elatior* and *F. pratensis. F. elatior,* when compared with *F. pratensis,* was taller, more drought and cold tolerant, and formed dense stands that were more competitive with weeds than *F. pratensis.* He referred to *F. elatior* as "tall English bluegrass" and to *F. pratensis* as "English bluegrass." Height, yields, and general vigor of *F. pratensis* were considerably less than that of *F. elatior.* From this report, it is obvious that *F. elatior* in the test was tall fescue (*F. arundinacea*) and *F. pratensis* was meadow fescue.

The Bureau of Plant Industry, USDA, organized an active testing program of all different kinds of grasses in the late 1800's (Kennedy, 1900). Among the grasses tested were the several species of fescues. In one test in 1900 (Ball, 1900), 19 different lots of fescues comprising 14 species were planted on an island in the Potomac River known as Potomac Flats near Washington, D.C. Two plots of tall meadow fescue from Russia with the numbers S.P.I. 1180 and 1337, and another plot of reed fescue (*F. arundinacea*) were noted for growth, stand, and height.

Reed fescue, known as *F. elatior* L., was also called tall fescue, tall meadow fescue, English bluegrass, Randall grass, and evergreengrass.

HISTORICAL DEVELOPMENT 3

Lamson-Scribner (1896) stated, "This grass has been widely cultivated in this country, having been introduced from Europe, and has become thoroughly naturalized. It is an exceedingly valuable grass either for mowing or pasture. It is productive on soils which are not too dry and being of long duration it is especially valuable for permanent pastures. It thrives best on moist soils rich in humus whether marls or clays. The variety *pratensis* is a common form, rather smaller than the species with a narrower and fewer flowered panicle. Variety *arundinacea* depicted in (Fig. No. 42, titled reed fescue) is a very vigorous tall form, 3 to 4 feet high, exceedingly hardy and yields a very large amount of hay of excellent quality, succeeding best on lands that are comparatively moist." Describing the various entry numbers of *F. elatior* strains in their 1908 test, Evans (M. W. Evans. 1908. Mimeographed Annual Field Report of the Division of Agrostology. USDA, Washington, D.C.) states, "Number 16960. This about as large and vigorous as any plot of *Festuca elatior* but is coarser and less leafy. It may be *Festuca arundinacea.*" Also, he states "Number 19803. This is somewhat smaller than the other members of *Festuca elatior*. It resembles *Festuca pratensis* in size and form and growth."

By 1908, the value of meadow fescue for the Central and Eastern States had been determined and its testing by USDA completed. The work on tall fescue was accelerated. Oakley (1908) stated, "Tall fescue on account of its rust resistance and vigorous growth will receive for the next few years the greatest part of the attention devoted to this project. A quantity has been sown at Pullman, Wash., with the idea of testing more thoroughly its seed production qualities." One-tenth-acre plots of tall fescue seedlings were planted in 1909 at Pullman, Wash., and Vinall (1909) recommended that rust resistant tall fescue be substituted for meadow fescue if rust made meadow fescue unprofitable to grow.

Although testing continued, not much prominence was given to tall fescue until Oregon and Kentucky released 'Alta' and 'Kentucky 31' tall fescue, respectively. The growth of these cultivars and the subsequent availability of good quality seed permitted a remarkable increase in utilization of this grass. The species became an important agricultural crop in the United States.

Kentucky 31 tall fescue is an ecotype found growing before 1890 on the Menifee County farm of W. M. Suiter. The grass was brought to the attention of E. N. Fergus, Professor of Agronomy, University of Kentucky, while visiting in the county during 1931 (Fergus and Buckner, 1972). Seed was obtained from the site and the grass was evaluated in small plots during the period 1931 to 1943 at the Kentucky Agricultural Experiment Station and at several locations in the state. Kentucky 31 was released by the University of Kentucky as a cultivar of tall fescue in 1943 (Anon., 1943; Fergus, 1952; Fergus and Buckner, 1972). The prominent features listed for the cultivar at the time of its release were as follows: 1) dependability; 2) adaptability to a wide range of soils; 3) affording grazing during most of the year; 4) palatability to livestock.

William C. Johnstone, Extension Specialist, University of Kentucky, quickly recognized the year-round grazing value and soil-conserving qualities of tall fescue and was instrumental in the phenomenal increase and acceptance of this grass by Kentucky farmers. Demonstrational plantings on the erodable, rolling loess silt loam soils of western Kentucky proved the ability of the tough root system and sod-forming quality of the grass to provide pasture for cattle and prevent erosion during the wet winter and early spring seasons (personal correspondence). Other cool-season grasses were unable to provide this kind of performance. The demonstrations created a great demand for seed of this "new" grass.

An indication of the rapid acceptance of Kentucky 31 tall fescue by farmers of Kentucky was the rapid increase in seed production in the state. In 1946, Kentucky produced about 34,020 kg of seed; in 1948, growers harvested about 1,814,400 kg, and average annual production during the period 1951 to 1960 was 5.7 million kg (USDA, 1946-1976). Thus, in a relatively short period, Kentucky had become an important tall fescue seed-producing area.

The USDA, Soil Conservation Service, recognized the value of tall fescue for soil conservation purposes, and beginning in 1943 established seed increase fields of Kentucky 31 tall fescue in districts throughout the southeastern states. Seed from these nurseries helped promote the rapid gain in hectarage of Kentucky 31 in the south central and southeastern U.S. (Tabor, 1952).

A book by Cope (1949), entitled *Front Porch Farmer,* undoubtedly had much to do with the rapid acceptance of Kentucky 31 tall fescue. This publication, highly recommended Kentucky 31 for use in a system of year-round grazing in combination with other forage species.

'Alta' tall fescue contributed substantially toward acceptance of tall fescue as a separate and distinct species. Alta is an ecotype selected over a number of years beginning in 1918. This cultivar was developed cooperatively by the Oregon Agricultural Experiment Station and the Forage and Range section of USDA (Cowan, 1956).

In 1916 H. A. Schoth, ARS, USDA, Corvallis, Oreg., observed a nursery of tall fescue on the farm of Max Heinricks at Pullman, Wash., that remained green during adverse summer conditions (Cowan, 1956). Seed of three plant introductions from this nursery was established during 1918 on the Oregon Experiment Station, Corvallis, Oreg. The introductions later became the progenitors of the Alta tall fescue cultivar. Two introductions from Germany and a line from the J. C. Peppard Seed Company, Kansas City, Mo., were selected on the bases of winterhardiness, ability of plants to remain green during long periods of drought, and persistence to form the cultivar. The ability of the cultivar to remain green during the dry summers in western Oregon resulted in its acceptance by farmers of the Pacific Northwest. The cultivar is characterized by wide adaptation and ability to produce high yields of nutritious forage. As an indication of its rapid acceptance, in 1940 the value of Alta tall fescue as a seed crop in Oregon was about $31,000, but in 1951 it had reached a peak of $2.5 million. Growing Alta tall fescue seed had become an important enterprise in Oregon and throughout the Pacific Northwest by 1949 (Cowan, 1956).

Alta tall fescue was named in 1940 and registered under that name by the Committee on Varietal Standardization and Registration of the American Society of Agronomy on 15 Nov. 1944. It has the distinction of being one of the first forage crop cultivars to receive such a certificate. Thus, Kentucky 31 and Alta cultivars were responsible for the rapid acceptance of tall fescue as an important species for forage, turf, and conservation purposes.

The confusion that existed in the early classification of meadow and tall fescue as a consequence of their morphological similarity could have been resolved by cytotaxonomic examination. Tall fescue was first reported to have the hexaploid number of $2n = 42$ chromosomes by Levitsky and Kuzmina (1927). They used this information to assist in their systematic and phylogenetic studies. Evans (1926) reported earlier that meadow fescue had $2n = 14$ chromosomes. Peto (1933) confirmed the findings of Evans (1926) and Levitsky and Kuzmina (1927) for meadow and tall fescue, respectively. Myers and Hill (1947) recognized that meadow and tall fescue were distinct crops, adapted to different conditions and uses. They collected both crops in central Pennsylvania and found that meadow fescue regularly had $2n = 14$ chromosomes, whereas tall fescue had $2n = 42$. Meiosis in the diploid was regular, while in the hexaploid there were found quadrivalents and, rarely, sexivalents at diakinesis, univalents at metaphase I, laggards at anaphase I, and micronuclei in the quartets. Crowder (1953) surveyed the meiotic behavior of the Kentucky 31, Alta, and 'S-170' tall fescue cultivars and native populations found growing in Georgia, North Carolina, Germany, and New Zealand. Generally, he found the hexaploid number of $2n = 42$ for all entries. Meiotic behavior was in general agreement with that observed by Myers and Hill (1947). More detailed cytogenetic information can be found in Chapter 6.

PRODUCTION AND UTILIZATION IN THE UNITED STATES

Tall fescue grew on approximately 16,000 ha in 1940 in the United States and by 1973 it had become the predominant cool-season perennial grass, occupying an estimated 12 to 14 million ha. Although tall fescue grows best in the transition zone, it is adapted to a wide range of climatic conditions and is grown from Florida to Canada. Tall fescue grows well on soils that vary from strongly acid (pH 4.7) to alkaline (pH 9.5) (Cowan, 1956). It thrives and conserves soil on thin, droughty slopes, yet forms dense sods and produces excellent growth on poorly drained soils where few other cool-season grasses survive. Hafenrichter et al. (1949) found that its adaptability and forage production on nonirrigated land is dependent on 45 cm or more of annual rainfall when growing on elevations under 1,524 m. Bailey (1952) did not recommend the grass in the southeastern U.S. for uplands in areas of less than 90 cm of rainfall. It is not too well adapted to sandy soils having long periods of drought.

For conservation purposes, the adaptability of tall fescue to a wide range of soils and climate, the coarse, deep root system and its ability to grow on low-lying sites on moist, heavy soils and withstand waterlogging and flooding make the species particularly valuable. Tall fescue provides excellent waterway protection and is ideal for cover in low wetlands and on steep, droughty slopes. It has stabilized eroded and disturbed land areas, and helped restore fertility to such land. This use is perhaps the species' greatest contribution to the transition region. Millions of hectarages that were gullied as a consequence of the plow have been healed and restored to lush, green, productive pastures by the establishment of tall fescue.

The Soil Conservation Service recognized early the value of tall fescue for conservation purposes and played an important role in promoting the crop in the transition zone. The contribution and utilization of tall fescue to conservation is discussed in Chapter 16 of this monograph.

Another indication of growth in popularity of tall fescue is derived from seed production figures. During the past 25 years, the production, cleaning, and distribution of tall fescue have become important agricultural enterprises. Before the release of Kentucky 31 and Alta (during the early 1940's), seed production of tall fescue in the United States was practically nil. In fact, Vinall (1909) stated that tall fescue had several desirable qualities that would probably permit it to replace meadow fescue as an important pasture grass in the east central states, but that poor seed production made usage of the grass prohibitive. He attributed poor seed yields to uneven ripening of seed stalks coupled with low production of culms.

Seed Crops (USDA, 1946-1976) reported the 1951 to 1960 average annual production of clean seed of tall fescue in the United States to be 14,400,000 kg, and during 1975 the production had increased to a record 56,824,367 kg. Average seed yields per hectare in the United States during 1974, 1975, and 1976 were 283, 295, and 258 kg, respectively. Early work by Rampton (1945), Spencer (1950), and Buckner and Burrus (1959) showed that proper cultural practices would approximately double production.

Most tall fescue seed in the United States is produced in the transition zone. Missouri presently produces approximately 65% of the seed grown in the United States. Carryover of seed supplies on farms and by dealers was approximately 13.6 million kg during 1974 to 1976 (USDA, 1946-1976). Thus, annual consumption of seed in the United States is approximately 43.2 million kg. The popularity of tall fescue for turf purposes, in addition to its value for forage, perhaps explains the heavy usage of tall fescue seed in this country. The specifics of tall fescue seed production are discussed in Chapter 8 of this monograph.

PRODUCTION AND UTILIZATION OUTSIDE THE UNITED STATES

Tall fescue has become naturalized in South Africa as well as Rhodesia where it grows well and is widely used in irrigated winter pastures in South Africa. Larin (1962) reported that it has occurred in pastures of Russia for many years and that although its hay is coarse, its yield exceeds that of all other cultivated grasses. The species is widely grown in the more temperate climates of South America. Chile recently tends to replace alfalfa (*Medicago sativa* L.) by irrigated pastures of Ladino clover (*Trifolium repens* L.) with tall fescue (Whyte et al., 1959). These pastures are strip-grazed behind electric fences on the rangelands of the south, and the cattle are then shipped north to fattening districts (Whyte et al., 1959). France has recently registered a tall fescue cultivar of improved palatability (Gillet and Huguet, 1977).

Although tall fescue seed production is one of the more specialized types of crop production, no pronounced interest in production in Great Britain and Western Europe has emerged. The increased demand of tall fescue as an important forage species worldwide has resulted in seed production becoming a specialized operation rather than an uncertain by-product of general farming operations in Great Britain, Australia, and Western Europe, as well as in the United States.

Tall fescue's worldwide distribution is determined more by winter temperature than by effects of historical geography or moisture factors (Whyte et al., 1959).

LITERATURE CITED

1. Anon. 1870. Descriptive catalogue of vegetable, agricultural and flower seed. 16th Ed. B. K. Bliss and Sons, New York.
2. ―――. 1943. Kentucky 31 fescue. Kentucky Agric. Exp. Stn. Circ. 390.
3. Ball, C. R. 1900. Grasses and fodder plants of the Potomac Flats. Circ. 28, USDA, Div. of Agrostology.
4. Bailey, R. Y. 1952. Tall fescue. Forages, 2nd Ed. Iowa State Univ. Press, Ames, Iowa. 724 p.
5. Borrill, M. 1976. Temperate grasses. p. 137-142. *In* N. W. Simmonds (ed.). Evolution of crop plants. Longman, London and New York.
6. Buckner, R. C., and P. B. Burrus, II. 1959. The effect of certain management and fertilization practices on seed production of tall fescue. Proc. 56th Assoc. South. Agric. Workers. Assoc. South. Agric. Workers, Memphis, Tenn.
7. Cope, Channing. 1949. Front porch farmer. Turner E. Smith and Co., Atlanta, Ga.
8. Cowan, J. R. 1956. Tall fescue. Adv. Agron. 8:283-320.
9. Crowder, L. V. 1953. Interspecific and intergeneric hybrids of *Festuca* and *Lolium*. J. Hered. 44:195-203.
10. Evans, G. 1926. Chromosome complements in grasses. Nature 118:841.
11. Fergus, E. N. 1952. Kentucky 31 fescue—culture and use. Kentucky Agric. Ext. Circ. 497.
12. ―――, and R. C. Buckner. 1972. Registration of Kentucky 31 tall fescue. (Reg. No. 7). Crop Sci. 12:714.
13. Garman, H. 1900. Kentucky forage plants—the grasses. Kentucky Agric. Exp. Stn. Bull. No. 87.
14. Gillet, M., and L. Huguet. 1977. Evaluation of the first tall fescue cultivar registered on the French list, selected for feeding value. [French, English summary]. Ann. Amélior. Plant. 27:331-339.
15. Hafenrichter, A. L., L. A. Mullen, and R. L. Brown. 1949. Grasses and legumes for soil conservation in the Pacific Northwest. USDA Misc. Publ. 678.
16. Hitchcock, A. S. 1935. Manual of the grasses of the United States. USDA Misc. Publ. 200.
17. ―――. 1950. Manual of the grasses of the United States. 2nd ed. USDA Misc. Publ. 200.
18. Kearney, T. H., Jr. 1895. Notes on grasses and forage plants collected or observed in the southeastern states. USDA Bull. 1.

19. Kennedy, P. B. 1900. Cooperative experiments with grasses and forage plants. USDA Bull. 22.
20. Lamson-Scribner, F. 1896. Useful and ornamental grasses. USDA Div. of Agrostology, Bull. 3.
21. Larin, I. V. 1962. Pasture economy and meadow cultivation [Russian Trans.]. Israel Program for Scientific Translations, Jerusalem. 600 p.
22. Levitsky, G. A., and N. E. Kuzmina. 1927. Karyological investigations on the systematics and phylogenetics of the genus *Festuca*. (English Summary). Bull. Appl. Bot. Genet. Plant Breed. 17:33-36.
23. Myers, W. M., and Helen D. Hill. 1947. Distribution and nature of polyploidy in *Festuca elatior* L. Bull. Torrey Bot. Club. 74(2):99-111.
24. Oakley, R. A. 1908. Extension of meadow fescue and tall fescue into sections where they are not commonly grown. USDA Bur. Plant Ind., Proj. No. 894.
25. Peto, F. H. 1933. The cytology of certain intergeneric hybrids between *Festuca* and *Lolium*. J. Genet. 28:113-156.
26. Rampton, H. H. 1945. Alta fescue production in Oregon. Oregon Agric. Exp. Stn. Bull. 427.
27. Spencer, J. T. 1950. seed production of Kentucky 31 fescue and orchardgrass as influenced by rate of planting, nitrogen fertilization, and management. Kentucky Agric. Exp. Stn. Bull. 554.
28. Tabor, Paul. 1952. Tall fescue grows up. Crops Soils 4(8):9-11.
29. Ten Eyck, A. M. 1903. Meadow fescue. Agric. Dep. Kansas Agric. Exp. Stn. Press Bull. 125.
30. USDA. 1946-1976. Annual summaries, seed crops. Stat. Rep. Serv., Crop Rep. Board, Washington, D.C.
31. Vinall, H. N. 1909. Meadow fescue; its culture and uses. USDA Farmers Bull. 361.
32. Whyte, R. O., T. R. G. Moir, and J. P. Cooper. 1959. Grasses in agriculture. FAO Agric. Stud. No. 42.

Chapter 2 Adaptation

JOSEPH C. BURNS
SEA-USDA
North Carolina State University
Raleigh, North Carolina

DOUGLAS S. CHAMBLEE
Crop Science Department
North Carolina State University
Raleigh, North Carolina

INTRODUCTION

Plant adaptation, in general, refers to the relationship between major environmental factors and the growth response of plants. The impact of environmental factors on altering plant response is limited at any given time by the genetic constitution of the plant (Daubenmire, 1959).

Geographic plant distribution as depicted by Good (1931, 1953) is primarily controlled by climatic conditions (temperature, air movement, light, and moisture) and secondarily by edaphic factors. Secondary placement of edaphic factors to climate is based on the large influence that climate has on altering edaphic factors. In essence, Good says that climate determines if a given plant is potentially adapted to an area, but that soil conditions determine if, and in what numbers, the plant is actually found.

In considering the adaptation of tall fescue (*Festuca arundinacea* Schreb.) consider the "theory of tolerance" set forth by Good in 1931 (Good, 1931). This theory states that each plant species is able to exist and survive only within a definite range of climatic and edaphic conditions; that the tolerance of a species is a specific character subject to the laws and processes of organic evolution in the same ways as its morphological characters, but that the two may be linked, and that wide differences in tolerance may not be correlated with morphological differences. In such cases, distribution of a species is determined by the result of competition between types.

Later, Cain (1944) expanded Good's theory of tolerance by noting the importance of biotic factors, indicating that the environment is holocoenatic (interdependent and interacting system) and that tolerance has a genetic base.

From the theory put forth by Good (1931) and Cain (1944), it is clearly evident that the adaptation of tall fescue in the United States, or throughout the world, will depend on the tolerances of the many ecotypes [a group of biotypes especially adapted to a specific environmental niche (Odum, 1959)] found within *Festuca arundinacea*.

Copyright © 1979 ASA-CSSA-SSSA, 677 South Segoe Road, Madison, WI 53711 USA. *Tall Fescue.*

A. General Adaptation of Festuca

Insight to the adaptation of tall fescue to the many environments of the world can be gained by considering the distribution map constructed by Hartley (1950) for the tribe Festuceae (Fig. 2-1). Hartley (1950, 1954) attributes the major factors controlling the distribution of the tribe to winter temperatures. He found that members of the Festuceae tribe occurred in greater abundance around the midwinter month (January for the northern hemisphere and July for the southern) 10 C isocheim (an imaginary line connecting places having the same mean winter temperature). This isocheim demarks, in a general way, those parts of the world in which severe winter frosts are normally experienced. Hartley concludes that the distribution of the Festuceae is mainly associated with temperature and to a minor extent with rainfall and historical factors.

Although edaphic and biotic factors influence the distribution of small vegetational units to the extent that the relative abundance of species is altered, rarely will they affect their presence or absence (Hartley, 1954). This statement assumes that the area selected is sufficiently large and diversified to include a wide range of soils and biotic influences.

Because of the wide range of adaptation of the Festuceae tribe (Fig. 2-1), it is not surprising that tall fescue also has a broad range of adaptation. Such a wide distribution suggests that the major factors limiting its presence are climatic (high and low temperatures and rainfall) and geographic (mainly altitude) with less influence from edaphic, pyric, or biotic factors. These aspects will be given further consideration.

A limited number of tall fescue cultivars has been developed in various environments throughout the world. In general, they have a similarly wide range of adaptation to both climatic and edaphic factors. The major cultivars developed in order of release, are 'Alta', 'Kentucky 31', 'Goar', 'Kenmont', 'Fawn', 'Kenwell', 'Aberystwyth S. 170', 'Demeter', and 'Melik' (see Chapter 7 for a detailed description of those commercially available).

Recognition of tall fescue's wide range in tolerance to climatic and edaphic factors has prompted interest in further evaluating this species as a hay or grazing forage. Proliferation of new cultivars is expected as new selections are evaluated and improved. Some other cultivars currently available, but not widely evaluated, and their countries of origin are 'Manade' (France), 'Ottawa Synthetic A' (Canada), 'Festal' (Netherlands), 'Yatesco' (United States), and 'Zapadnaya' (Russia).

CLIMATIC AND EDAPHIC FACTORS THAT INFLUENCE GROWTH AND PERSISTENCE

Climatic (rainfall and temperature), edaphic (soil texture and moisture), and geographic (latitude and elevation) factors are primarily instrumental in determining the distribution of tall fescue. The significance of any

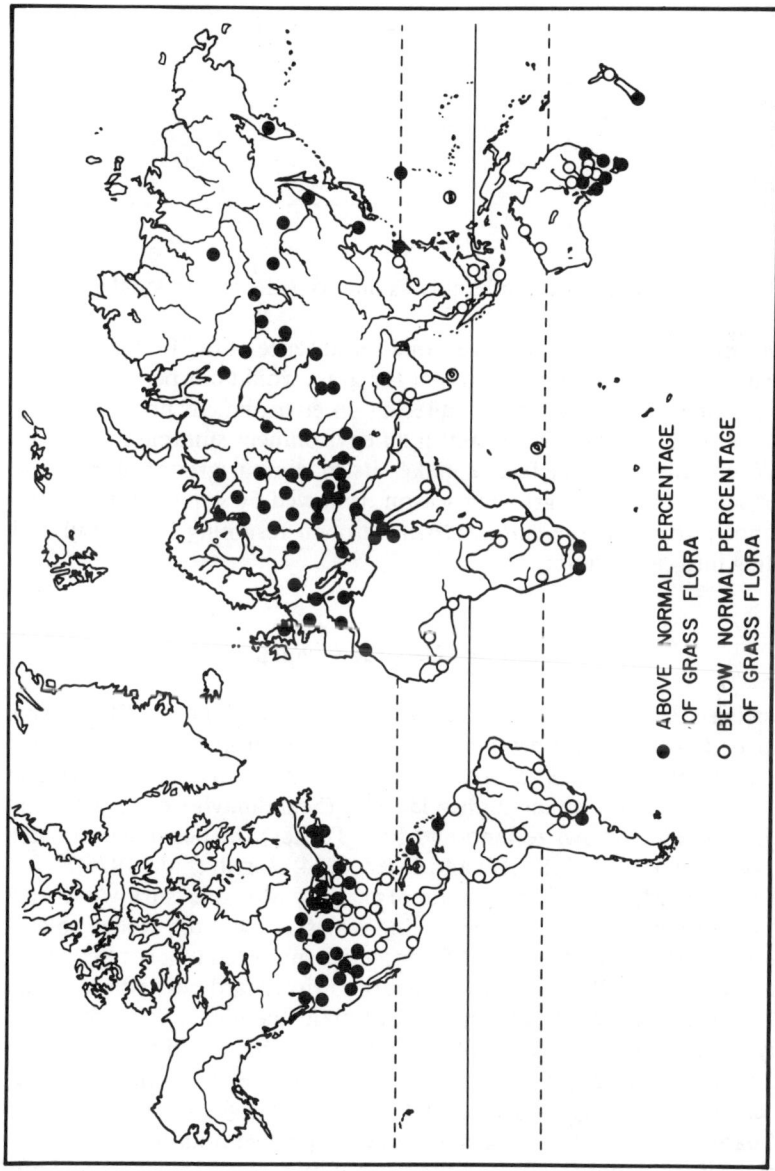

Fig. 2-1—Geographic distribution of the Festucae tribe. The circles (of equal radius for convenience) are placed approximately in the geographic center of the area considered (Hartley, 1950).

Table 2-1—Survival percentages of tall fescue tillers introduced from the Atlas Mountains of Morocco and exposed to the natural environmental conditions existing at Aberystwyth, Wales, during the winter of 1962-63.†

Altitude of collection	Origin				
	Mid Atlas			High Atlas	
	North	South	West	West	Central
——m——					
914–1,219	4	--	27	13	--
1,219–1,523	3	4	49	14	--
1,523–1,828	23	40	59	47	--
1,828–2,133	93	73	--	89	71
2,133–2,437	--	--	--	100	94
2,437–2,742	--	--	--	--	94

† From Breese (1964).

one of these factors varies considerably as one considers more local environments.

It is important to note, however, that the absence of tall fescue in a particular environment does not mean that the grass would not persist if introduced into that climate. Likewise, altering an environment previously uninhabitable to tall fescue may result in an environment suitable for its survival. Where adapted, tall fescue makes its maximum growth during the cold spring portion of the growing season, followed by semidormancy during the hot portion of the summer, with growth resuming in the fall and continuing into early winter.

A. Climatic and Geographic Factors

Hartley (1950) suggests that temperature extremes limit the distribution of the tribe Festuceae and probably tall fescue [see distribution map (Fig. 2-1) for the tribe Festuceae]. More specifically, as one moves into the colder region associated with either higher latitude (Scandinavia) or higher altitudes (Alpine zone of Switzerland and East Turkey) the occurrence of tall fescue is greatly restricted (Borrill and Tyler, 1974; Borrill et al., 1976).

The relationship between altitude of origin and winter survival of tall fescue was demonstrated at Aberystwyth, Wales, when a tall fescue collection from the Atlas Mountains of Morocco was subjected to a severe winter without snow cover (Breese, 1964). Superior winterhardiness was demonstrated by collections obtained within a region at the higher altitudes, particularly those above 1,828 m (Table 2-1). The higher survival rate of fescue from the lower altitude of the western, mid-Atlas region compared with fescue collected from the north and south regions was attributed to its natural habitat of lower winter temperatures and higher humidity.

Frost injury to tall fescue and the variability associated with cold temperature injury was noted in studies at Brandon, Manitoba, Canada. Stand survival of Alta tall fescue for 1969, 1970, and 1972 was 67, 94, and 73%, respectively (H. Gross, unpublished data). Data from the university campus of Saskatoon, Saskatchewan, Canada, (R. P. Knowles, personal communication) showed complete winter killing of tall fescue in the 1st year of establishment while 282 km (175 miles) southeast (Indian Head, Canada) survival has been moderately good. This finding illustrates the northern limit at which tall fescue can be grown. However, its survival under cold temperature conditions is greatly modified by factors such as humidity and snow cover.

Although the distribution of tall fescue into the more northern latitudes is restricted by cold winter temperature, it does make appreciable growth at rather low temperatures. At Knoxville, Tenn., tall fescue grew whenever the mean weekly temperature was above 4.4 C (Leasure, 1952). Tall fescue was not completely dormant until the mean weekly temperature dropped to 1.1 C. In contrast, orchardgrass (*Dactylis glomerata* L.) grew very little unless mean weekly temperature was above 10 C and was completely dormant around 4.4 C. Similarly, Templeton et al. (1961) found that tillers of tall fescue continued to emerge throughout the winter at Lexington, Ky., although mean weekly temperature exceeded 4.4 C only once during the period 22 December to 12 March. The occurrence of high temperatures after 12 March increased tillering markedly with nearly twice the tillers present by 1 April and 3.5 times as many tillers present by 19 April. Mean weekly temperatures during the period (12 March to 19 April) ranged from 6.1 to 16.3 C. Furthermore, tillering of tall fescue was greater under short days than long days and at an exposure of 7.2 compared with 22.2 C.

The response of tall fescue to hot temperatures is more difficult to assess because of the confounding influence of associated water stress (insufficent or poor distribution of rainfall, high evapotranspiration rates, and droughty soils). Most temperate grass growth is maximum when temperatures are between 20 and 25 C with growth nearly ceasing when temperatures rise above 30 to 35 C (Cooper and Tainton, 1968). Such a response is shown by 6 week-old meadow fescue (*Festuca elatior* L.) grown at four day/night temperature regimes (Sprague, 1943). Regimes of 13/4, 21/13, 29/21, and 38/29 C gave dry weight yields of 12.0, 17.4, 26.6, and 0.0 mg/plant, respectively. Similar results are reported (Hoveland et al., 1974) for tall fescue (expressed as leaf area in cm^2) where a day/night temperature regime of 24/18 C produced plants totaling approximately 10 cm^2 of leaf area in 10 days. Plants grown in a 30/24 C regime required 20 days to produce a similar leaf area. For the first 28 days after defoliation, new leaf appearance required 20 days under the warm regime compared with 13 days for the cool regime. During the second 28-day period, new leaf appearance was similar (16 days for the warm regime vs. 15 days for the cool regime).

Although cool-season grasses generally exhibited reduced growth above 25 C, even when ample water is supplied (Cooper and Tainton, 1968), the persistence of tall fescue under hot conditions appears to be mainly associated with the soil moisture status. When soil moisture is adequate, tall fescue remains green and continues some growth while stand thinning occurs if moisture stress becomes severe (Dennis, 1969; Chamblee, 1960). Net C exchange was not influenced much until water stress was below −13 bars

(G. L. Horst. 1973. Physiological and growth response of tall fescue genotypes to drought and irrigation. Ph.D. Thesis. Univ. of Missouri, Columbia, Mo). This topic is considered further in Chapter 5.

Rainfall is also a factor in determining distribution. Tall fescue is not considered drought tolerant in the major portion of the arid U.S., but has been grown under irrigation and is recommended as a forage in both Nevada (Jensen et al., 1971) and Arizona (Dennis, 1969). The rainfall required for its survival is dependent on both temperature and soil conditions. Under dry land conditions, Kentucky 31 tall fescue survival averaged 49% at Red Bluff, Mont. (Joppa and Roath, 1964) where rainfall averaged 354 mm/year (4-year average). Survival in Idaho required in excess of 457 to 559 mm (Ensign and Harris, 1975), compared to 762 mm in Kansas (F. L. Barnett, personal communication), and above 889 mm in Texas (J. Neal Pratt, personal communication). Although dormancy in tall fescue is associated with high temperatures and moisture stress, this condition is alleviated when the water stress is removed. Persistence and growth under such conditions should not be confused with high productivity.

B. Edaphic Factors

For best growth, tall fescue requires good moist soils that are heavy to medium in texture and have considerable humus (Buckner and Cowan, 1973). Tall fescue will survive and persist much better than many forage species on soils that are fine textured with moderate to slow permeability; moderately coarse to medium textured soils underlain by a clay pan (25 to 91 cm deep); moderately coarse to fine textured soils that are poorly drained; soils over 51 cm deep, but moderately to strongly saline alkali and usually imperfectly to poorly drained; shallow to moderately deep, well drained soils over hardpan, bedrock, or other material that prevents root penetration and retains less than 13 cm of available water in the root zone; and soils more than 51 cm deep, but medium to extremely acid (Ensign and Harris, 1975).

The area of major use for tall fescue, designated in Fig. 2-2, demonstrates its tolerance to the adverse soil conditions noted above, compared with the tolerance of other cool-season species. This region represents a transition zone between the temperate North and mild South and receives an annual average precipitation of 1,016 to 1,168 mm (Heath, 1969, 1970). The soils are low in organic matter, with a pH of 5.3 to 5.5, and of low fertility. They range from heavy, poorly drained clays, to deep droughty sand and from being shallow (46 to 85 cm) and underlain with bedrock to fragipans. Since the area receives much of its winter precipitation as rainfall, the soils remain wet and are subjected to considerable winter heaving. In addition, long cold and dry or long hot and dry periods (30 days or more 80% of the years) are not unusual during the growing season. Yet, tall fescue has been found to be one of the best adapted cool-season forage species in both growth and persistence on all of these soils, except the coarse, deep, droughty sands.

ADAPTATION

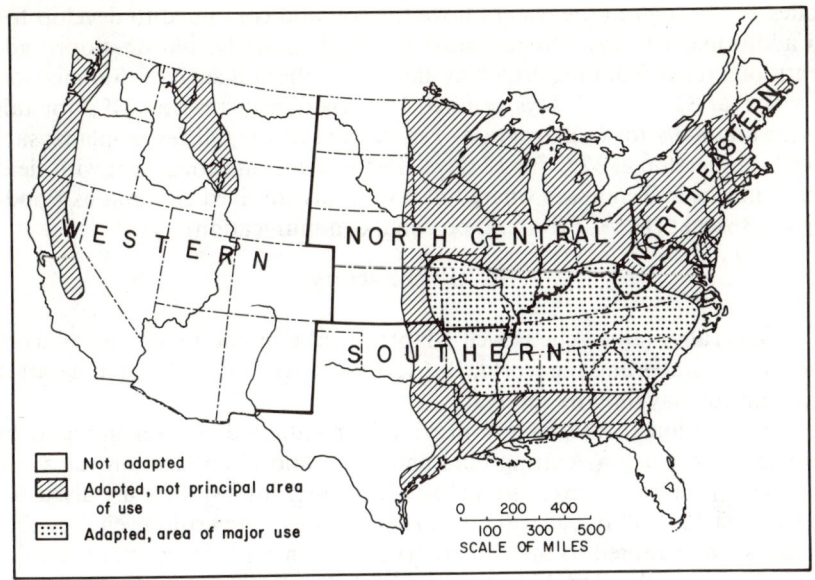

Fig. 2-2—Adaptation and use of tall fescue in the United States.

Soil characteristics not suited to the persistence of tall fescue within its temperature tolerance are coarse to very gravelly, medium-textured soils that are excessively drained (dry land conditions with less than 13 cm of available moisture in the root zone). Also, low pH soils with high Al availability may not be a suitable medium for tall fescue (Shoop et al., 1961).

Tall fescue's massive root structure is frequently attributed to its wide adaptation to many soil types. Hafenrichter et al. (1968) reports yields of more than 7,844 kg/ha of roots in the top 20.3 cm of a deep prairie soil after 6 years of establishment. Yields of 5,603 kg/ha were obtained when it was clipped at 3-week intervals. Its roots are credited with decreasing soil density, improving soil structure, and reducing soil erosion.

Water table fluctuations and periodic flooding of plants can greatly influence root growth and plant persistence. Maintaining the water table at 15 cm revealed traces of tall fescue roots at a depth of 45 to 60 cm compared with complete absence of roots for orchardgrass (Gilbert and Chamblee, 1959). At a water table of 30 cm, the percentage distributions of tall fescue roots at 0 to 15, 15 to 30, 30 to 45, and 45 to 60 cm were 71.6, 22.3, 5.3, and 0.8, respectively. Values for orchardgrass were 85.6, 13.4, 1.0, and a trace. Fluctuating the water table (15 cm 1 week and 50 cm for 3 weeks) compared with a constant 50-cm water table reduced roots at the 15- to 30-cm depth from 29.2 to 9.3% for orchardgrass compared with a reduction from 23.4 to 17.2% for tall fescue. These results partially explain why tall fescue is better adapted than orchardgrass to low-lying areas.

Partial submersion of mature tall fescue plants (2.5 cm of forage above the water) for 40 days at 18 C reduced its yield 31% (Gilbert and Chamblee, 1965). After 30 days, well-developed adventitious roots had formed at the nodes of the stem in the water above the soil and continued to develop for an additional 10 days. Orchardgrass behaved similarly, but developed adventitious roots from the crown at and below the soil surface. Submersion in water at 32 C for 2 days resulted in a yield reduction of 64% for tall fescue and 84% for orchardgrass. Ten percent of the tall fescue plants survived submersion at 32 C for 18 days. These results are consistent with field observations in Tennessee where tall fescue has survived continuous flooding for 3 weeks (H. A. Fribourg, personal communication).

C. Other Factors

Generally, tall fescue is well adapted to the humid temperate areas of the United States and the world. But several factors other than temperature and rainfall may limit its survival.

In addition to the factors previously mentioned stresses imposed by management (such as cutting date and time, and N application rates and frequency) can cause severe stand losses or winter injury; these are discussed in Chapter 10. Tall fescue will not persist on deep sandy soils even though it appears well adapted to the climate when grown on heavier textured soils. An example is the Sandhill area of North Carolina (Typic Quartzipsamment, Lee County) where tall fescue died during the summer months in spite of several reseeding efforts (Chamblee, 1960). Apparently the root system of tall fescue does not reach suitable depths to extract moisture for survival during prolonged periods of drought. Warm season species such as bermudagrass (*Cynodon dactylon* L. Pers.) are much better adapted to these deep sandy soils, yielding from 8,000 to 14,000 kg of dry matter/ha when adequately fertilized.

Aluminum concentrations of low pH soils may influence the survival of tall fescue (Mirasol, 1920). Recent studies by Virginia workers (Shoop et al., 1961) showed a linear increase in the growth of tall fescue when lime was applied at rates up to 1,121 kg/ha and a concomitant linear decrease in exchangeable Al. Examination of the root system showed that roots of tall fescue failed to develop in the unlimed and unfertilized layer of soil (lower 5.1 cm of soil in the pot).

Pests, such as nematodes (Hoveland et al., 1975; McGlohon et al., 1961), and leaf diseases (Berry and Gudauskas, 1972) can also be important factors, particularly in the southern U.S., in limiting the adaptation of tall fescue (see Chapter 15 for additional information). Data from Alabama indicate that nematodes may be more of a problem in limiting survival of tall fescue in the extreme Deep South than are high temperatures (C. S. Hoveland, personal communication).

ADAPTATION OF TALL FESCUE IN THE UNITED STATES COMPARED WITH THAT OF OTHER COOL-SEASON GRASSES

The distribution of tall fescue in the United States under natural rainfall conditions is shown in Fig. 2-2. Specific comparisons in production and persistence with other cool-season grasses are considered below for each region.

A. Adaptation in the Northeast Region

The Northeast region (Fig. 2-2) is characterized by a frost zone in the northern portions of the states of New York, Vermont, New Hampshire, and Maine, with the rest of the region enveloped by a temperate, humid climate. Although tall fescue is not widely grown in this region because of better alternatives, it is well adapted in most of the area. Frost injury has been reported in northern Vermont (Burlington) where cold exposure studies (Wood, 1974) with several species seeded in pots showed tall fescue survival to be 38% compared with 27% for orchardgrass and 64% for Kenhy tall fescue (a tall fescue × rye-grass hybrid). The lowest soil temperature recorded was −9.8 C for a gravel medium and −16.3 C for a vermiculite medium. Winter injury was associated with no snow cover.

Tall fescue is grown throughout Maine on road cuts and is well adapted to the coastal area with snow cover being an important factor in its survival (C. S. Brown, personal communication). Lack of winter-hardiness has been noted as far south as Rhode Island (C. R. Skogley, personal communication).

Well adapted to the other states in the region, tall fescue has been successfully established and grown in many parts of Massachusetts, ranging from roadside sites to sand dune stabilization at Provincetown (Zak, 1967). In the latter situation, a stand established in 1962 still retained a fair rating by 1966.

New Jersey data showed tall fescue to be as productive as orchardgrass on fertile, well-drained soils throughout the state and generally superior where soil drainage is poor. Yields of dry matter ranged from 8,068 to 10,085 kg/ha. Tall fescue has established better than either orchardgrass or bromegrass (*Bromus inermis* Leyss.) on the edge of brackish tidewaters and further onto the estuary where the exposure to salt is greater (M. A. Sprague, personal communication). Tall fescue is adapted throughout Pennsylvania (J. B. Washko, personal communication), Maryland (A. M. Decker, personal communication), and West Virginia with its distribution not limited by any particular soil type, drainage condition, or fertility level.

B. Adaptation in the North Central Region

The North Central region (Fig. 2-2) is composed of three major climatic zones: 1) The arid zone which occupies 70 to 90% of the states along the western edge of the region; 2) the frost zone of the northern states; 3) the humid, temperate zone of the remaining states.

Tall fescue is not well adapted to dryland conditions of the arid zone (C. R. Krueger and J. T. Nichols, personal communications). This limitation appears to be mainly associated with insufficient rainfall because it has been found to be quite persistent at North Platte, Neb., when additional water is supplied (J. T. Nichols, personal communication). Tall fescue does persist well and is productive in the eastern one-third of Kansas where the

Table 2-2—Summary of dry matter yields for three cool-season species for three locations in Minnesota (1972 to 1974).†

Grasses‡	Location		
	Waseca (Southern)	Pine City (Central)	Roseau (Northern)
	kg/ha		
Tall fescue	8,883	6,343	5,678
Bromegrass	6,813	4,572	7,098
Orchardgrass	7,217	6,963	5,812

† Adapted from A. R. Schmid, unpublished data.
‡ Forages were topdressed with 224 kg/ha of elemental N.

annual average rainfall is 762 mm or more (F. L. Barnett, personal communication; Hyde, 1969; Hyde and Kilgore, 1973).

Although well adapted to the states in the frost zone of the North Central region (Fig. 2-2), tall fescue is subject to considerable winter injury. Such injury has been associated with a lack of snow cover as well as cold temperatures. Adaptation studies in Minnesota showed a noted reduction in dry matter yields of well-fertilized tall fescue from southern to northern locations (Table 2-2). Similar results occurred for orchardgrass, while bromegrass yields were generally higher at the northern locations. These yield differences appear to be associated with cold temperature injury (A. R. Schmid, unpublished data) even though good snow cover occurred at the northern location. Both tall fescue and orchardgrass showed winter injury, with tall fescue being more severely injured than orchardgrass.

Severe winter injury to tall fescue has also been noted in space planting at Arlington, Wis. Winter survival for tall fescue averaged 40%, compared with 100% for bromegrass and 5% for orchardgrass (D. A. Rohweder, personal communication). Although the frequency of winter injury was high for tall fescue at both Arlington and Lancaster, Wis., its survival further north was better, aided by snow cover.

Tall fescue was productive and yielded similarly to bromegrass and orchardgrass in northern Michigan when well fertilized (Tesar, 1974). Yields for the last year (1973) of a 6-year study were 11,699, 11,587, and 10,579 kg/ha of dry matter, respectively. Mean yields for the 6-year study were 10,220 kg/ha for tall fescue and bromegrass compared with 9,906 kg/ha for orchardgrass.

Winter injury of tall fescue has not been observed in northern Ohio (R. W. Van Keuren, personal communication) and it is well adapted throughout the remainder of the mild, humid area of the region.

C. Adaptation in the Southern Region

The southern region (Fig. 2-2) may be divided into the upper South (north of a line from the southern edge of North Carolina through Little Rock, Ark. and along the Oklahoma-Texas border) and lower South (south of the arbitrary line noted above).

Table 2-3—Field yield of three cool-season grasses grown at three locations in North Carolina (second year following seeding) 1947, and greenhouse yields from soils from each location.†

Grass	Location and soil type					
	Coast, Typic Ochraquult		Piedmont, Typic Hapludult		Mountains, Typic Rhodudult	
	Field‡	Greenhouse§	Field‡	Greenhouse§	Field‡	Greenhouse§
	kg/ha	g/pot	kg/ha	g/pot	kg/ha	g/pot
Tall fescue	2,438	3.62	1,728	4.96	4,356	4.34
Kentucky bluegrass	1,593	1.74	439	3.42	2,004	3.79
Orchardgrass	3,546	2.65	1,681	4.06	4,398	3.90

† From S. H. Dobson (1947). The adaptation of five species of permanent pasture grasses to three soil types. Unpublished M.S. Thesis. North Carolina State College, Raleigh, N.C. L.S.D. (0.05) for comparing among grasses on any one soil of the greenhouse study = 0.46.   ‡ Forages received 108 kg/ha of elemental N applied as a split application in the spring and summer.   § Yields are the average of six harvests taken at 30-day intervals and two levels of N (18 and 54 kg/ha of elemental N) topdressed every 60 days.

Tall fescue is well adapted to the upper South. In Virginia (4-year study) tall fescue was more productive than orchardgrass or bromegrass under good fertilization (224 kg/ha of elemental N). Dry matter yields (kg/plot) when frequently defoliated (five cuts) averaged 737 for tall fescue, 672 for orchardgrass, and 674 for bromegrass (Wolf, 1973).

Tall fescue is well adapted from the coast to the mountains in North Carolina. A comparison of field yields from tall fescue with orchardgrass and Kentucky bluegrass for the geographical regions of the state (Coast, Piedmont, and Mountains) is shown in Table 2-3 (Dobson, S. H. 1947. The adaptation of five species of permanent pasture grasses to three soil types. Unpublished M.S. Thesis. North Carolina State College, Raleigh, N.C.). When the same species were evaluated on the same soil types in the greenhouse (constant environment), tall fescue produced higher yields than the other grasses on all soil types.

Tall fescue is well adapted throughout Tennessee. Stands are known to persist 20 years or more even on shallow soils underlain with bedrock or fragipan (H. A. Fribourg, personal communication). A comparison of tall fescue and orchardgrass in Kentucky (Templeton et al., 1965) revealed that both species yielded similarly whether grown alone or in combination with ladino clover (*Trifolium repens* L.). The decision of which grass to grow with a legume was in favor of orchardgrass for pastures and meadows because of its quick establishment, leafiness, and compatibility with legumes. However, tall fescue was favored where conditions of stress or poor management might occur; in permanent type grasslands; and on land subject to excessive water erosion.

Tall fescue is well adapted to most soils north of Little Rock, Ark. (K. L. Smith, personal communication) extending into eastern Oklahoma where rainfall averages 889 mm (Fuller et al., 1971; Rommann and McMurphy, 1973). Its distribution can be extended further west on sites that receive extra moisture, or under very careful management (L. M. Rommann, personal communication). Tall fescue is presently being recommended to improve the native ranges in the Owachita region of Oklahoma (McMurphy et al., 1975).

Table 2-4—Survival of cool-season perennial grasses in Alabama after 3 years when harvested monthly during the growing season.†

Species and cultivars	Locations		
	Northern Alabama	Central Alabama	Southern Alabama
	Survival %		
Tall fescue (Kentucky 31)	100‡	100	82
Orchardgrass (Potomac)	91	60	0
Bromegrass (Lincoln)	82	61	33

† Adapted from Hoveland et al. (1970).   ‡ Values represent the mean of three, five, and three different sites for the northern, central, and southern locations, respectively.

In the lower South (Fig. 2-2) tall fescue makes considerable growth during the winter period and is valued for this production. It is well adapted to most sites throughout South Carolina, Georgia, Alabama, Louisiana, Mississippi, and into eastern Texas; however, its distribution in these states is restricted by deficiencies in soil water (M. W. Jutras, personal communication; Woodle and Turner, 1949). As mentioned previously, pests such as nematodes and leaf diseases are also important factors limiting its adaptation in this region.

Tall fescue is productive and has persisted in northern Florida (West Florida Experiment Station, Jay, Fla.). Three years after establishment, stands yielded 6,173 kg/ha of dry matter from two harvests taken in November 1960 and March 1961 (Dunavin, 1961). However, it did not persist well at Ona, Fla. (Hodges et al., 1953) nor has it made satisfactory growth on acid flatwoods soils. Although tall fescue is generally considered adapted to only northern Florida, Bair and Kidder (1949) have maintained stands for 3 years on the peat and muck soils of the Everglades.

The persistence of tall fescue compared with orchardgrass and bromegrass has been superior at northern, central, and southern Alabama locations (Table 2-4). Only tall fescue survived more than 1 year on the sandy soils at the Southern Alabama Brewton experimental field (Hoveland et al., 1970).

Tall fescue is the major perennial winter grass in Louisiana for the poorly drained, heavy clay soils (C. L. Mondart, Jr., personal communication), and its distribution extends to the Colorado River in the gulf coastal region of east Texas (J. N. Pratt, personal communication). It also has been found to persist in north central Texas on bottom land or deep upland clay soils of the Blackland or Grand Prairie types where annual rainfall approaches 870 mm.

When irrigated, tall fescue is adapted to the trans-Pecas area and to the high plains of west Texas. It is considered the best perennial cool-season grass for the Wichita Valley area (Brooks and Holt, 1954). Dry matter yields and stand survival for several species compared with orchardgrass and bromegrass are shown in Table 2-5. Tall fescue produced higher yields than either orchardgrass or bromegrass. Growth continued later in the spring and began earlier in the fall than either of the other species. Tall fescue exhibited only a 2-month dormancy period.

Table 2-5—Yield and plant survival of several cool-season grasses under irrigation at Iowa Park, Tex. (1949 to 1951).†

Species	Yield			Stand		
	1949‡	1950	1951	1949	1950	1951
	kg/ha			%		
Tall fescue (Kentucky 31)	2,734	6,466	5,513	99	99	99
Bromegrass (Lincoln)	1,927	3,429	2,678	94	94	94
Orchardgrass (Cultivar not identified)	235	818	1,871	40	32	30

† Adapted from Brooks and Holt (1954).
‡ Yields taken on 30 May and 23 June were low because of a late fall planting.

Table 2-6—Rainfall data, mean yields (12% moisture), and percent occupancy of five tall fescue types at seven locations in Montana.†

Location	Rainfall		Conditions‡			
			Dry land		Irrigated	
	Average 1959-61	Long term average	Yield	Occupancy§	Yield	Occupancy§
	mm		kg/ha	%	kg/ha	%
Sidney (Northeast)	274	329	Died	0	4,801	97
Huntley (Southeast)	251	295	Died	0	8,951	95
Moccasin (Central)	321	357	1,215	85	--	--
Harve (North Central)	236	290	Died	0	--	--
Creston (Northwest)	532	551	5,724	96	4,236	95
Bozeman (Southwest)	426	447	--¶	--	3,546	89
Norris (Southwest)	352	334	883	46	--	--

† Adapted from Joppa and Roath (1964). ‡ Dry land plots harvested once each year. Yields for Creston are an average for 1960 and 1961 and for Norris and Moccasin, 1959, 1960, and 1961. Irrigated plots were harvested at least three times each year to simulate grazing. Yields are an average for 1959, 1960, and 1961. § Occupancy scores taken in 1961. ¶ Test not evaluated.

D. Adaptation in the Western Region

In the western region (Fig. 2-2), tall fescue is limited under dry land conditions primarily by inadequate rainfall. Evaluation of five tall fescue cultivars at seven locations in Montana (Joppa and Roath, 1964) showed that they lacked sufficient drought tolerance to persist when the mean annual rainfall was less than 305 mm (Table 2-6). Yet, irrigation at the Huntley (southeast) and Sidney (northeast) locations resulted in nearly 100% plant occupancy scores after 3 years compared with complete stand failures at the same sites under dry land conditions.

Tall fescue has good winterhardiness in Idaho and is adapted to dry land conditions where annual rainfall exceeds 508 mm (R. D. Ensign, personal communication). It is also adapted to the subhumid irrigated areas

Table 2-7—Adaptation of irrigated cool season forages for Nevada.†

Factors	Regions of Nevada							
	Northern			Western			Southern		
	TF‡	OG	SB	TF	OG	SB	TF	OG	SB
Winterhardiness	F§	P	E	E	G	E	--	NA	NA
Seedling vigor	E	G	G	E	G	G	E		
Yield potential	M	M	H	H	H	M	M		
Recovery rate	R	R	M	R	R	M	R		
Longevity	M	M	L	L	M	M	M		
Drought tolerance	F	F	G	F	F	G	F		
Wetness and flood tolerance	G	P	F	G	P	F	G		
Salt and alkali tolerance	G	P	F	G	P	F	G		

† Adapted from Jensen et al. (1970a, b, c).   ‡ TF = tall fescue, OG = orchardgrass, and SB = smooth bromegrass.   § E E excellent, G = good, F = fair, and P = poor; H = high and M = medium; L = Long; R = rapid; NA = not adapted.

of the Pacific Northwest and the Great Basin States (Washington, Oregon, Idaho, Utah, Nevada, and California), particularly in the wet, alkali areas of the basin, as well as on the lighter soils. It remains green during the summer months, but grows only when soil moisture is adequate (Turner, 1966; Van Keuren et al., 1960).

Tall fescue was more productive, more winterhardy, and persisted better through the hot summers of Wyoming than did orchardgrass (Moyer and Seamands, 1975). Yields (12% moisture) of 12 tall fescue cultivars grown under irrigation averaged (1973 and 1974 seasonal totals) 5,334 and 7,799 kg/ha at Laramie and Torrington, Wyo., respectively.

Tall fescue was found more tolerant to poorly drained alkaline soils than bromegrass, with stand failure occurring with the latter after the year of establishment (Seamands and Kolp, 1975). Supplemental water permits the growth of tall fescue throughout Nevada, while orchardgrass and bromegrass can be grown only in the north and west sections of the state (Table 2-7).

Tall fescue will persist under irrigation above 609 m in Arizona (Dennis, 1969). Higher elevations are required for orchardgrass (above 914 m) and bromegrass (above 1,219 m). However, at the higher elevations, tall fescue will grow all summer while at the lower elevations, growth is restricted mainly to spring and fall.

ADAPTATION VS. PRODUCTION

The pressure on the land to produce food for human and animal consumption makes it difficult to determine if the presence of a species is due to its natural genetic makeup to compete and survive in that environment or if its occurrence is due to man's manipulation caused by a species' economic impact on society. Tall fescue is a species whose area of major production should not be confused with its area of adaptation (Fig. 2-2). It is well adapted to large areas in the north central and northeastern U.S. However, its general absence throughout these two regions is attributed to the adaptation of more desirable species such as orchardgrass, bromegrass, timothy (*Phleum pratense* L.), and Kentucky bluegrass (*Poa pratensis* L.).

The region designated in Fig. 2-2 as the area of major use of tall fescue is actually a transition zone (see paragraph 2 of section on Edaphic Factors for a description). The tolerance that tall fescue exhibits in terms of its perennial growth habit and the lack of persistence exhibited by other cool-season grasses to such a heterogeneous environment has resulted in reliance by producers upon tall fescue as the major forage grass. This reliability compared with other cool-season grasses primarily accounts for its predominance.

ADAPTATION OF TALL FESCUE IN OTHER REGIONS OF THE WORLD

Tall fescue is currently receiving consideration as a forage species in many parts of the world, even where other species have historically been considered of more economic importance. Such emphasis has probably resulted from the realization that the genus *Festuca* is very diverse (Frame et al., 1970) as suggested by the distribution map of the tribe Festuceae (Fig. 2-1). Through careful selection, the potential exists of obtaining tall fescue cultivars that are very tolerant to adverse environmental (climate and soil) and management conditions, yet are productive, persistent, and of adequate nutritive value for many classes of livestock.

A. Adaptation in Europe and Asia

Tall fescue is adapted throughout Europe, extending over a broad region between northern Italy on the south, to Scandinavia on the north, and Hungary to the east, with less frequent appearance in Turkey. Its apparent ubiquitous distribution is restricted as one moves into the colder regions characteristic of higher latitudes and altitudes or as one moves south into the drier regions of north Italy (Borrill et al., 1976). In general, meadow fescue (*Festuca pratensis*) is adapted further north into Scandinavia than is tall fescue and is a major constituent of grasses that have been collected from the mountain uplands of East Turkey. Meadow fescue is naturally adapted to the cold or very cold winters of eastern Anatolia (a portion of Turkey) where January temperatures of -15 C and over 100 days of snow cover occur. Expeditions from 1961 to 1972 yielded tall fescue introductions from Norway, Sweden, Denmark, Belgium, France, Italy, Switzerland, Turkey, Spain, Portugal, Tunisia, and Morocco (Davies et al., 1972). Except for alterations associated with altitude and localized climates or management practices, growth in the Mediterranean environments is limited by summer drought, with winter being the active growing season. Selections of Mediterranean types generally do not show winter dormancy and grow well at moderately low temperatures and/or short nights, but are not winterhardy. In the north and central region of Europe the cold winter temperatures and short days and associated low input of light energy limit winter production (a high degree of cold tolerance and winter dormancy is required). There, growth occurs mainly in late spring, summer, and early autumn. Selections from this environment frequently exhibit winter dormancy and considerable winterhardiness. In the maritime region (from Portugal to Scotland) both winter cold and summer droughts are less severe. Consequently, fewer winter of summer dormancy mechanisms have developed and production is more closely correlated with seasonal input of light energy (Breese, 1964; Davies et al., 1972).

In long-time mowing experiments at Wageningen, Netherlands, tall fescue yielded about 20% more dry matter than perennial ryegrass (*Lolium perenne* L.). Tall fescues occurrence in perennial pastures is estimated at about 10%, being most frequently found on sand and clay soils and less frequently on peat soils. Interest has increased in recent years in sowing tall fescue for zero grazing and in seeding on lands adjacent to rivers (M. L. 't Hart, personal communication).

According to Kirillov (1974), tall fescue is found in the wild state along the Baltic coasts, on the coasts of the Island of Ihmen and Ladojsk, on the river banks and mountains of the midwest and the northern Caucasus as well as in the Caucasus in the southern part of European USSR, and in western Siberia extending into China. He describes three subspecies and their area of adaptation. The first, pre-Atlantic [*F. arundinacea* ssp. *arundinacea* (Schreb.) Tzvel 1972], is found on the coasts of the Baltic and in northern Europe. The second, eastern continental [*F. arundinacea* ssp. *orientalis* (Hack) Tzvel, 1970) is adapted to the southern part of European USSR, the Caucasus, southern and western Siberia, central Asia, southern Europe, the Near East, Asia, and China. The third subspecies, Mediterranean [*F. arundinacea* ssp. *interrupta* (Desf) Tzvel, 1971], is found in the Caucasus of USSR, central Asia, and occasionally in the southern parts of European USSR, southern Europe, the Near East, and North Africa.

Tall fescue is adapted to very different soils and varied climatic conditions in Russia and Asia, ranging from forest soils (non-chernozem) to the steppes and semiarid areas when irrigated. Forms from both the mild and severe winter latitudes have sufficient winterhardiness for the area, with the southern forms having a higher salt tolerance. It is a long-lived perennial throughout this vast region, remaining in pure stands from 12 to 15 years. Because of its conservation characteristics on slopes and road embankments, its high yielding potential, and its excellent disease resistance, Kirillov places tall fescue among the best of the traditional perennial grasses in the USSR.

B. Adaptation in the United Kingdom

Tall fescue is generally well adapted to the cool, moist climate of the United Kingdom. Recent emphasis in the United Kingdom on special purpose, as opposed to general purpose, grasses has highlighted the potential of tall fescue as a winter growing species. Annual dry matter yields (3-year average) of well fertilized Kentucky 31 tall fescue at Auchincruive, Ayr, Scotland averaged 13,500 kg/ha (Frame et al., 1970). The Mediterranean ecotypes were frequently found to lack winterhardiness, especially when intensively defoliated.

C. Adaptation in Canada

Tall fescue is apparently limited to the extreme southern portion of Canada (Fig. 2-1). Results from Saskatoon, Saskatchewan showed tall fescue not to be adapted to that area as stands were winter killed the first winter (1950) of seeding (R. P. Knowles, unpublished data). Tall fescue has persisted better at Brandon, Manitoba (located south and east of Saskatoon), but is still subject to considerable winter injury at that location (H. Gross, unpublished data). These results indicate that temperatures at Saskatoon were too cold for the survival of tall fescue, while Brandon, Manitoba demarks its northern limit of survival. This demarcation agrees well with Hartley's distribution map (Fig. 2-1).

D. Adaptation in Japan

Kentucky 31 tall fescue is widely adapted throughout Japan (Suzuki, S. 1971. Studies on the adaptability of cross fertilizing forage crops. JIBP/UM Symp. Tokyo, Japan. Mimeo. Rep. p. 35–63). In general, production increases from north to south, with 1.4 and 2.2 times more production occurring in central (Nasu) and southern (Kagoshima) Japan, respectively, than in northern (Sapporo) Japan. Several new cultivars being evaluated were more productive than Kentucky 31 in the northern and southern locations, while all cultivars appeared to yield similarly at the central location (Nasu).

E. Adaptation in Australia and New Zealand

Sown pastures in Australia are limited by rainfall, temperature, and day length (Bank of New South Wales, 1971). In general, sown pastures are limited to an area receiving 305 mm of annual rainfall in western Australia; 381 mm in southeastern Australia; 508 mm in Queensland; 762 mm in the northern territory and northern, western Australia.

A Morocco selection of tall fescue was introduced into Australia through France in 1931. This selection has since been released as the cultivar 'Demeter'. In early trials at Canberra (southern tablelands), it showed good fall production (Neal-Smith and Wright, 1969), but survival was poor (67%) compared with bulbous canarygrass (*Phalaris aquatica* L.) (formerly *tuberosa* L.) (87%) and orchardgrass (80%). The potential of tall fescue was not realized until its evaluation on the tablelands in northern New

Table 2-8—Elevation and climatic conditions for five locations in Mexico and subsequent adaptation scores for Kentucky 31 tall fescue.†

Location	Elevation	Temperature		Rainfall	Adaptation‡ score
	m	Max.	Min.	mm	
		— C —			
Mexico Valley	2,240	32.2	−6.0	596	1
Toluca Valley	2,675	26.8	−3.0	792	1
El Bajío	1,680–1,809	37.0	−2.0	730	2
Saltillo, Coahuila	1,609	38.0	−7.3	339	2
Llanuras de Ver.	61	41.2	7.2	1,526	3

† Adapted from Buller et al. (1955).
‡ 1 = very good; 2 = good; 3 = fair; 4 = not adapted.

South Wales where trials at Armidale (characterized by a mean maximum and minimum temperature of 27.4 and 13.3 C in January and 12.0 and 0.2 C in July and an average annual rainfall of 800 mm) showed that Demeter fescue had a longer effective growing season than either a commercial cultivar of Phalaris or Victoria perennial ryegrass (Hilder, 1963).

Both temperate and Mediterranean ecotypes have survived at Kojonup (western Australia), Canberra, and Armidale (New South Wales). comparisons at Canberra showed greater persistence for the Mediterranean ecotypes (Neal-Smith, 1969) while at Armidale, the Mediterranean types yield less total production, but gave higher production during the winter (Schiller and Lazenby, 1975). The Mediterranean types maintained their greenness and actually grew during the winter period and have not been observed to be severely injured by the winter cold (Schiller and Lazenby, 1975).

Tall fescue is adapted to much of New Zealand's environment (Levy, 1970). It is well adapted on soils of average fertility and tolerates tidal salt mud flats, swampy and flood areas, as well as very wet and waterlogged sites. However, it is less tolerant to New Zealand's dry land region. It is considered of moderate value on the extremely wet sites, but termed a weed on first and second class lands. Tall fescue is not adapted to the friable loams, pumice, and sand or gravel sites of New Zealand.

F. Adaptation in Mexico

Of the introduced cool-season species evaluated under Mexico conditions, tall fescue (both Kentucky 31 and Alta) has been found to be the best adapted. They have been successfully grown on soils with poor drainage, even when irrigated, and have shown tolerance to both alkalinity and salinity. These two conditions are prevalent in the soils of Mexico. Of the many climatic regions in Mexico (Table 2-8), best results with tall fescue have been obtained in the more temperate areas such as the Valley of Mexico and the Valley of Toluca. The Toluca Valley is a semi-cold region with dry winters. Tall fescue's adaptation has been rated as "good" in the Bajío and Saltillo regions. The former is a medium-hot climate with the latter being more temperate with dry winters. Tall fescue showed only fair survival in the tropical environment of Llanuras de Ver. where a winter dry period is experienced (Buller et al., 1955).

G. Adaptation in South America

The distribution map (Fig. 2-1) of Hartley (1950) shows the Festuceae tribe distributed across South America, but in less than average frequency compared to other species. Tall fescue has been found productive at the higher elevations in Colombia (Crowder et al., 1959) with recent interest being expressed in its use in Chile for irrigated pastures in mixture (Whyte et al., 1965). It has been found well adapted throughout the pampas of Argentina when seeded as improved pasture (J. Maddaloni, personal communication).

Some insight to the adaptation of tall fescue in the hot (0 to 762 m), warm (762 to 1,980 m), cool (1,980 to 3,047 m), and cold (above 3,047 m) regions of South America can be obtained from the studies in Colombia by Crowder et al. (1959) and Crowder (1967). They found adaptation to be good in the cool climate (Bogotá, 2,590 m and Las Palmas, 2,437 m), only fair in the cold (Usme, 3,351 m) and the higher elevations of the warm climate (Manizales, 1,980 m), but not adapted to either the lower elevations of the warm climate (Medellín, 1,371 m and Palmira, 914 m) or in the hot climate (Villavicencio, 457 m and Montería, 46 m).

H. Adaptation in Africa

Tall fescue is adapted throughout the Atlas Mountain region in Morocco and also in Tunisia, Table 2-1 (Breese, 1963; Borrill et al., 1971; Davies et al., 1972). The Festuceae tribe comprises less than the normal percentage of flora across Central Africa (Fig. 2-1) with higher percentages occurring in South Africa. The tribe's lack of occurrence in Central Africa is probably associated with the tropic or subtropic environment. Such conditions have been found in Australia to restrict the occurrence of tall fescue to either irrigated or high altitude sites (Humphreys, 1969). Kentucky 31 tall fescue has been found to be a productive winter grass in South Africa where rainfall exceeds 750 mm (Scott, 1967). It grows in the spring and fall in the summer rainfall areas (receiving over 1,000 mm) where temperatures are cool (eastern coast area from Elizabeth to just below the Drakensbery Mountains) and in the fall, winter, and spring where annual rainfall is 762 to 1,016 mm (mist belt area of Eastern Transvaal and the eastern cape region). Tall fescue is not adapted to the northern portion of South Africa (rainfall is less than 508 mm), but its adaptation can be extended into this dry region when supplemental water is provided. It will not survive in regions characterized by high temperature or where soil moisture is limiting in the winter (Scott, 1967).

ACKNOWLEDGMENT

Grateful acknowledgment is made to all contributors who supplied information through personal communications or publications or both. The authors express special appreciation to Dr. R. P.Knowles, Head Plant Breeding Section, Research Station, Research Branch, Agriculture, University Campus, Saskatoon, Saskatchewan, Canada, for his English translations of portions of the article by Kirillov (1974).

LITERATURE CITED

1. Bair, Roy A., and R. W. Kidder. 1949. Pasture investigations on the peat and muck soils of the Everglades. Florida Agric. Exp. Stn. Annu. Rep. p. 195-197.
2. Bank of New South Wales. 1965. Pasture legumes and grasses. 2nd Ed. (revised). Waite and Bull Proprietary Ltd., Sydney, Australia. 76 p.
3. Berry, Charles D., and Robert T. Gudauskas. 1972. Races of *Puccinia coronata* var. *coronata* on tall fescue grass. Plant Dis. Rep. 56:614-615.
4. Borrill, M., and B. F. Tyler. 1974. Distribution of fescue chromosome races. Welsh Plant Breed. Stn. Annu. Rep. p. 22-23.
5. ─────, ─────, and M. Lloyd-Jones. 1971. Studies in Festuca 1. A chromosome atlas of bovinae and scariosae. Cytologia 36:1-14.
6. ─────, ─────, and W. G. Morgan. 1976. Studies in Festuca 7. Chromosome atlas (part 2). An appraisal of chromosome race distribution and ecology, including *F. pratensis* var. *apennina* (De Not.) Hack.-tetraploid. Cytologia 41:219-236.
7. Breese, E. L. 1964. Herbage plant breeding. Welsh Plant Breed. Stn. Annu. Rep. 1963. p. 26-29.
8. Brooks, Lester E., and Ethan C. Holt. 1954. Cool-season grasses in the Wichita Valley. Texas Agric. Exp. Stn. Prog. Rep. 1716. 4 p.
9. Buckner, Robert C., and J. Ritchie Cowan. 1973. The fescues. p. 297-306. *In* M. E. Heath, D. S. Metcalfe, and R. F. Barnes. Forages. 3rd Ed. The Iowa State Univ. Press, Ames, Iowa.
10. Buller, R. E., J. B. Pitner, and Hector Porras M. 1955. Adaptacion de zacates y leguminosas para forraje, conservacion y mejoramiento del suelo en Mexico. Foll. Tec. 18. 75 p.
11. Cain, Stanley A. 1944. Foundations of plant geography. Harper and Brothers, New York. 556 p.
12. Chamblee, Douglas S. 1960. Adaptation and performance of forage species. North Carolina Agric. Exp. Stn. Bull. 411. 47 p.
13. Cooper, J. P., and N. M. Tainton. 1968. Light and temperature requirements for the growth of tropical and temperate grasses. Herb. Abstr. 38:167-176.
14. Crowder, L. V. 1967. Grasslands of Colombia. Herb. Abstr. 37:239-245.
15. ─────, Jaime Vanegas A., Jaime Lotero C., and Angelo Michelin. 1959. The adaptation and production of species and selection of grasses and clover in Colombia. J. Range Manage. 12:225-230.
16. Daubenmire, R. F. 1959. Plants and environment. 2nd Ed. John Wiley and Sons, Inc., New York. 422 p.
17. Davies, W. Ellis, B. F. Tyler, M. Borrill, J. P. Cooper, Hugh Thomas, and E. L. Breese. 1972. Plant introduction at the Welsh Plant Breeding Station. Welsh Plant Breed. Stn. Annu. Rep. p. 143-162.
18. Dennis, R. E. 1969. Establishment and management of irrigated pastures in Arizona. Arizona Agric. Ext. Serv. Exp. Stn. Bull. A-49. 35 p.
19. Dunavin, L. S., Jr. 1961. Another look at tall fescue for northwest Florida. Soil Crop Sci. Soc. Fla. Proc. 21:128-136.
20. Ensign, R. D., and H. L. Harris. 1975. Idaho forage crop handbook. Idaho Agric. Exp. Stn. Bull. No. 547. 54 p.

21. Frame, J., I. V. Hunt, and R. D. Harkess. 1970. Potentiality studies of tall fescue (*Festuca arundinacea* Schreb.). Int. Grassl. Congr. Proc. 11th (Surfers Paradise, Australia). p. 210–214.
22. Fuller, William W., William C. Elder, Billy B. Tucker, and Wilfred E. McMurphy. 1971. Tall fescue in Oklahoma. Oklahoma Agric. Exp. Stn. Prog. Rep. 650. 22 p.
23. Gilbert, W. B., and D. S. Chamblee. 1959. Effect of depth of water table on yield of ladino clover, orchardgrass and tall fescue. Agron. J. 51:547–550.
24. ———, and ———. 1965. Effect of submersion in water on tall fescue, orchardgrass and ladino clover. Agron. J. 57:502–504.
25. Good, Ronald. 1931. A theory of plant geography. New Phytol. 30:149–171.
26. ———. 1953. The geography of the flowering plants. 2nd Ed. Longmans, Green and Co., New York. 452 p.
27. Hafenrichter, A. L., John L. Schwendiman, Harold L. Harris, Robert S. MacLauchlan, and Harold W. Miller. 1968. Grasses and legumes for soil conservation in the Pacific Northwest and Great Basin States. USDA Agric. Handb. 339. 69 p.
28. Hartley, W. 1950. The global distribution of tribes of the Gramineae in relation to historical and environmental factors. Aust. J. Agric. Res. 1:355–373.
29. ———. 1954. The agrostological index: A phytogeographical approach to the problems of pasture plant introduction. Aust. J. Bot. 2:1–21.
30. Heath, Maurice E. 1969. Fitting plants to fragipan soils in Southern Indiana. Proc. Indiana Acad. Sci., 1968. 78:429–434.
31. ———. 1970. Naturalized big trefoil (*Lotus pedunculatus* Cav.) ecotypes discovered in Crawford County, Indiana. Proc. Indiana Acad. Sci., 1969. 79:193–197.
32. Hilder, E. J. 1963. Growth curves of three grass species at Armidale N.S.W. CSIRO Division of Plant Industries. Field Stn. Rec. Vol. 2. No. 1:25–28.
33. Hodges, E. M., D. W. Jones, and W. G. Kirk. 1953. Forage variety trials. Florida Agric. Exp. Stn. Annu. Rep. p. 312.
34. Hoveland, C. S., E. M. Evans, and D. A. Mays. 1970. Cool-season perennial grass species for forage in Alabama. Auburn Agric. Exp. Stn. Bull. 397. Auburn, Ala. 19 p.
35. ———, H. W. Foutch, and G. A. Bachanan. 1974. Response of phalaris genotypes and other cool-season grasses to temperature. Agron. J. 66:686–690.
36. ———, R. Rodriguez-Kabana, and C. D. Berry. 1975. Phalaris and tall fescue forage production as affected by nematodes in the field. Agron. J. 67:714–717.
37. Humphreys, L. R. 1969. A guide to better pastures for the tropics and subtropics. 2nd Ed. CSIRO Trop. Pasture Div., Brisbane, Australia. 79 p.
38. Hyde, Robert. 1969. Cool season grasses in Kansas. Kans. State Ext. Serv. Circ. 257. 11 p.
39. ———, and Gary L. Kilgore. 1973. Tall fescue production. Kansas State Ext. Serv. Circ. 470. 11 p.
40. Jensen, E. H., Frank L. Brooks, Jr., Gayland D. Robison, J. Boyd Price, Edmund R. Barmettler, Norman R. Ritter, and B. Brooks Taylor. 1971. Management of irrigated forages in Nevada. Nevada Agric. Exp. Stn. Bull. 29. Reno, Nev. 39 p.
41. ———, J. Boyd Price, Fredrick F. Peterson, Harold Harris, Norman Ritter, and Clark Torell. 1970a. Irrigated forages for northern Nevada type climate. Nevada Agric. Ext. Serv. Circ. 105. 18 p.
42. ———, ———, ———, Robert S. MacLauchlan, and Norman Ritter. 1970b. Irrigated forages for western Nevada type climate. Nevada Agric. Ext. Serv. Circ. 106. 20 p.
43. ———, ———, Gayland D. Robison, Clinton Renney, Fredrick F. Peterson, and Norman Ritter. 1970c. Irrigated forages for southern Nevada type climate. Nevada Agric. Ext. Serv. Circ. 107. 13 p.
44. Joppa, L. R., and C. W. Roath. 1964. A comparison of Kenmont, Alta, and other tall fescue varieties in Montana. Montana Agric. Exp. Stn. Bull. 582. 11 p.
45. Kirillov, Yu. I. 1974. Tall fescue (*Festuca arundinacea* Schreb.). Tr. Prikl. Bot. Genet. Sel. 52:146–168.
46. Leasure, J. K. 1952. The growth pattern of mixtures of orchardgrass and tall fescue with ladino clover in relation to temperature. Proc. Assoc. South. Agric. Workers. 49:177–178.
47. Levy, E. Bruce. 1970. Grasslands of New Zealand. 3rd Ed. A. R. Shearer, Government Printer, Wellington, New Zealand. 374 p.
48. McGlohon, Norman E., J. N. Sasser, and R. T. Sherwood. 1961. Investigations of plant-parasitic nematodes associated with forage crops in North Carolina. North Carolina Agric. Exp. Stn. Tech. Bull. 148. 39 p.

49. McMurphy, W. E., Jr., L. M. Rommann, and B. B. Webb. 1975. Sarkeys research and development report. Oklahoma State Agric. Exp. Stn. Misc. Publ. 95. 31 p.
50. Mirasol, Jose Jison. 1920. Aluminum as a factor in soil acidity. Soil Sci. 10:153-217.
51. Moyer, J. L., and W. J. Seamands. 1975. Tall fescue. Wyoming Agric. Ext. Serv. Leaflet B-626. 2 p.
52. Neal-Smith, C. A., and L. G. Wright. 1969. Cool season performance of some tall fescue (*Festuca arundinacea*) lines at Canberra, A.C.T. Aust. J. Exp. Agric. Anim. Husb. 9:304-309.
53. Odum, Eugene P. 1959. Fundamentals of ecology. 2nd Ed. W. B. Saunders Co., Philadelphia. 546 p.
54. Rommann, L. M., and W. E. McMurphy. 1973. Tall fescue establishment and management. Oklahoma State Ext. Facts No. 2559. 4 p.
55. Schiller, J. M. A., and Alec Lazenby. 1975. Yield performance of tall fescue (*Festuca arundinacea*) populations on the northern tablelands of New South Wales. Aust. J. Exp. Agric. Anim. Husb. 15:391-399.
56. Scott, J. D. 1967. Advances in pasture work in South Africa. Herb. Abstr. 37:159-167.
57. Seamands, Wesley J., and B. J. Kolp. 1975. Regar bromegrass. Wyoming Exp. Stn. Rep. B-625. 2 p.
58. Shoop, G. J., C. R. Brooks, R. E. Blaser, and G. W. Thomas. 1961. Differential responses of grasses and legumes to liming and phosphorus fertilization. Agron. J. 53:111-115.
59. Sprague, V. G. 1943. The effects of temperature and day length on seedling emergence and early growth of several pasture species. Soil Sci. Soc. Am. Proc. 8:287-294.
60. Templeton, W. C., Jr., G. O. Mott, and R. J. Bula. 1961. Some effects of temperature and light on growth and flowering of tall fescue, *Festuca arundinacea* Schreb. I. Vegetative development. Crop Sci. 1:216-219.
61. ———, T. H. Taylor, and J. R. Todd. 1965. Comparative ecological and agronomic behavior of orchardgrass and tall fescue. Kentucky Agric. Exp. Stn. Bull. 699. 18 p.
62. Tesar, M. B. 1974. Nitrogen on grasses compared to alfalfa-grass mixtures in northern Michigan. Michigan State Agric. Exp. Stn. Res. Rep. 256. p. 101-104.
63. Turner, Darrell O. 1966. Yield, longevity, and response to lime of certain grass and legume varieties on Nisqually loamy sand. Washington Agric. Exp. Stn. Circ. 460. 13 p.
64. Van Keuren, R. W., V. F. Bruns, and D. D. Suggs. 1960. Forages for grazing and weed control in wet areas of the Columbia Basin. Washington Agric. Exp. Stn. Circ. 374. 12 p.
65. Whyte, R. O., T. R. G. Moir, and J. P. Cooper. 1965. Grasses in agriculture. FAO, UN. FAO Agric. Stud. No. 42 (3rd printing) 417 p.
66. Wolf, Dale D. 1973. Report of research with forage crops. Dep. Agron. Virginia Polytech. Inst. State Univ. 15 p.
67. Woodle, H. A., and E. C. Turner. 1949. Tall fescue. Clemson Agric. Coll. Ext. Circ. 345. 12 p.
68. Wood, Glen M. 1974. Evaluation of perennial ryegrass-tall fescue hybrids as forage for northern areas. Vermont Agric. Exp. Stn. 38th Annu. Rep. of Forage Research in the Northeastern U.S.
69. Zak, John M. 1967. Controlling drifting sand dunes on Cape Cod. Massachusetts Agric. Exp. Stn. Bull. 563. 15 p.

Chapter 3 Taxonomy, Morphology, and Phylogeny

EDWARD E. TERRELL

SEA-USDA
Beltsville Research Center
Beltsville, Maryland

INTRODUCTION

The genus *Festuca* includes approximately 80 species (Willis, 1973) occurring in temperate and cool regions of the world. In the United States there are about 20 species, all perennial (excluding *Vulpia*). Hitchcock's Manual (Hitchcock, 1951, rev. by Chase) included 13 additional annual species in the Section *Vulpia* of *Festuca*; however, recent taxonomic and floristic treatments have recognized *Vulpia* as a distinct genus (North American species were revised by Lonard and Gould, 1974). *Festuca, Vulpia,* and the related genus, *Lolium* (Terrell, 1968a), are members of the tribe Festuceae in the subfamily Festucoideae. Some earlier writers mistakenly placed *Lolium* in the tribe Triticeae (Hordeae). Although the contrast between a spike and a panicle makes *Lolium* and *Festuca* look superficially very different, there are close basic resemblances when florets and caryopses are compared; for example, Musil (1963) illustrated how florets of *L. perenne* and *F. arundinacea* differ only by minor morphological characteristics. Hubbard (1948) noted that *Lolium* has compound starch grains like other Festuceae instead of simple grains like Triticeae.

The outstanding classical taxonomic treatment of European species of *Festuca* was provided by Hackel (1882), who recognized 28 European species placed in six sections. In section *Ovinae* (now called Section *Festuca* because it includes the type species of the genus, *Festuca ovina*) he included *F. ovina* L. (sheep fescue) and *F. rubra* L. (red fescue), among other species. Section *Bovinae* comprised *F. gigantea* (L.) Vill. (giant fescue) and *F. elatior* L. (including tall fescue and meadow fescue).

Other taxonomic treatments of portions of *Festuca* were done by Piper (1906), Howarth (1928, 1948), and St. Yves (1922, 1927, 1929a, 1929b). There have been other published treatments for certain countries or regions by other authors. A number of the European taxonomic studies of the *Festuca ovina* and *F. rubra* species-groups have "split" species into varieties, subvarieties, and forms on the basis of differences in leaf anatomy in cross section. It is questionable whether this proliferation of minor taxa was justified, as many of these studies did not clarify the possible role of environment in leaf modification (see comments by Bor, 1970, p. 74–75).

Copyright © 1979 ASA-CSSA-SSSA, 677 South Segoe Road, Madison, WI 53711 USA. *Tall Fescue.*

NOMENCLATURE

The earliest Linnaean name applied to tall fescue and meadow fescue was *Festuca elatior,* described by Linnaeus in his *Species Plantarum* (1753). Subsequently *F. elatior* L. either included both tall and meadow fescues or only meadow fescue. The latter usage became common in the United States due to the influence of Hitchcock's *Manual of Grasses* (Hitchcock, 1935, 1951, rev. by Chase). Another name, *F. pratensis* Hudson, has been used in Europe for meadow fescue. A detailed study of Linnaeus's original description of *F. elatior* showed that it included both tall and meadow fescues and in addition one or two other taxa (Terrell, 1967). The Linnaean type specimen is actually tall fescue; therefore, if the type concept of the *International Code of Botanical Nomenclature* (Stafleu et al., 1972) is followed strictly, *F. elatior* would apply to tall fescue, and much more confusion would result. It seems best to reject the name *F. elatior* L. as an ambiguous name, as European taxonomists have been doing for some years. The next earliest name is *F. pratensis* (Hudson, 1762), which clearly referred only to meadow fescue.

The earliest name for tall fescue as a distinct species is *F. arundinacea*, described in 1771 by J. D. C. Schreber in *Spicilegium Florae Lipsicae,* p. 57, a local flora of the Leipsig area in Germany. Some later synonyms are: *Bromus arundinaceus* (Schreber) Roth (*Tentamen Florae Germanicae* 2: 141. 1789); *Festuca elatior* var. *arundinacea* (Schreber) Wimmer (Flora von Schlesien, ed. 3: 59. 1857); and *F. elatior* subsp. *arundinacea* (Schreber) Hackel (*Monographia Festucarum Europaearum,* 152. 1882).

In addition to the common name tall fescue, at least two other common names are mentioned in the American botanical literature—reed fescue and alta fescue—but the latter more properly refers to the cultivar 'Alta'. Names for tall fescue in other languages include: French—fétuque élevée; German—Rohrschwingel; Spanish—cañuela alta, festuca alta.

GEOGRAPHIC DISTRIBUTION

Festuca arundinacea Schreber has a wide native distribution in temperate and cool climates throughout Europe (north to 62° in Scandinavia), North Africa, and in west and central Asia and Siberia. It has been introduced in North and South America, Australia, New Zealand, and in south and east (Kenya) Africa.

The time of introduction of tall fescue to the United States is unknown. Cowan (1956) noted that the earliest collection of tall fescue in the United States was dated 1886. A recent perusal of the U.S. National Herbarium by the present writer turned up a still earlier collection: one by Isaac C. Martindale on ballast in Camden, N.J., in 1879. After 1890 there were several collections from various parts of the United States, apparently mostly from

TAXONOMY, MORPHOLOGY & PHYLOGENY

Fig. 3-1—*Festuca arundinacea*. Plant × 0.2. Caryopses (immature) × 2.4; florets including lemma (awned) and palea × 2.4; spikelet × 1.6.

cultivated plots. The first official plant introduction of *F. arundinacea* by the USDA was P.I. 5835 from Sweden in 1901 (P.I. records began in 1898); however, as Cowan noted, early collections were usually recorded as *F. elatior* and may have been either tall fescue or meadow fescue.

MORPHOLOGY AND ANATOMY

A brief description of *F. arundinacea* follows (see also details in key and Fig. 3-1, 3-2).

Perennial bunchgrass, tufted, with or without short rhizomes; culms usually erect, stout, smooth (or rough below panicle), to 2 m tall; ligules membranous, to 2 mm long; leaf sheaths smooth; leaf blades minutely scabrous or smooth, scabrous on margin, to 60 cm long, 3 to 12

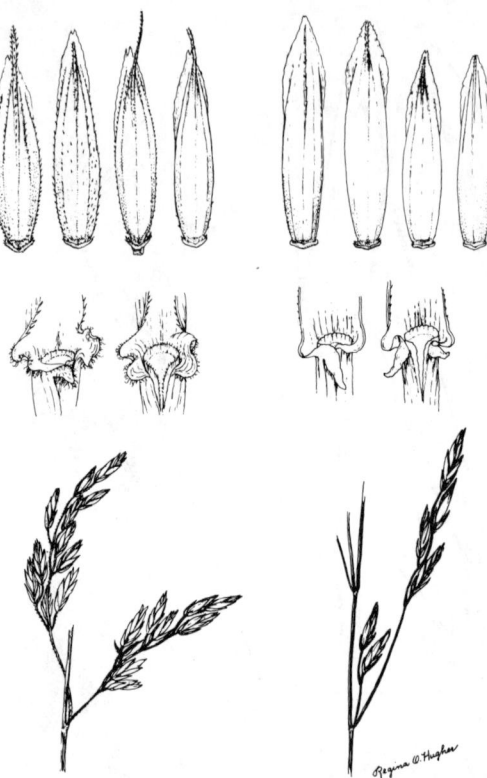

Fig. 3-2—Three key characters in *Festuca arundinacea* (left) and *F. pratensis* (right) comparing (top to bottom) scabrous, awned vs. glabrous, unawned lemmas; auricles and collars, ciliate vs. glabrous; shorter of a pair of branches bearing several vs. two spikelets. Lemmas × 2.4; auricles × 1.2; branches × 0.4.

mm wide; auricles absent or short, auricles and collars ciliate or sometimes glabrous; panicles 10 to 50 cm long, broad and loosely branched varying to rather narrow with short branches; spikelets elliptical to oblong, 10 to 18 mm long, each with 3 to 10 florets, rachilla internodes scabrous; lower glumes narrow-lanceolate, 3 to 6 mm long, one-nerved; upper glumes lanceolate to narrow-oblong, 4.5 to 7 mm long, three-nerved; lemmas narrowly elliptic or lanceolate, 6 to 10 mm long, five-nerved; often scabrous with minute prickles, or rarely glabrate, awns to 4 mm long or absent; paleas as long as lemmas; stamens three, with anthers 3 to 4.5 mm long; caryopses (grains) oblong, tightly enclosed by lemma and palea.

Anatomical data for *Festuca* species were summarized by Metcalfe (1960) from his own studies and from previous literature. The leaf blades of *Festuca arundinacea* are described by Metcalfe, as follows.

The lower epidermis has short cells paired or solitary; silica bodies roundish and projecting into adjacent cork cells; macrohairs and prickles present on upper epidermal surface; microhairs absent; stomata have low, dome-shaped subsidiary cells. In cross section the leaf blades have festucoid vascular bundles with double bundle sheaths and bulliform cells in fan-shaped groups. Culms have a large, central cavity with three circles of vascular bundles, and a well-defined peripheral ring of sclerenchyma. All of these characteristics are typically festucoid.

Additional data on root and development anatomy are also reported in Metcalfe (1960). Barnard (1964) illustrated a culm section and the lower epidermis of *F. arundinacea*. Konstantinova (1968) devised a key for determining Ukrainian fescues by anatomical characters.

The leaf of *F. pratensis* shows only minor differences from *F. arundinacea* (Metcalfe, 1960; Huon, 1965). Crowder (1956) compared certain characteristics of the leaf epidermis in *F. arundinacea* and *F. pratensis* (as *F. elatior*). He found *F. arundinacea* to have larger stomata than *F. pratensis*. On the lower epidermis the stomata were more numerous in tall fescue than in meadow fescue; however, on the upper epidermis they were about equally numerous. There were two to three rows of stomata adjoining the sclerenchymatous tissue in tall fescue but usually only one in meadow fescue. In both species the distribution pattern of the stomata on the upper epidermis was different from that on the lower. Badoux (1971) compared the leaf epidermis of *F. arundinacea, F. pratensis, Lolium multiflorum,* and their hybrids. He stated that *F. pratensis, L. multiflorum,* and their hybrid have long cells with straight cell walls and lack cork cells in the intercostal epidermis; *F. arundinacea* and the hybrid with *L. multiflorum* have sinuous cell walls and cork cells.

IDENTIFICATION OF TALL AND MEADOW FESCUES

Comparison of tall and meadow fescues reveals much variation and overlap in such characteristics as leaf width, number of spikelets per panicle, and length of panicle branches. A study was made of the distinguishing characters suggested by Crowder (1953, 1956) and other writers, and these were tested on many specimens. The resulting key (from Terrell, 1968b) is presented here, along with illustrations of three important characteristics (Fig. 3-2). The more significant differences are in italics in the key. Intermediates and hybrids will not be clearly identifiable and may have shrunken anthers or sterile pollen. Sulinowski (1968) stated that he and previous workers found ciliate auricles to be dominant in F_1 hybrids. *F. pratensis* var. *apennina* will not fit the key given here because it has awns up to 4 mm long. However, it is of restricted geographic distribution in Europe, and has only recently appeared among agricultural accessions (of the Welsh Plant Breeding Station).

Key to *Festuca pratensis* and *F. arundinacea*

Lemmas *(Fig. 3-2) glabrous, glabrate, or scabrous only at apex,* awnless or rarely short-awned; spikelets cylindrical to narrow-oblong; rachilla (axis of spikelet) internodes glabrous or nearly so; *shorter branch of each pair of panicle branches bearing only 1 to 2 spikelets (Fig. 3-2);* panicles 10 to 35 cm long; *basal leaf sheaths breaking up and often decaying into irregular dark-brown fibers;* leaf blades usually 2 to 6 (−8) mm wide; *auricles often long, curved; auricles and collars glabrous (Fig. 3-2);* plants 30 to 120 cm high. . . . . . *F. pratensis* Huds.

Lemmas (Fig. 3-2) scabrous or short-hispid with minute teeth all over or only on nerves or keels (visible at 10× or more magnification) or rarely glabrate, short-awned (to 4 mm long) or awnless; spikelets elliptic to oblong; rachilla internodes scabrous; *shorter branch of each pair of panicle branches bearing three or more spikelets (Fig. 3-2);* panicles 10 to 50 cm long; *basal sheaths tough, whitish to darkish, persistent;* leaf blades three to 12 (usually 4 to 6) mm wide, usually stiffer and more heavily nerved than blades of *F. pratensis; auricles absent or short; auricles* and *collars ciliate (Fig. 3-2) or sometimes glabrous;* plants 45 to 200 cm high, usually taller, more robust, and with thicker culms than *F. pratensis.* . . . .*F. arundinacea* Schreb.

TAXONOMIC RELATIONSHIPS

Taxonomic opinion has been divided about whether tall and meadow fescues are two separate species or one broad species. As treated by Hackel (1882), *F. elatior* was a broad, inclusive species containing subsp. *pratensis* (Hudson) Hackel and subsp. *arundinacea* (Schreber) Hackel. These, in turn, were each subdivided into three varieties and certain subvarieties.

Recent floras in the United States have had varied treatments. For example, Grays Manual (Fernald, 1950) listed them as one species, simply *F. elatior.* In contrast, Britton and Brown's Illustrated Flora (Gleason, 1952) recognized *F. elatior* var. *pratensis* (meadow fescue) and var. *arundinacea* (tall fescue). Hitchcock's Manual of Grasses (1935) recognized *F. elatior* (meadow fescue) and *F. elatior* var. *arundinacea* (tall fescue), but later Hitchcock (1951), rev. by Chase) elevated tall fescue to the rank of species.

Recent European floras have often considered them as two separate species, seemingly the best course to follow until much more is known about them. They differ in several partly overlapping characters (see key) and in chromosome numbers. The most widely distributed, typical race of *F. arundinacea* is hexaploid (2n=42), while *F. pratensis* is diploid (2n=14). That there are other infraspecific taxa is well-known; however, treatments of these in various floras is often incomplete and contradictory. Maire (1955) in his flora of North Africa recognized several infraspecific taxa, tending toward "splitting." A list of the vascular plants of central Europe (Ehrendorfer, 1973) records the following under *F. arundinacea:* subsp. *arundinacea;* subsp. *fenas* (Lag.) Arc. (*F. fenas* Lag.; *F. arundinacea* var. *glaucescens* Boiss.), and subsp. *uechtritziana* (Wiesb.) Hegi. Under *F. pratensis* he lists subsp. *pratensis* and subsp. *apennina* (De Not.) Hegi (var. *apennina* (De Not.) Hack.). To what extent these taxa overlap in morphology and in geography is not known at present.

TAXONOMIC AND PHYLOGENETIC IMPLICATIONS OF CYTOGENETICS

The results of breeding experiments within *Festuca* and between *Festuca* and *Lolium* are discussed elsewhere in this volume. F_1 hybrids have been produced artificially or naturally among: 1) species within *Festuca* Section *Bovinae*; 2) Section *Bovinae* × Section *Festuca (Ovinae)*, e.g., *F. arundinacea* × *F. rubra*; 3) Section *Bovinae* × Section *Scariosae*; 4) *Festuca* × *Vulpia*; 5) *Festuca* × *Lolium*.

Taxonomic implications of crosses between *Festuca* and *Lolium* were reviewed by Terrell (1966; see also Connor, 1968). The closest relationships are among the four agricultural taxa, *F. pratensis, F. arundinacea, L. perenne,* and *L. multiflorum,* as proved by numerous plant breeding experiments over a period of many years. This close relationship was further confirmed recently in a serological study of seed proteins (Jaworski et al., 1975). In a comprehensive review of crosses in various combinations and on various ploidy levels, Jauhar (1975) concluded regarding *F. pratensis, L. perenne,* and *L. multiflorum* that there is little structural differentiation among chromosomes of the three taxa and no effective barriers to gene exchange. On a genetic basis certain workers have advocated the union of *Festuca* and *Lolium.* This question was reviewed by Terrell (1966, 1968a) and discussed more recently by Jauhar (1975, 1976). The present writer's opinion is (essentially as advocated in Terrell, 1968a) that other species of *Lolium* and the several sections of *Festuca* need to be assessed as a whole on the basis of all available evidence before any definite conclusions should be reached.

Recent field collections and cytogenetic research by the staff of the Welsh Plant Breeding Station have contributed significantly to knowledge of the *F. pratensis-F. arundinacea* complex. The report of a 2n = 70 race of *F. arundinacea* (Levitsky and Kuzmina, 1927) was confirmed and additional ploidy levels were found (Borrill et al., 1971, 1976). *Festuca pratensis* var. *apennina* (2n = 28) was collected at several localities in Switzerland, and herbarium specimens were seen from Romania, Germany, Austria, Italy, and France. A tetraploid race (2n = 28) of *F. arundinacea* in France and Spain agreed in morphology with var. *glaucescens* Boiss. In the High Atlas Mountain region of Morocco two polyploid races were found: (1) 2n = 56 plants conforming to *F. arundinacea* var. *atlantigena* St. Yves forma *pseudomairei* Lit. & Maire; (2) 2n = 70 plants of *F. arundinacea* var. *letourneuxiana* St. Yves and var. *cirtensis* St. Yves, Thus, for *F. pratensis* a tetraploid variety is known in addition to the widespread diploid. For *F. arundinacea* there are 2n = 28, 56, and 70 races in addition to the widely distributed hexaploid (2n = 42). Cytogenetic studies show a close relationship between the Section *Bovinae* and the Section *Scariosae,* the latter represented by the diploid *F. scariosa* from Spain, and the tetraploid *F. mairei* St. Yves, from Morocco (Malik, 1967; Chandrasekharan and Thomas, 1971a, 1971b; Chandrasekharan et al., 1972; Borrill, 1972).

The polyploid series in the *F. pratensis-F. arundinacea* complex could have arisen by natural hybridization and chromosome doubling, according to Chandrasekharan et al. (1972). Chandrasekharan and Thomas (1971a) made the logical proposal that *F. pratensis* and *F. arundinacea* var. *glaucescens* could be the progenitors of the hexaploid *F. arundinacea.*

Future study of the taxonomic relationships of the *F. pratensis-F. arundinacea* complex should include detailed plant exploration in order to reveal exact geographic distributions and natural hybridizations among the component taxa. This could be followed by further cytogenetic experiments and taxonomic work using modern methods.

ACKNOWLEDGMENT

Thanks are expressed to Dr. Regina Hughes for preparing the illustrations used in this chapter.

LITERATURE CITED

1. Badoux, S. 1971. Sur l'anatomie de la feuille de *Festuca arundinacea* Schreb., *Festuca pratensis* Huds., *Lolium multiflorum* Lam. et leurs hybrides. Bull. Soc. Vaud. Sci. Nat. 71:15-22.
2. Barnard, C. (ed.). 1964. Grasses and grasslands. Macmillan and Co. Ltd., London. 269 p.
3. Bor, N. L. 1970. Gramineae (Lfg. 70). 573 p. *In* K. H. Rechinger (ed.) Flora Iranica. Akad. Druck-u. Verlagsansalt, Graz, Austria.
4. Borrill, M. 1972. Studies in *Festuca*. III. The contribution of *F. scariosa* to the evolution of polyploids in sections *Bovinae* and *Scariosae*. New Phytol. 71:523-532.
5. ———, B. Tyler, and M. Lloyd-Jones. 1971. Studies in *Festuca*. I. A chromosome atlas of *Bovinae* and *Scariosae*. Cytologia 36:1-14.
6. ———, ———, and W. G. Morgan. 1976. Studies in *Festuca* 7. Chromosome atlas (Part 2). An appraisal of chromosome race distribution and ecology, including *F. pratensis* var. *apennina* (De Not.) Hack.,—tetraploid. Cytologia 41:219-236.
7. Chandrasekharan, P., E. J. Lewis, and M. Borrill. 1972. Studies in *Festuca*. II. Fertility relationships between species of sections *Bovinae* & *Scariosae* and their affinities with *Lolium*. Genetica 43:375-386.
8. ———, and H. Thomas. 1971a. Studies in *Festuca*. 5. Cytogenetic relationships between species of *Bovinae* and *Scariosae*. Z. Pflanzenzucht. 65:345-354.
9. ———, and ———. 1971b. Studies in *Festuca*. 6. Chromosome relationships between *Bovinae* and *Scariosae*. Z. Pflanzenzucht. 66:76-86.
10. Connor, H. E. 1968. Interspecific hybrids in hexaploid New Zealand *Festuca*. N.Z. J. Bot. 6:295-308.
11. Cowan, J. R. 1956. Tall fescue. Adv. Agron. 8:283-320.
12. Crowder, L. V. 1953. A simple method for distinguishing tall and meadow fescue. Agron. J. 25:453-454.
13. ———. 1956. Morphological and cytological studies in tall fescue (*Festuca arundinacea* Schreb.) and meadow fescue (*F. elatior* L.). Bot. Gaz. 117:214-223.
14. Ehrendorfer, F. 1973. Liste der Gefässpflanzen Mitteleuropas. Gustav Fischer Verlag, Stuttgart. 318 p.
15. Fernald, M. L. 1950. Gray's manual of botany. American Book Co., New York. 1632 p.
16. Gleason, H. A. 1952. The new Britton and Brown illustrated flora of the Northeastern United States and adjacent Canada. Vol. 1. Lancaster Press, Lancaster, Pa. 482 p.
17. Hackel, E. 1882. Monographia Festucarum Europaearum. Th. Fischer, Kassel and Berlin. Reprint 1964, Johnson Reprint Corp. 216 p.
18. Hitchcock, A. S. 1935. Manual of the grasses of the United States. USDA Misc. Publ. 200. 1040 p.
19. ———. (rev. by A. Chase). 1951. Manual of the grasses of the United States. 2nd Ed. USDA Misc. Publ. 200. 1051 p.
20. Howarth, W. O. 1928. The genus *Festuca* in New Zealand. J. Linn. Soc. Bot. 48:57-77.
21. ———. 1948. A synopsis of the British fescues. Bot. Soc. Exch. Club Br. Isles Rep. 13: 338-346.
22. Hubbard, C. E. 1948. Gramineae. p. 284-348. *In* J. Hutchinson (ed.) British flowering plants. P. R. Gawthorn Ltd., London.

23. Hudson, W. 1762. Flora Anglica. London.
24. Huon, A. 1965. Caracteres epidermiques distinctifs des ssp. *arundinacea* (Schreb.) Hack. et *pratense* (L.) Hack. du *Festuca elatior* (L.) Hack. Bull. Soc. Bot. France 112:37-42.
25. Jauhar, P. P. 1975. Chromosome relationships between *Lolium* and *Festuca* (Gramineae). Chromosoma 52:103-121.
26. ―――. 1976. Chromosome pairing in some triploid and trispecific hybrids in *Lolium-Festuca* and its phylogenetic implications. p. 165-177. *In* P. L. Pearson and K. R. Lewis (eds.) Chromosomes today 5. John Wiley and Sons, New York.
27. Jaworski, A., S. Sulinowski, and E. Nowacki. 1975. Seed proteins of the *Lolium* and *Festuca* genera and the ability to produce alloploid hybrids. Genet. Pol. 16:271-275.
28. Kontstantinova, A. H. 1968. (Key for determining Ukrainian fescue species by anatomical characters). (Russ.). Ukr. Bot. Zh. 25:35-42.
29. Levitsky, G. A., and N. E. Kuzmina. 1927. Karyological investigations on the genus *Festuca*. Bull. Appl. Bot. Genet. Plant Breed. 17:3-36. (English summary p. 33-36.)
30. Linnaeus, C. 1753. Species plantarum. Vol. 1. Stockholm. 560 p.
31. Lonard, R. I., and F. W. Gould. 1974. The North American species of *Vulpia* (Gramineae). Madrono 22:217-230.
32. Maire, R. 1955. Flore de l'Afrique de Nord. Vol. 3. Paul Lechavalier, Paris.
33. Malik, C. P. 1967. Hybridization of *Festuca* species. Can. J. Bot. 45:1025-1029.
34. Metcalfe, C. R. 1960. Anatomy of the monocotyledons. I. Gramineae. Clarendon Press, Oxford. 731 p.
35. Musil, A. F. 1963. Identification of crop and weed seeds. USDA Agric. Handb. 219. 171 p.
36. Piper, C. V. 1906. North American species of *Festuca*. Contrib. U.S. Nat. Herb. 10:1-48.
37. Saint-Yves, A. 1922. Les *Festuca* (sub. Eu-Festuca) de l'Afrique de Nord et des Iles Atlantiques. Candollea 1:1-63.
38. ―――. 1927. Tentamen clavis analyticae Festucarum veteris orbis. Rev. Bret. Bot. 1927-1939:1-124.
39. ―――. 1929a. Contribution a l'étude des *Festuca* (Subgen. Eu-Festuca) de l'Amerique de Sud. Candollea 3:151-315.
40. ―――. 1929b. Contribution a l'étude des *Festuca* (Subgen. Eu-Festuca) de l'orient, Asie, et region mediterraneene voisine. Candollea 3:321-466.
41. Stafleu, F. A. (ed.). 1972. International code of botanical nomenclature. Int. Bur. Plant Taxon. Nomen. Utrecht, Netherlands. 426 p.
42. Sulinowski, S. 1968. Results of interspecific crosses *Festuca pratensis* Huds. (2n=14) × *Festuca arundinacea* Schreb. (2n=42) performed without applying emasculation. Genet. Pol. 9:77-85.
43. Terrell, E. E. 1966. Taxonomic implications of genetics in ryegrasses (*Lolium*). Bot. Rev. 32:138-164.
44. ―――. 1967. Meadow fescue: *Festuca elatior* L. or *F. pratensis* Hudson? Brittonia 19:129-132.
45. ―――. 1968a. A taxonomic revision of the genus *Lolium*. USDA Tech. Bull. 1392. 65 p.
46. ―――. 1968b. Notes on *Festuca arundinacea* and *F. pratensis* in the United States. Rhodora 70:564-568.
47. Willis, J. A. (rev. by H. K. Airy Shaw). 1973. A dictionary of the flowering plants and ferns. 8th Ed. Cambridge Univ. Press, London. 1245 p.

Chapter 4 Mineral Nutrition

S. R. WILKINSON
*SEA-USDA, Southern Piedmont
Conservation Research Center
Watkinsville, Georgia*

D. A. Mays
*TVA Soils and Fertilizers Branch
National Fertilizer Development Center
Muscle Shoals, Alabama*

INTRODUCTION

Tall fescue (*Festuca arundinacea* Schreb.) grows under a wide variety of soil, environmental, and management conditions. At times it has a reputation for unpalatability and low feeding value for livestock. The possibility of fescue toxicity has further tarnished its image. However, farmers have demonstrated that tall fescue is reliable, persistent, and productive when managed properly. Consequently, its contributions to the beef industry are extremely valuable. Tall fescue is widely used for conservation, recreation, and turf purposes, and; more recently, as vegetative cover on land used for application of liquid and solid waste.

The adaptability of tall fescue is demonstrated by its tolerance of poorly drained, cool soils, its ability to survive flooding as well as its high productivity on a variety of well-drained mineral and organic soils (Tesar and Shepard, 1963; Chamblee and Gilbert, 1965; Waddington and Zimmerman, 1972). Rogers and Davis (1973) found tall fescue to be less sensitive to waterlogging than orchardgrass (*Dactylis glomerata*), timothy (*Phleum pratense*), or perennial ryegrass (*Lolium perenne*). However, dry matter yields and tissue contents of K, N, and Mg were correlated positively with soil O_2 concentrations in the field and pot studies with waterlogged and well-drained soils. Tall fescue also has moderate salt tolerance (U.S. Salinity Laboratory Staff).

In this chapter, we review research relating to mineral nutrient requirements of tall fescue as well as effects of mineral nutrition on its utilization by livestock. Since most tall fescue is grazed, salient features of mineral cycling in grazed pasture ecosystems are also reviewed.

Copyright © 1979 ASA-CSSA-SSSA, 677 South Segoe Road, Madison, WI 53711 USA. *Tall Fescue.*

NITROGEN REQUIREMENT

A. Nitrogen Nutrition

The basic functions of N in higher plants have been known for many years. Briefly, N occurs primarily as a constituent of the various protein compounds in plants, accounting for 16 to 18% of the protein molecule (Meyer and Anderson, 1952). Protoplasmic proteins are part of the basic structural system and are necessary for cell division, whereas storage proteins are often found in stored foods, particularly seeds. Nucleoproteins occur in chromosomes and thus are involved in genetic reproduction and also are found in the cytoplasm. Nitrogen is also found in alkaloids and many of the intermediate metabolic products which are the building blocks of proteins. Besides the organic compounds listed above, N can be found as NO_3 in rather large amounts in plants (as much as 3%); small amounts of NO_2 or NH_4 are sometimes found.

Although most N reactions in tall fescue are similar to those in other green plants, three N-related phenomena—NO_3 and alkaloid accumulation and the effect of N fertilization on nonstructural reserve carbohydrate concentrations—deserve special attention.

B. Total Nitrogen and Nitrate Accumulation

Total N concentrations in cool-season grasses are strongly dependent on stage of maturity at harvest and on the N supply available to the plant. The available N supply depends on N fertilization, soil organic matter content, and rate of NO_3 release from the soil organic matter. Total N in foliage can range from less than 1.5% in mature tall fescue to greater than 5% in highly fertilized and frequently clipped immature tall fescue. Martin and Matocha (1973) indicated that N contents of less than 2.5% are deficient for tall fescue growth harvested at 5- to 6-week intervals. They suggest that critical N contents of tall fescue range from 2.8 to 3.4%. The adequate range is from 3.4 to 3.8%, and values greater than 4.0% are considered high.

Estimating protein content of forages by total N × 6.25 assumes that all tissue N is protein. This may be so for grasses grown with well-balanced fertility, but under high N fertilization or with a growth limiting deficiency of some other nutrient, tall fescue commonly contains significant amounts of nonprotein N (NPN). Teel (1966) found that applying 168 kg of N/ha increased NPN by 35%. The organic acids succinate and malate were increased sixfold, while fumarate was decreased 66%.

Increasing N rates commonly increases forage yield, with a modest increase in total N and true protein, until yields are near maximum, after which NPN forms, particularly NO_3, accumulate. Hojjati et al. (1972) reported this type of accumulation in May, July, and August when forage was

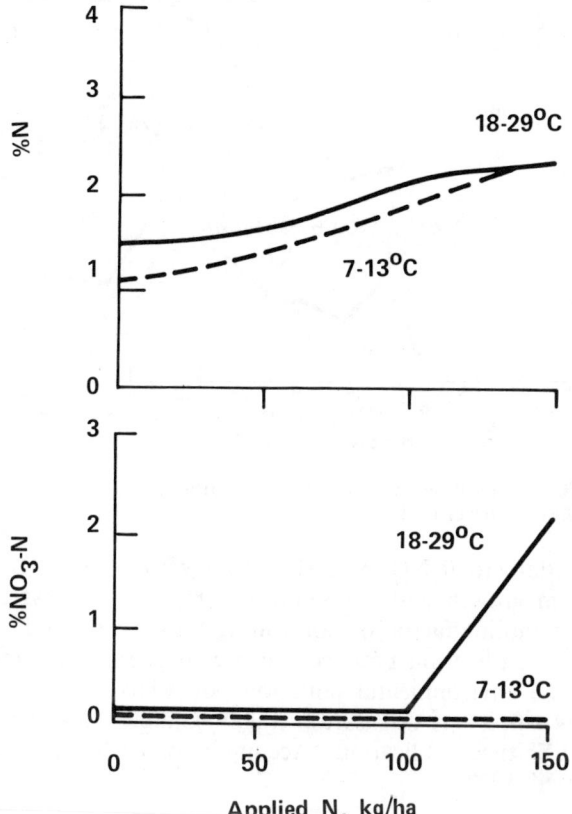

Fig. 4-1—Nitrogen and nitrate-N concentrations in tall fescue in response to four rates of applied N (Duncan et al., 1969).

harvested 30 days after N applications, while that which received no N during the growth cycle had uniformly low NO_3 levels. Nitrate-N accumulations of 0.86% have been observed in 'Kenwell' tall fescue fertilized with 18 metric tons/ha of broiler litter (S. R. Wilkinson, unpublished data). Similar levels of NO_3-N were observed in 'Kentucky 31' tall fescue.

Duncan et al. (1969) showed that total N increased with increased N fertilizer application rates. Nitrate increased sharply when the application rate was increased from 100 to 150 kg/ha at high temperatures, but did not increase under the low temperatures (Fig. 4-1). Maximum growth of tall fescue was obtained at 0.09 to 0.11% NO_3-N. Duncan et al. (1969) also cites research where 0.05 to 0.10% was sufficient for maximum growth of Italian ryegrass (*Lolium multiflorum* L.) and Van Burg (1962) reported no increase in dry matter yields when herbage exceeded 0.07 to 0.14% NO_3-N. Terman et al. (1976) observed concentrations of NO_3-N less than 0.1% up to 3.0% total N in tall fescue forage with NO_3-N increasing linearly thereafter up to 2% NO_3-N when total N levels reached 6% in the forage. Normally ac-

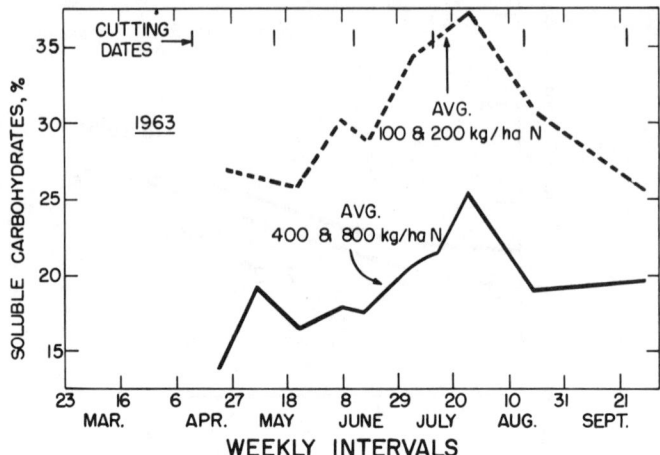

Fig. 4-2—Effect of N fertilizer on the mean soluble carbohydrate content of Kentucky 31 tall fescue stubble (Hallock et al., 1965).

cepted potentially toxic NO_3-N levels in forage are above NO_3 levels needed for maximum growth and, consequently; represent excess N supply, or some growth limiting factor(s). Controlling N supply to tall fescue swards is important in fertilization practice because of economics, animal toxicity problems, and environmental pollution potentials. Nitrate accumulation and its many aspects are discussed further in Chapter 13 and in the National Academy of Science publication "Accumulation of Nitrate" (National Research Council, 1972).

C. Nitrogen Effects on Soluble Carbohydrates

Nonstructural soluble carbohydrate relationships are particularly important in forage plants subjected to frequent defoliations. Carbohydrates supply the energy needed for initial regrowth when the leaf area is insufficient to provide the necessary photosynthetic activity. Reserve carbohydrates are also important in early spring growth of perennial plants.

Lechtenberg et al. (1972) found that N fertilization decreased sucrose and fructosan contents of tall fescue but did not affect the diurnal trends. They proposed that N deficiency results in accumulation of nonstructural carbohydrates because growth is restricted and photosynthate is poorly used.

Hallock et al. (1965, 1973) found that high N fertilization, 448 kg/ha or higher, resulted in lowered soluble carbohydrate levels during the growing season (Fig. 4-2), and severe stand thinning. McKee et al. (1967) found tall fescue stands markedly decreased by high N under a no-clipping pretreatment, but stands did not decrease on plots frequently clipped as a pretreatment. At Muscle Shoals, Ala., tall fescue cut biweekly throughout the summer tolerated higher N rates without stand loss than did swards cut

MINERAL NUTRITION

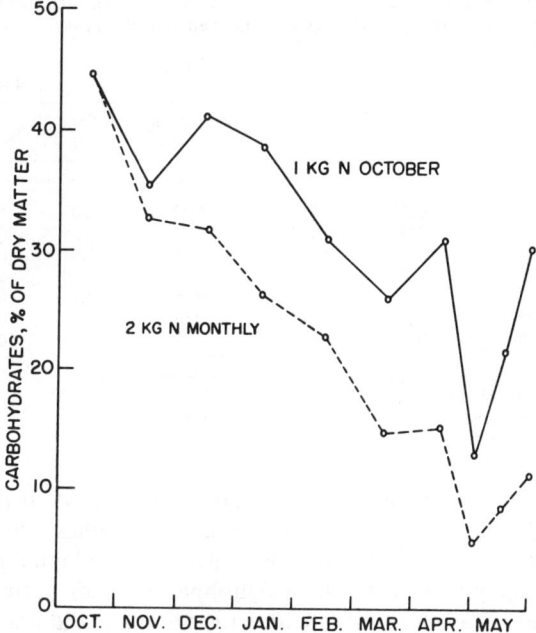

Fig. 4-3—Carbohydrate content of tall fescue stems as affected by low and high N fertilization (Powell et al., 1967).

three times annually (D. A. Mays, unpublished data). Powell et al. (1967) found that N in tall fescue managed as turf showed the same effects on soluble carbohydrates in winter as in summer (Fig. 4-3), and that adequate N kept tall fescue green throughout the winter.

D. Nitrogen Effects on Alkaloids

Gentry et al. (1969) found that perloline, the predominant alkaloid in tall fescue, can be increased in concentration from four to eightfold by N fertilization. Conversely, perloline levels were decreased by P and K fertilization. Perloline content of both Kenwell and Kentucky 31 tall fescue increased more with increased N fertilization in August than in November in the Southern Piedmont (Table 4-1). Perloline and NO_3-N both accumulated at the August sampling, but NO_3-N accumulated more so than perloline at the November sampling. These data suggest an interaction between perloline accumulation, N fertilization, and season.

Two other alkaloids found in tall fescue, N-acetylloline and N-formylloline did not increase in response to increasing levels of N fertilization in field plot studies (Table 4-1, and Robbins et al., 1973). However, in grazed tall fescue pastures receiving 224 kg N/ha/year as NH_4NO_3 or 789 kg N/ha/year as broiler manure, perloline, and N-acetylloline plus N-formylloline

Table 4-1—Interaction between date of sampling and fertility treatment on nitrogenous compounds in tall fescue (average of Kentucky 31 and Kenwell) Watkinsville, Ga.

Fertility treatment	Date of sampling						
	26 Aug. 1969				29 Nov. 1969		
	Total N[†]	NO$_3$-N[†]	Perloline	N-acetylloline[‡] N-formylloline	Total N	NO$_3$-N	Perloline
kg N/ha	———————————————— % of dry matter ————————————————						
112	2.24	0.007	0.024	0.12	2.03	0.009	0.002
224	2.40	0.034	0.029	0.11	2.25	0.008	0.012
448	2.67	0.020	0.059	0.09	3.04	0.028	0.029
896	3.28	0.128	0.135	0.07	3.59	0.072	0.023
298§	2.61	0.008	0.055	0.09	2.78	0.013	0.016
597§	2.96	0.029	0.075	0.09	3.63	0.146	0.018
1,195§	3.46	0.074	0.123	0.06	3.79	0.382	0.012

[†] Total N and NO$_3$-N unpublished data, W. A. Jackson and S. R. Wilkinson.
[‡] Alkaloid levels from Robbins et al., 1973.
§ Actual N input from 9, 18, and 36 metric tons broiler manure/ha/year.

both increased at the higher N-fertilization level. Levels of these alkaloids were lowest from January through April, and highest during July and August (Robbins et al., 1973). Nitrate levels were also high during late summer. That N-acetylloline and N-formylloline alkaloids accumulated in response to higher N application rate under grazing and did not accumulate in harvested field plots with increasing N fertilization suggests that alkaloid levels are also affected by management conditions.

E. Nitrogen Effects on Other Nutrients

Nitrogen availability to plants may affect the concentration of other macro- and micronutrients. The actual nutrient concentrations depend on soil availability relative to total plant need of N and other nutrients. Nutrient concentrations are affected by NO$_3$ or NH$_4$ ions in the soil solution by dilution from increased growth response to adequate N, or by an insufficiency of N for maximum growth and use of other nutrients in growth. Effects of N on mineral concentration will depend on the soil's mineral status and the amounts of applied nutrients. For example, Duell (1965) found that NH$_4$NO$_3$ applied alone lowered K concentrations in tall fescue, while applying a complete fertilizer (10–4.4–8.3) did not. However, Reid et al. (1970) found that N fertilization of tall fescue increased tissue K concentrations in tall fescue when soil K was adequate. Analyses of tall fescue clipped every 2 weeks in April and May showed that N fertilization increased K concentration but had little effect on Ca and Mg levels (D. A. Mays and G. L. Terman, unpublished data).

The effects of three N application rates on concentrations of nine minerals in tall fescue grown in West Virginia are shown in Table 4-2 (Reid and Jung, 1965). They found that K concentrations were increased 50% or more by N fertilization, while P, Ca, and Mg levels remained constant or in-

Table 4-2—Concentrations of nutrients in tall fescue as affected by applied N.†

Mineral element	N applied (kg/ha)							
	0	56	168	504	0	56	168	504
	First cutting†				Aftermath†			
	% of dry matter							
K	2.29	3.02	3.32	3.49	2.18	3.28	3.24	3.40
P	0.34	0.35	0.34	0.32	0.37	0.30	--	0.26
Ca	0.50	0.57	0.59	0.64	0.62	0.65	0.62	0.61
Mg	0.17	0.21	0.21	0.19	0.23	0.24	0.26	0.26
	µg/g/dry matter							
Mn	75	82	74	69	93	95	81	90
Cu	17	13	15	15	15	14	14	14
Zn	28	30	30	29	22	26	25	25
Mo	0.9	1.0	1.3	0.4	0.6	0.8	1.2	1.1
Co	1.1	1.2	1.2	0.2	0.5	0.5	1.4	1.8

† Reid and Jung (1965).

creased only slightly in the first cutting and subsequent aftermath forage. Applied N had little effect on Mn and Zn; effects on Cu, Mo, and Co were fairly large, but inconsistent among N rates and growth stages.

Rinne et al. (1974) found that N fertilization initially increased the K contents of orchardgrass and meadow fescue (*Festuca pratensis*, Huds.). This increase in K diminished until the third growing season when increasing N fertilization resulted in decreased K concentration in association with declining soil K. Calcium and Mg content increased with increased N fertilization, while P content decreased. Nitrogen fertilization decreased soil pH resulting in probable soil fixation losses of P at the lower soil pH. Whitehead (1972) summarizes this effect of N on K content as depending greatly on the K supply since N tends to increase herbage K content when K supplies are plentiful and to depress it when K supplies are limited. Therefore, it is difficult to generalize the effect of N fertilization on other element composition unless supplies of other nutrients are also assessed. Nutrient interactions between soils and plants have been recognized and are briefly discussed by Allaway (1971). Extreme care is necessary in designing experiments and collecting and interpreting experimental results when studying effect of N on uptake of other nutrients.

F. Nitrogen Effects on Feeding Value

Nitrogen effects on NO_3, crude protein, amino acids, and true protein, all of which may affect feeding value, were discussed earlier. Several other organic constituents are also important to forage quality. Miaki (1969) found that doubling the N rate increased the various N and protein fractions but did not affect crude fiber, lignin, or digestible organic matter concentrations. Reid and Jung (1965) found that cellulose, acid detergent fiber, lignin, and cell wall constituents were not greatly different at 56, 168, or 504 kg N/ha; however, with no added N, cellulose, fiber, and cell wall constituents were somewhat higher. Protein digestibility was improved significantly by N fertilization, while cellulose and total dry matter digestibility were changed only slightly. Grimes (1967) found that increasing the N application from 58 to 580 kg/ha decreased the cellulose and water-soluble carbohydrates, but did not affect lamb weight gains. George et al. (1972) reported that increasing increments of N fertilizer up to 168 kg/ha did not influence average daily gain of steers grazing tall fescue.

G. Yield Responses

Nitrogen effects on yield are interrelated with forage management systems and these effects are discussed in Chapter 10. However, the effects of various N-application schedules, sources, and rates on yield and N-use efficiency receive some emphasis here.

In practice, tall fescue is usually fertilized with low rates of N (100 kg of N/ha or less annually) for several reasons:
1. Farmers have traditionally considered pasture or hay as a low-value crop; thus, they cannot justify heavy fertilization;
2. They often rely on legumes present in the sward to provide N;
3. Utilization of the high spring yields from heavy fertilization requires better management than most farmers are able to provide because of lack of time, resources, or understanding.

The prediction of tall fescue response to N is complicated by the presence of a legume in the sward. Typically, small N applications merely substitute for the N contribution of the clover and cause no net increase in yield, while heavy N applications tend to suppress or eliminate legumes and may result in an eventual decrease in yield, unless heavy N fertilization is continued. Frame (1973) found that N response of a mixed tall fescue-clover sward was rectilinear at N rates above the contribution of the clover. Templeton and Taylor (1966) reported that spring N applications increased the spring yield and decreased the fall yield of a tall fescue-clover sward.

Practices which eliminate legumes from a mixed grass-legume sward may decrease overall forage quality and average daily gains or milk production of grazing cattle. Gains per hectare, however, may be increased if fertilization and management increases carrying capacity. Fertilizer N applications on grazed tall fescue swards can increase carrying capacity and greatly increase yields of livestock products if management systems effectively utilize the increased forage yield (Mott et al., 1971).

Many N rate studies have involved maximum N rates of 150 to 250 kg/ha (Leamer, 1963; Mays and Terman, 1969; Duell and Trout, 1972). In this range, response to N is almost always rectilinear unless other environmental factors severely limit yield. Hallock et al. (1965), Planquaert (1973), and Cooper et al. (1962) have reported positive yield responses to annual N applications of 400 kg/ha or more, although yield responses began to level off at the highest N rates and stand reductions were sometimes reported.

MINERAL NUTRITION

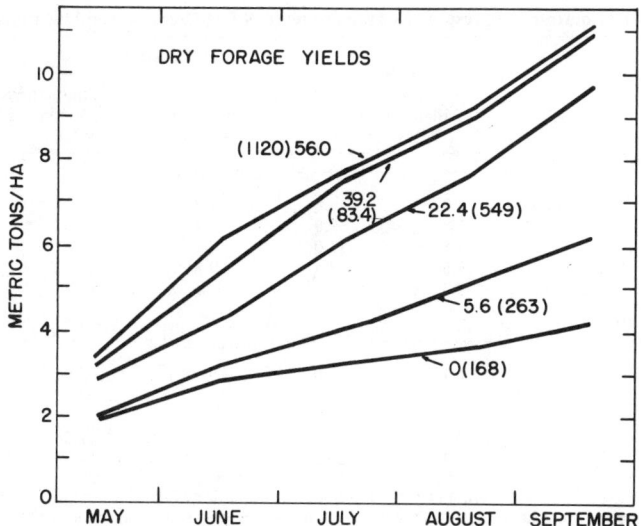

Fig. 4-4—Cumulative yields from five harvests of tall fescue fertilized annually with 168 kg/ha of N in March plus several rates in 17 weekly applications May through August, Holland, Va., 1966 to 1970. Values in parentheses are total rates (in kg/ha) of N applied annually (Hallock et al., 1973).

The data from Hallock et al. (1973) (Fig. 4-4) are an example of response to extremely high N rates. In the Georgia Mountain region yields were maximum at 448 kg N/ha/year as NH_4NO_3, but yields increased 2 metric tons/ha when N was increased from 448 to 896 kg N/ha/year in the Southern Piedmont region (Table 4-3). Tall fescue fertilized with broiler manure did not respond above 27 metric tons/ha when plots were cut three to five times per year at 5 to 8 cm height. When these same plots were cut six times per year at 2.5 to 4 cm height, yields were appreciably higher and an increased yield response was observed with application of broiler manure above 27 metric tons/ha. Kentucky 31 and Kenwell tall fescue had the highest yields at 36 metric tons/ha broiler manure at Watkinsville, Ga. Stand reductions were found at 896 kg N/ha/year as NH_4NO_3 treatments in both the Southern Piedmont and Mountain locations. Stand reductions did not occur at the high manure rates in the Mountain region, and were only slight in the Southern Piedmont. These high yields of tall fescue under broiler manure fertilization with six close clippings per year have been confirmed in other studies with and without irrigation at the Southern Piedmont location (S. R. Wilkinson and R. N. Dawson, unpublished data).

The magnitude of the N response, as well as the seasonal distribution of forage may depend strongly on the N application schedule. Late summer and fall yields were annually highest when N was applied in several split applications or when delayed-release N source was used (Mays and Terman, 1969). However, total annual yields were frequently better when all N was applied in early spring, particularly if soil moisture and other growing conditions are more favorable in spring and early summer than in fall. Results

Table 4-3—Dry matter yield response of tall fescue to N fertilizer at several locations.[†]

Fertilizer or N treatment	Experimental location				
	Mountain, Blairsville, Ga. Ky 31[†]		Slat Belt, Saluda, S.C. Ky 31[§]	Southern Piedmont, Watkinsville, Ga.	
				Ky 31[¶]	Kenwell[¶]
kg N/ha/year	metric tons/ha				
Control	4.0	6.8	1.7	1.2	1.3
112	5.8	9.9	4.4	4.6	3.3
224	7.8	12.0	7.1	7.4	6.7
448	8.6	13.4	9.8	10.3	9.6
896	8.7	11.4	--	12.2	11.5
Poultry manure tons/ha/year	metric tons/ha				
9	6.1	11.0	--	5.0	4.6
18	--	--	--	8.5	7.7
27	9.5	18.1	--	--	--
36	0	--	--	11.3	11.4
54	9.1	20.2	--	--	--

[†] S. R. Wilkinson, R. N. Dawson, and J. W. Dobson, Jr., unpublished data.   [‡] Hayesville sandy clay loam, (a clayey, oxidic, mesic Typic Hapludult) uniform application of about 45 kg P/ha/year and 86 kg K/ha/year to all plots. The first column represents total of three or four harvests per year with a sickle-bar mower at 5- to 8-cm height (1969, 1970). The second column represents yields of six harvests per year with a flail type harvester at 2.5- to 4-cm height (1971, 1972). Poultry manure rates are on a fresh basis.
[§] Georgeville silt loam (1967–1970). All N treatments also received 98.6 and 186 kg/ha/year P and K, respectively.   [¶] Cecil sandy clay loam (1968–1970). All N treatments received 98.6 and 186 kg/ha/year P and K per year, respectively, except the 896 kg/ha/year treatment which received 336 and 561 kg/ha P and K, respectively. Poultry manure rates are on dry matter basis.

in Table 4-4 show greater fall yields with split application even though total yields were higher with spring application of N.

Maximizing yields with heavy spring N applications causes difficulties in effectively utilizing the growth by grazing, and problems in making hay when haymaking weather is often poor. Because of uncertainty in many areas of good late summer and fall rains, high fall growth yields from split fertilizer applications are not always obtained.

Henderlong and Street (1974) indicate that tall fescue grown in a modified continuous flow culture system grew better with NO_3-N than with NH_4-N at equivalent rates[1]. Shoot N content did not differ greatly, but total N uptake was increased two to threefold for the plants supplied with NO_3-N. Studies where NH_4 and NO_3 source were switched indicated a reversible effect of NH_4 toxicity on plant growth. These results are interesting, but we do not know of field comparisons of NH_4-N and NO_3-N sources that substantiate this apparent greater efficiency. However, Zagallo and Bollen (1962) in a study of the rhizosphere of tall fescue found the densities of bacteria responsible for ammonification and nitrification in association with tall fescue roots to be lower than with other plants. This may partially explain the results of Henderlong and Street.[1]

Mays and Terman (1969) found that N uptake by tall fescue from NH_4NO_3 was not significantly different than from nitric phosphate, but was

---
[1] P. R. Henderlong and J. R. Street. 1974. Comparative effects of nitrogen source and rates on the development and chemical composition of tall fescue. Plant Physiol. Suppl. 32 (Abstr.).

Table 4-4—Yields of tall fescue as affected by N application rate and schedule.†

N treatment	Total yield (kg/ha)	
	Spring applied N	N split five times
kg/ha		
122	5,006 cd*	4,371 d
243	7,139 b	6,462 bc
487	9,136 a	7,412 b
Harvest date	Daily yield at highest N rate (kg/ha)	
5 Apr.	86	42
6 May	57	32
14 June	58	54
12 July	52	29
19 Aug.	32	15
20 Sept.	19	31
31 Oct.	9	26
15 Nov.	7	16

* Duncan multiple range test at 5% level of significance, means of the column that do not have a letter in common differ significantly. † Mays, 1974, unpublished data.

Table 4-5—Effect of N-P-K fertilization on mass and distribution of Kentucky 31 tall fescue roots in Georgeville silt loam.†

N-P-K treatment			Depth (cm)				Total
			0 to 15	15 to 31	31 to 61	61 to 152	
kg/ha/year			dry matter, kg/ha				
0	0	0	1,966	291	188	126	2,572
112	25	93	2,452	406	442	298	3,598
112	99	186	2,504	426	597	513	4,040
224	25	93	3,191	344	328	238	4,101
224	99	186	3,341	369	555	372	4,637
448	25	93	3,562	312	368	373	4,615
448	99	186	4,161	248	296	395	5,100

† Root cores were taken September 1971, after 4 years of treatment. All plots received a complete micronutrient mix. (S. R. Wilkinson, unpublished data).

higher than from sulfur coated urea (SCU), urea, and urea ammonium phosphate. Yields of tall fescue from SCU were similar to those from NH_4NO_3 at 84 and 168 kg N/ha/year in two of these experiments. Apparent N recoveries for 84 and 168 kg N/ha were 68 and 62% for SCU, 66 and 62% for urea, and 79 and 80% for NH_4NO_3. Kaempffe and Lunt (1967) found that N recovery reached a maximum for 'alta' tall fescue harvested 4 to 6 weeks after fertilization. Without conclusive information to the contrary, tall fescue appears to respond to different N sources like other cool-season grasses.

Tall fescue root mass was almost doubled in plots receiving 448-99-186 kg N-P-K/ha/year as compared with unfertilized tall fescue growing in a slate belt soil in South Carolina (S. R. Wilkinson, unpublished data, Table 4-5). There were also increases in root mass with increments of 112 and 224 kg N/ha/year. Nitrogen fertilization with P and K fertilization increased root mass over unfertilized treatments at all depths sampled. Fertilization with 112 kg N/ha slightly increased the proportion of total roots below 15 cm, but at the highest fertilization level this proportion was decreased.

H. Nitrogen Cycling

The N cycle describes the transfers, transformations, accumulations, and N losses within the ecosystem under different managements. Whitehead (1970) reviews transformations important to understanding and promoting effective N use in grassland ecosystems. Ecosystems are communities of organisms which have defined boundaries which are usually open with respect to one or more properties and are characterized by influxes and outfluxes of energy, nutrients, and matter. In improved pasture ecosystems, we are concerned with young, productive growth systems, and how man, as a dominant organism, must manage them to maintain productivity.

Important features of the N cycle are the amounts of N in each compartment and the rate of movement, or flux, between compartments. Nitrogen turnover rates between compartments are important in the maintenance of pasture productivity. Nitrogen becomes deficient when flux between compartments does not meet the minimum for growth. An era where nutrients are not conserved or used more than once is coming to a close because of plant nutrient costs and concern over environmental quality. We are also gaining a better understanding of N and other nutrient cycles. The soil contains various sources, or pools of N, including mineral N, as well as soil organic matter and plant and animal residues. These sources vary in availability of N for plant uptake. Organic N sources must be mineralized to either NH_4 or NO_3-N forms, before the N becomes available for plant uptake. The rate at which this mineralization takes place depends on C/N ratios. Residues containing more than 1.8% N mineralize rapidly, while those containing less than 1.2% N mineralize slowly. Therefore, N fertilization may increase turnover rates of organic N. The nutrients in the ungrazed plant recycle to the soil. Grazed swards seldom have above 60% utilization efficiency (Minderhoud et al., 1974). This finding implies a residue's return of 40% to soil pools directly from plant sources in grazed situations. In harvesting situations, utilization efficiencies are much higher and the N recycled is reduced. Nitrogen also may be redistributed and reused (internal recycling) in ungrazed plants with increasing maturity.

The animal consumes N in grazed herbage, converts it to protein, and excretes in the urine, feces, and in belched gases what is not incorporated into its tissues. Nitrogen distribution between feces and urine was shown to be about 0.8 g N/100 g dry matter consumed in the feces with the remainder of the N excreted in the urine (Barrow and Lambourne, 1962). This distribution suggests that the proportion and amount of N excreted in urine increases as the N content of the grass increases. Since the N excreted in urine is mostly in urea and amino N forms, it is readily available for plant uptake. However, much of the N excreted in feces must undergo mineralization before it is available for plant uptake. In terms of pasture fertilization, N excreted in urine is readily recycled, with a higher proportion of the N excreted in a readily available form as N status of the sward is increased. However, this form of N excretion may result in greater losses through NH_3

volatilization. Stewart (1970) indicated up to 90% N loss under laboratory conditions when urine is placed on dry soil and water was allowed to evaporate. Australian workers (Denmead et al., 1974) have found an average daily flux of N to the atmosphere of 0.26 kg N/ha in a 3-week test on a pasture stocked with 50 sheep/ha. These losses, as well as denitrification losses should be studied in tall fescue pastures. Such N losses are weather and soil type dependent as are losses of N from the pasture by leaching.

The distribution of excreta over space and time in the grazed sward is very important in assessing the effectiveness of recycled N. Wilkinson and Lowrey (1973) suggested that N return in urine is much more effective than N return in feces because of the greater effective area per excretion of urine. Residual effects of N from feces would be expected to be greater than for urine. Lotero et al. (1966) reported that the tall fescue yield response to each urination decreases quite rapidly. The importance of earthworms and dung beetles in increasing plant and animal residue mineralization has been recognized (Mott, 1974).

Losses of N in livestock products in grazed pastures are relatively small. For example, a livestock weight gain of 500 kg/ha means a harvested product loss of about 14 kg N/ha, whereas 8,000 kg/ha of herbage (containing 2.5% N) removed means a loss of 200 kg N/ha.

Nitrogen does not recycle completely because of losses from volatilization, leaching, runoff, immobilization of organic N in residues, nutrient removal in livestock, and uneven distribution of excreta over the pasture area. The magnitude of these losses must be assessed for each location. However, appreciable N may be recycled in grazed pasture systems. Some experiment stations have recognized this recycling of N and are making lower N recommendations for grazed situations than for haymaking or forage harvesting situations.

Studies involving N fertilization of Kentucky 31 tall fescue pastures stocked at 0.4 ha/cow-calf pair indicate similar levels of calf weight production per hectare at 77, 224, 789 kg N/ha/year rates with no advantage to rates greater than 224 kg N/ha/year (Stuedemann et al., 1975).

The question of fertilizer N, or biologically fixed N is important for tall fescue pastures. Templeton and Taylor (1966), Templeton et al. (1965) indicate how effective clovers can be in supplying N for tall fescue/ladino clover associations. Literature surveys suggest that a mixture containing 40% or more clovers will result in dry matter production equivalent to that obtained with grass alone fertilized with 67 to 224 kg N/ha/year. When legumes supply the N through symbiotic fixation, the N cycle most nearly resembles the ideal closed cycle. The question of fertilizer N vs. legume N becomes one of legume adaptability, persistence, and reliability in the different regions where tall fescue is grown.

The ability of legume-free Kentucky 31 tall fescue sods to accumulate N has been demonstrated by Giddens et al. (1971), who found an increase of 0.041% N in topsoil after 5 years of unharvested tall fescue, fertilized annually with 34 kg N/ha. After 5 years, accumulated tall fescue tops yield was 10,928 kg/ha and root yield was 11,631 kg/ha containing 1.4 and

0.99% N, respectively. Total N in the tops and roots was 273 kg N/ha. Reddy and Giddens (1975) found that the increase in total N in the soil crust after 4 years of tall fescue (0.06 to 0.18%) was 32 kg/ha/year, assuming that the 1-cm depth crust with 1.0 bulk density contains most of the accumulated N. In grasslands in warm, humid regions with low fertilizer N application where the soil is not completely shaded and maintained at pH 6.5 or above the N accumulation by blue-green algae may be significant. These observations support the probable importance of nonsymbiotic N fixation in the practical N nutrition of tall fescue pastures.

Smith (1974) reported evidence for appreciably more soil absorption of atmospheric N in Florida than that contributed directly in the rainfall. Leachates, from rains passing through fine sands contained 45 to 56 kg N/ha/year, suggesting this much N was absorbed from the atmosphere. Certainly, if NH_4 and related gaseous N compounds exist in the atmosphere, acid soils are effective absorbers. Hutchinson et al. (1972) demonstrated that plants can absorb significant amounts of NH_3 from the air, even at naturally occurring low atmospheric concentrations.

SULFUR REQUIREMENT

A. Nutrition

The literature on S nutrition and yield responses contains very little information concerning tall fescue. Observed S deficiency is infrequent in areas where much tall fescue is grown. The common tall fescue management systems that include grazing put a minimum stress on S supplies because of recycling. Thus, there has been little pressure for S research with tall fescue. Metson (1973) reviewed 125 papers on the S nutrition of forage crops. Much of the information which he summarizes on the S status of other cool-season grasses should also apply to tall fescue.

Sulfur in plants occurs in protein as sulfate and in low molecular weight organic compounds such as free amino acids, glycosides, mercaptans, and sulfides. When S is more limiting to plant growth than N, almost all the S is in proteins. The nonprotein forms occur in abundance only when N or another growth factor is limiting and S is surplus. When sulfate is present it has no apparent physiological function except to serve as a reserve of available S for assimilation. Grasses are more vigorous sulfate accumulators than clovers and may often have an appreciable sulfate content when clovers growing in the same location have almost none. Most of the organic S is contained in the amino acids, cystine and methionine, which are components of the protein fraction.

Much has been written about N:S ratios in forages, both from a plant growth standpoint and for animal nutrition. Metson (1973) stated that within a given plant species the protein fraction has an almost constant N:S ratio. Thus, a low N:S ratio in the plant suggests a surplus of S with insufficient N for maximum protein production. Conversely, the more common-

Table 4-6—Effect of N rate on N:S ratio in tall fescue forage, average of three locations in Kentucky.†

	Rate of spring applied N (kg/ha)			
	0	45	90	135
	N:S ratio			
Cut 1	7.6	9.4	11.2	13.6
Cut 2	6.5	5.3	6.1	7.1

† K. L. Wells, 1975, unpublished data.

ly occurring high N:S ratio may indicate an over abundance of various non-protein N compounds with insufficient S for maximum protein. A N:S ratio of about 14:16 seems the most desirable range from a plant productivity standpoint, and approximates the ratio of 10:1 to 15:1 reportedly effective for N utilization by cattle and sheep (Allaway and Thompson, 1966). Under conditions of low S supply, N accumulation can increase this ratio considerably.

Analysis of Kentucky 31 tall fescue samples from three locations in Kentucky (Table 4-6) showed that the N:S ratio at the first cutting increased from 7.6 to 13.6 as the spring N application increased from 0 to 136 kg/ha with no S applied. The lower N:S ratio at the second cutting shows the effect of the decreased N supply caused by crop removal. Nitrogen fertilization after the second cutting widened the ratio again in the third cutting forage (data not shown), (K. L. Wells, Univ. of Kentucky, unpublished data, 1975).

B. Yield Responses

Dry matter yield responses to S fertilization have been shown with a wide variety of crops in many parts of the world. Sulfur behaves like most other plant nutrients in producing a major crop response only when it is the sole nutrient which is in short supply. Plant uptake is mostly in the SO_4 form.

Among forage crops, the greatest number of responses to S fertilization have been reported with alfalfa and the various clovers, since they are less aggressive accumulators of sulfate than the grasses. Orchardgrass, the ryegrasses (*Lolium* spp.), and various annual grasses have produced higher yields with S fertilization at a number of locations. Soils very low in available S are more common in the Pacific Northwest than in other regions of the United States (Allaway, 1975).

The only S fertilization experiment with tall fescue which has come to our attention was conducted at Prosser, Wash. (A. I. Dow. 1975. Washington State Univ., personal communication). Applying 56 kg/ha S more than doubled the average yield of the second, third, and fourth cuttings of both Alta and 'Fawn' tall fescue and halved the N:S ratio (Table 4-7). Similar results might be obtained at other locations where soil S supplies are limited.

Table 4-7—Effect of S fertilization on yield and N:S ratio of tall fescue at Prosser, Wash. Average yield of second, third, and fourth cuttings.†

S rate	Yield	N:S ratio
kg/ha	kg/ha	
0	1,194	20.0
56	2,421	10.3

† A. I. Dow, 1975, unpublished data.

The lack of apparent need for S fertilization of tall fescue relates to the favorable contribution of S in precipitation, and the ability of soil to absorb S. Bertramson et al. (1950) estimated that Indiana soils may absorb as much as 30 kg S/ha/year. When 11.2 kg/ha/year are carried down from the atmosphere with precipitation, the occurrence of S deficiency is unlikely (Whitehead, 1964).

The importance of S as a needed plant nutrient for tall fescue may increase in the future because of 1) increased usage of high-analysis, low-sulfur content fertilizer materials; 2) increased crop yields where harvested for hay; 3) greater emphasis on control of air pollution and reduction of S emissions into the air.

C. Sulfur Cycling

The critical S concentration for tall fescue has apparently not been established. Metson's review (1973) however, indicated that 0.30% in plant tissue is the level needed for near maximum yields of several grasses. If one assumes a tall fescue yield of 8,000 kg/ha, this S concentration would result in the removal of 24 kg/ha of S if the crop were removed. Live-weight of steers, calves, and sheep contain about 0.14% S, unwashed wool about 3.5% S, and cows's milk about 0.42% S (from Wilkinson and Lowrey, 1973). Therefore, a calf weight gain of 500 kg/ha means a harvested product loss of about 0.8 kg/ha. Animals can utilize inorganic S in synthesizing protein. Apparently there is no detrimental effect on animal health or performance of excess S within wide ranges.

Barrow and Lambourne (1962) found that about 0.1 g of S was excreted in the feces with each 100 g of feed eaten, with the remainder of the S excreted in urine (after retention by the animal). The proportion of consumed S excreted in the urine then depends on the S content of the grass. As the S content of the diet increases, the proportion and amount excreted in the urine increases.

Wilkinson and Lowrey (1973) describe S cycling in a hypothetical tall fescue pasture ecosystem. A series of papers by Till, May, and co-workers at the Pastoral Research Laboratory in Armidale, Australia describe field studies of ^{35}S kinetics in pastures grazed by sheep (May et al., 1973).

PHOSPHORUS REQUIREMENT

A. Nutrition

Black (1957) discussed the functions of various organic and inorganic forms of P in green plants. Organic forms are found in structural and storage compounds and are important constituents in some intermediate metabolic compounds. The P containing storage compounds include phytin, the principle P form in seeds and phospholipids which also occur in seeds and nucleoprotein. Nucleoproteins are present in chromosomes and change during mitosis; thus, by inference P is important in cellular reproduction.

In metabolic processes, P plays a direct role as a carrier of energy. The compound adenosine triphosphate, as well as some other P-containing groups of lesser importance, contain high energy phosphate bonds which, when broken by hydrolysis, supply the energy needed for various carbohydrate transformations. Phosphorus is also involved in the photosynthetic reaction of CO_2 with five-C sugars containing P to form phosphoglyceric acid.

The inorganic P in plant tissue serves as a pool to supply phosphate ions that enter the metabolic cycle and is a repository for those ions that leave the cycle. Thus, a deficiency of inorganic P in the plant reduces the rate of metabolic processes.

B. Animal Requirements

Reid and Jung (1974) who discussed the importance of adequate P for herd health, reproductive efficiency, and meat or milk production indicated a marked lack of agreement in research findings on the minimum P concentrations needed in the ration. The National Research Council requirements are given as 0.18% of the ration for pregnant cows and 0.18 to 0.23% for lactating beef cows; however, there were instances where less than one-half of these P levels in forage had little apparent adverse effect on grazing animals. Phosphorus concentrations, which are above the "critical" level for maximum plant growth, should meet the animal requirement. Because of the simplicity of mineral box supplementation of animals, fertilization to increase P concentrations in forage above sufficiency levels for plant growth is usually not economical.

C. Yield Responses

Areas of P deficiency in the United States were delineated by Beeson (1945). Since then a Federal cost sharing program for more than 20 years has provided fertilizers for pasture and hay establishment. Federal and

Table 4-8—Yield of Alta tall fescue as affected by level of soil and applied P.†

HCO₃ extractable P in soil	P added (ppm)		
	0	6.6	66
ppm	g dry matter/pot		
2.8	2.65	3.25	4.76
4.2	4.35	5.26	6.28
4.6	2.72	4.03	5.46
8.2	5.21	5.20	6.18
13.8	5.14	5.30	6.08
17.5	5.47	5.92	6.33

† Lunt et al. (1966).

Table 4-9—Concentration of P in Alta tall fescue forage as affected by soil and applied P.†

Soil P status	P added (ppm)		
	0	6.6	66
	% P		
Average low P soils	0.26	0.30	0.39
Average of high P soils	0.35	0.36	0.46

† Lunt et al. (1966).

state agencies and fertilizer companies have also conducted extensive educational programs to promote the use of fertilizers on all crops. The net result is that P deficiency is now extremely localized. Many fields are now so well supplied with P that only maintenance applications may be necessary for many years. Thus, P should be applied only on the basis of soil test results and fertilization history on a field-by-field basis.

Lunt et al. (1966) grew Alta tall fescue on six soils with HCO_3-extractable P levels ranging from 2.8 to 17.5 ppm. Their data (Table 4-8) clearly showed the relationship between soil test P and plant response to applied P. Yield response to the highest rate of applied P (66 ppm) was less than 19% when soil test P was 8.2 ppm or greater, but was 80% at the lowest soil P level. They found that 66 ppm of applied P increased tissue P 50% on low P soils, but only 31% on high P soils (Table 4-9).

Lunt et al. (1966) estimated that about 50% of maximum yield was obtained in Alta tall fescue when tissue P levels were 0.19% P; for maximum yields 0.35% P was needed. Martin and Matocha (1973) list the following P ranges for tall fescue cut at 4- to 5-week intervals: Deficient, less than 0.24%; critical, 0.26 to 0.32%; adequate, 0.34 to 0.40%; high, more than 0.45% P.

Usually P deficiency is extreme only in disturbed soils such as strip mines and highway cuts where there is no topsoil. Mays and Bengston (1974) grew several cool-season forages on a coal mine spoil in northeast Alabama, a site so P-deficient that an application of 56 kg/ha of P at planting was necessary for winter survival and subsequent growth. Application of 28 kg P/ha was insufficient to allow development and winter survival of satisfactory cover, even though no forage was removed. Phosphorus responses of this magnitude are now extremely rare on U.S. agricultural land.

Table 4-10—Effect of P and K fertilizer on percent clover, yield, and N removal of tall fescue/ladino clover associations in Kentucky (Tilsit silt loam) (1955-58 average).†

Fertility treatment N-P-K	Clover content (average for 12 harvests over 4-year period)	Total herbage production	Total N obtained over 4 years from		
			Fescue	Clover	Total
kg/ha/year	%	———————— kg/ha ————————			
0- 0- 0	27.9	4,260	289	198	489
0-98-130	38.8	7,399	497	472	969
Increase from P & K	10.9	3,139	208	274	480

† Templeton et al. (1965).

Neller and Hutton (1957) found that on Marlboro fine sandy loam (a clayey, Kaolinitic, thermic Typic Paleudult) Kentucky 31 tall fescue obtained more P from superphosphate placed at the soil surface at 15-cm spacings than from applications placed at 7.5 and 15 cm below the soil surface at 30.5-cm spacings.

If P is applied to mixed grass-legume swards, the grass may benefit indirectly. When Wolfe and Lazenby (1973) applied superphosphate to a tall fescue-clover sward at rates up to 375 kg/ha only the clover responded the first year. However, in subsequent years, tall fescue responded as much as the clover to additional increments of P. Apparently, the P increased the clover sward's vigor which in turn supplied greater amounts of N to the grass.

Templeton et al. (1965) found that annual applications of P and K to a tall fescue/ladino clover association on Tilsit silt loam (a fine-silty, mixed, mesic Typic Fragiudult) increased clover content, total herbage production, and total N obtained in the tall fescue and clover over the 4-year period (Table 4-10). Total N obtained in grasses and legumes increased 120 kg/ha/year from the application of 98 kg/ha of P and 130 kg/ha of K. The increase in N obtained in tall fescue was 52 kg/ha/year. Thus, their data aptly demonstrates the importance of adequate P and K supply for legumes and N fixation. Some of the response shown in Table 4-10 was probably due to P since the authors state that the Tilsit soil was very low in available P.

There was also a 58% increase in dry matter yield from P and K when the tall fescue-clover association was fertilized with 135 kg N/ha/year. Peters and Lowance (1974) reported that mowing treatments and application of P and K nearly eliminated broomsedge (*Andropogon virginicus* L.) from a heavily broomsedge-infested tall fescue-legume pasture. Their results confirm farmer observation that broomsedge invasion in tall fescue pastures was caused by poor soil fertility and grass management.

D. Phosphorus Effects on Other Nutrients

Reports on the effects of P fertilization on inorganic and organic constituents of tall fescue are not common. Reid and Jung (1965) show that 220 kg of applied P/ha on a high P soil had relatively little effect on the concen-

Table 4-11—Mineral and organic constituents in tall fescue as affected by P fertilization.†

Constituent	First cutting		Aftermath	
	No P	220 kg P/ha	No P	220 kg P/ha
	% dry matter			
K	2.29	2.50	2.18	2.08
P	0.34	0.38	0.37	0.37
Ca	0.50	0.46	0.62	0.62
Mg	0.17	0.17	0.23	0.19
	µg/g dry matter			
Mn	75.00	62.00	93.00	113.00
Cu	17.00	14.00	4.00	3.00
Zn	28.00	26.00	22.00	22.00
Mo	0.90	0.70	0.60	0.80
Co	1.10	1.70	0.50	0.40
	% dry matter			
Cellulose	31.10	31.90	29.90	30.40
Acid detergent fiber	33.60	34.70	36.50	36.80
Acid insoluble lignin	3.20	3.00	4.00	4.60
Cell wall constituents	65.20	66.10	65.10	64.70
Soluble carbohydrates	8.80	13.00	11.70	10.50

† Reid and Jung (1965).

tration of any inorganic constituent in either first cutting or aftermath forage (Table 4-11). However, Cu and Co concentrations were much lower in aftermath forage. Of the organic constituents only soluble carbohydrates in first cutting forage were markedly increased in concentration, considered a quality improvement due to P fertilization.

E. Phosphorus Cycling

Phosphorus is considered an immobile nutrient and leaching losses are neglible except on very coarse, sandy soils. Phosphorus may be immobilized in organic materials containing less than 0.2%, as well as fixed in acid soil as Fe and Al compounds, or in calcareous soils as calcium phosphates. Neither P or K are volatilized and except for the action of plants are not recycled.

Phosphorus removal is a function of the yield level, P concentration in the harvested forage, and disposition of the crop (hay or grazing). A yield of 8,000 kg/ha containing 0.30% P would remove 24 kg P/ha under a hay management system. However, with 50 to 60% of available forage consumed, about 14 kg P/ha would actually be ingested. Phosphorus removal in 500-kg live-weight gain would be about 3.5 kg P/ha. Production of 13,451 kg milk/ha/year would remove about 14 kg P/ha/year. Phosphorus not contained in the animal body is mainly excreted in feces with about 0.06 g organic P excreted per 100 g feed eaten; the remainder is excreted as inorganic P (Barrow and Lambourne, 1962). This suggests that the higher the P content of the grass, the greater the inorganic P content of the feces. Generally, only trace amounts of P are excreted in the urine. Inefficiency in P recycling may occur because of uneven return of feces over the pasture area.

Nutrients contained in the feces in general are not as readily available for plant uptake as those in urine, and less pasture area may be directly affected by dung pats (Wilkinson and Lowrey, 1973).

POTASSIUM REQUIREMENT

A. Nutrition

The primary role of K in plants is metabolic. Potassium is not a constituent of intermediate organic compounds within plants and is not found in the structural components (Black, 1957). *The Role of Potassium in Agriculture* (Kilmer et al., 1968), contains an extensive review of the functions of K in plants.

Jackson and Volk (1968) reported that photosynthetic CO_2 fixation is repressed by inadequate K, while both light and dark respiration rates are increased. These effects can decrease dry matter production before visual deficiency symptoms appear. Potassium appears to be necessary for the transformation of simple sugars to complex carbohydrates in transphosphorylation reactions, in activity of enzyme systems, and in translocations of carbohydrates (Liebhardt, 1968). Sufficiency of K seems necessary to maintain good water relations in plants. Transpiration is higher from K deficient plants in sunlight but the opposite is true under cloudy conditions. The leaf cuticle layer is reduced in thickness or entirely missing with severe K deficiency. This may be the reason for the transpiration changes and this may predispose plant tissue to fungal attacks. Potassium may be the major cation involved in the ion influx which raises guard cell osmotic concentration (Fischer, 1968).

Several symptoms of K deficiency appear to be the direct result of its effect on carbohydrate relationships in the plant. Poor stalk strength in corn and small grains, as well as parenchyma cell breakdown with concurrent disease organism entry, are due to thin cell walls, and are probably a direct result of K deficiency-induced failure in converting simple sugars to long chain structural carbohydrates.

Important K-deficiency effects in corn (*Zea mays* L.) or small grain production may not be equally important in tall fescue, but decreased net CO_2 fixation rate and failure to convert simple sugars to complex carbohydrates will adversely affect forage yield and, thus, be important economically. Blaser and Kimbrough (1968) reported that Kentucky 31 tall fescue under K deficiency stress became dark green, grew slowly, wilted, and leaves later turned brown along edges and the apex. Stands seemed depleted during winter.

B. Animal Health

Forage K levels associated with good plant growth exceed the animal dietary requirements of 0.5% in the ration. Recently, however, Pfander and Rubic (1972) found that the K content of tall fescue harvested after January in Missouri contained 0.4% K, indicating deficiency for ruminants. Dead leaf tissue contained 0.47% K, while new growth tall fescue contained 4.48% K on a dry matter basis in late February and early March pasture samplings at Watkinsville, Ga. (S. R. Wilkinson, unpublished data). Potassium content of stockpiled, or deferred tall fescue, which has been frozen and weathered, may be deficient for ruminants.

High K has been implicated in grass tetany through an effect on availability of herbage Mg to ruminants (Fontenot, 1973). The capability of tall fescue to absorb K in excess of needs has been generally accepted with levels in excess of 6% reported (Miller, R. W. 1974. Effect of deficient nutrient solutions on the concentrations of 16 mineral elements in clippings of Kentucky 31 tall fescue. p. 209. *In*: Proc. of the 2nd Int. Turfgrass Res. Conf. Abstr.). Some nutritionists feel that excessive K uptake, leading to a K/(Ca+Mg) ratio (meq basis) greater than 2.2 has been associated with an increased incidence of grass tetany. Attempts to relate grass tetany incidence to this ratio have met with limited success. More recently, high N and high K have been implicated in the etiology of grass tetany. Stuedemann et al. (1975) report seasonal distribution of K in tall fescue from a high in early March, associated with the onset of spring growth, to a low associated with frozen, weathered tissue in January and February at Watkinsville, Ga. Considerable variation in K content exists with soil type, season, climate, soil water, and temperature of soil and air.

C. Yield Responses

Crop response to applied K depends on the supply of available K at a given soil site. This includes exchangeable K measured by soil test and in most soils includes some of the reserve K (non-exchangeable K). Sources of soil K include past fertilization, native soil minerals, return of plant residues, and animal excreta.

Different soils with similar exchangeable K levels may vary greatly in capacity to support plant growth without K fertilization. For example, a fertile Sango silt loam (a coarse-silty, siliceous, thermic Glossic Fragiudult) at Muscle Shoals, Ala. supported only 1 year of alfalfa production without deficiency symptoms and stand loss when K was not applied annually (D. A. Mays, unpublished TVA data, 1965). In contrast, R. E. Blaser and H. T. Bryant (personal communication, 1962) grew alfalfa for 5 years on Chester silt loam (a fine-loamy, mixed, mesic Typic Hapludult) in northern Virginia without yield response to applied K because an adequate supply was continually being released by the soil minerals. Joy et al. (1973) found that tall fescue took up 520 kg K/ha over 3 years with only 270 kg of applied K/ha. They attributed the difference to reserve K and possible uptake from deeper depths.

Potassium contents of tall fescue, listed as deficient, critical, adequate and high by Martin and Matocha (1973), were less than 2.2%, 2.5 to 2.8%, 3.0 to 3.5%, and greater than 4.0%, respectively. Data from Blairsville, Ga.

suggests that about 2% K was necessary for near maximum growth of Kentucky 31 tall fescue (Wilkinson, unpublished data, 1975). Clement and Hopper (1968), working with various cool-season grasses, found they could obtain 90% of maximum yield by harvesting at either grazing or silage stages when herbage contained no more than 2% K in the dry matter. Concentrations of K less than 1.5% have been associated with moderate to severe K deficiency. Potassium concentration in excess of 2.5% probably represents luxury consumption.

Duell and Trout (1972) showed the variation in K concentration that could exist with only minor effects on yield. In this experiment the major yield response was to N, but K uptake tripled from zero to the highest K application rate. Duell (1965) reported that yield response to applied K decreased N concentrations in grasses; however, this may have been the result of N dilution caused by increased growth.

Soils with relatively low available K, such as coarse textured soils, can be depleted rapidly by a fast growing crop of heavily N-fertilized tall fescue. An 8,000 kg/ha yield of tall fescue containing 2% K removes 160 kg K/ha. If the soil has a cation exchange capacity (CEC) of 2 meq/100 g and 5% of this exchange capacity is K, the exchangeable K level is 87 kg/ha. Unless this soil has a high reserve K with a rapid release rate, K deficiency will be likely, unless K fertilizer is frequently applied. This illustrates the importance of frequent K applications on soils with low CEC, or slow reserve K release rates. Phosphorus and K are considered immobile in the soil (Bray, 1954). For ions whose movement in soil is relatively rapid, like NO_3, the soil requirement would approximate the amount removed by the crop, or the crop and its production level would dictate its requirement. However, for immobile nutrients with low tendencies to move in soil, soil requirements are almost independent of crop removal, and are determined by the crop's ability to absorb sufficient nutrient to satisfy the growth demand. Consequently, the concept has been formulated to applying N to meet production levels, while P and K are applied to the soil as indicated by soil test to supply intensity—replenishment capacity.

Phosphorus and K are often applied together in mixed fertilizers. This is often justified by a need for both. Fuller et al. (1971) found yields of tall fescue on Taloka loam (a fine, mixed, thermic Mollic Albaqualf) fertilized at 717 kg N/ha/year to be increased from 4,459 to 8,195 kg/ha by fertilizing annually with 79 kg P/ha and 149 kg K/ha. They concluded that 79 and 74 kg P and K were adequate. There was a complete tall fescue stand loss at 717 kg N/ha rate without P and K fertilization. Fuller et al. (1971) found that tall fescue yields increased as N was increased to 717 kg N/ha whenever P and K were adequate.

Higher soil K levels are needed to support a legume-grass sward than for grass alone because clovers and alfalfa are less competitive for K, particularly under conditions of low fertility (Blaser and Kimbrough, 1968). Under conditions of K stress, legumes are invariably lower in K concentration than the associated grasses. Blaser and Kimbrough (1968) also reported that there are differences among grasses in competitiveness for soil K. Tall fescue was more aggressive than Kentucky bluegrass at low K fertility levels; at high K levels of orchardgrass was the most aggressive.

D. Potassium Cycling

The amount of fertilizer K needed for maximum tall fescue yield appears to be dependent on the amount and rate at which soil K can be supplied to the crop and on the amount of K recycled. A yield level of 8,000 kg/ha containing 2.0% K would require 160 kg of K/ha/year from soil and fertilizer sources. Approximately 50 to 60% of the available forage may be ingested under grazing. Seventy to 90% of the K may be returned to the soil in the urine. The K concentrations of calves, steers, lambs, sheep, unwashed wool, and cow's milk are about 0.17, 0.15, 0.14, 0.13, 4.66, and 0.11%, respectively. Therefore, removal of 500 kg calf gain would constitute a loss of 0.9 kg/ha, or a neglible amount. Losses to drainage water, soil, and imperfect redistribution of excreta by livestock would likely be appreciably greater. Mott (1974) and Wilkinson and Lowrey (1973) contain a fuller discussion of K cycling in grazed pastures.

LIME, CALCIUM, AND MAGNESIUM REQUIREMENT

A. Lime

Lime requirement relates to the control of soil pH (hydrogen ion concentration) and to the tolerance of tall fescue to acidic or basic soil conditions. Spurway (1941) indicated that tall fescue grew best when soil pH ranged from 6.5 to 8.0. Rampton (1950) observed that the lower limit of soil pH for growth of Alta was 4.7, and the upper limit was 9.5. Vogel and Berg (1968) recommend tall fescue for reclamation of acid strip-mine spoils only when the soil pH is greater than 4.5. Fleming et al. (1974) found that tall fescue tolerated low nutrient solution pH (4 to 5) and Mn (4 to 64 ppm Mn), but that 4 ppm Al in the nutrient solution severely inhibited tall fescue top and root growth. The Al sensitivity of tall fescue was associated with relatively large decreases in Ca and Mg concentrations and increases of K and P concentrations in tall fescue.

In spite of tall fescue's tolerance to a wide range of soil pH, there are several reasons for maintaining soil pH values in the 6.0 to 7.0 range. Among these are the high soil pH requirements for possible legume companion crops, P compounds are less soluble at low pH, and CEC of soils may decrease and reduce soil capacity to hold cations from leaching. Perhaps more importantly, the rate of nitrification decreases greatly below a soil pH of 6.0 and becomes neglible below pH 5.0. An acid soil may have low N mineralization rates from plant and animal residues, and low nutrient turnover rates. Consequently, N requirements may be increased. Tall fescue swards, even without legumes, should have soil pH's maintained above 6.2

for these reasons. The actual amounts of lime required to maintain this soil pH is best determined by a soil test correlated for the specific soils and localities.

B. Calcium and Magnesium

No specific tall fescue growth response to Ca or Mg is known except for some unpublished data obtained at Blairsville, Ga. In this study, small, but statistically significant yield responses (about 0.7 metric ton/ha were obtained from annual liming of tall fescue with 2.24 metric tons/ha of dolomitic limestone (S. R. Wilkinson, R. N. Dawson, and J. W. Dobson, Jr., unpublished information, 1974). Upon examination of individual clipping yields, the statistically significant responses were usually found early or late in the growing season. Whether the response was caused directly by improved Ca or Mg nutrition, or indirectly by improved soil conditions resulting from liming is not known. There was a trend toward a yield response to dressings of 112 kg/ha as MgO.

Embleton (1966) indicated that a shoot Mg concentration of 0.1% was sufficient for maximum grass growth. DeWit et al. (1963) reported critical levels for Ca and Mg in grasses to be about 0.1 and 0.06%, respectively. Experimentally determined critical levels of Ca and Mg for tall fescue are not known.

Many tall fescue pastures are used for cow-calf operations. As a consequence of this usage pattern and the climates where tall fescue is used, grass tetany is an economic problem. The problem is brought into perspective by the relatively low Mg requirements for plant growth and the much higher concentration in the forage generally required to prevent grass tetany (0.2 to 0.25% Mg). Wilkinson et al. (1973) and Grunes et al. (1970) reviewed the prospects of preventing grass tetany by Mg fertilization.

Magnesium concentrations in tall fescue vary greatly with season (Stuedemann et al., 1975). Not only does season of year affect Mg concentrations, but so may stage of maturity, available Mg and K supplies in the soil, soil balance between K, Ca, and Mg, soil texture and mineral content, soil temperature, and soil moisture. Elkins et al. (1978) also found variation in Mg concentration of tall fescue ecotypes grown at 2% and 21% soil oxygen. Magnesium levels of all genotypes were greatly reduced at the 2% level of soil oxygen.

Fertilization of tall fescue with NH_4NO_3 increased Mg concentrations when K inputs were held low and constant across all N levels. Fertilization with broiler litter had little effect on Mg concentrations, but increased K concentrations considerably (S. R. Wilkinson, unpublished data, 1974). When available soil K levels exceeded 160 kg K/ha (by dilute double acid extraction methods) Mg concentrations were relatively constant at about 0.17% Mg without Mg fertilization, and about 0.20% with annual additions of 112 kg Mg/ha/year as either MgO or $MgSO_4$. Annual surface applications of dolomitic limestone at 2.24 tons/ha for 4 years did not increase

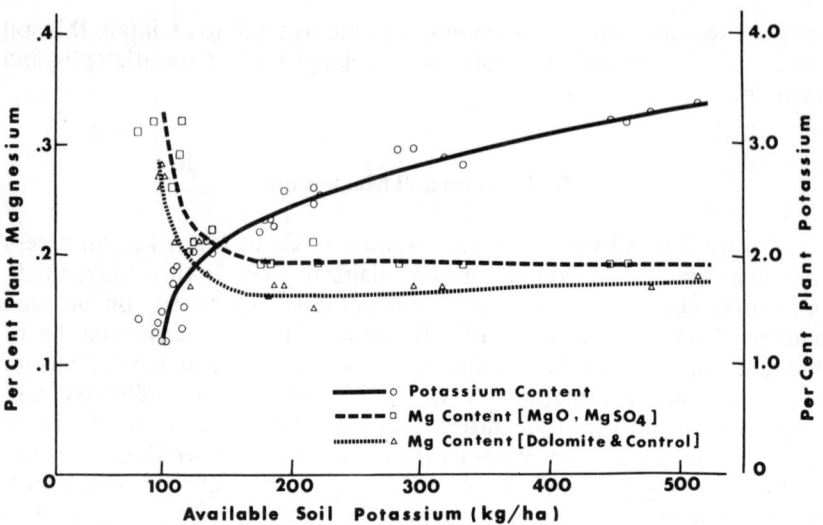

Fig. 4-5—Association between K and Mg concentrations in early spring growth of Kentucky 31 tall fescue and available soil K (dilute-double acid extraction). (S. R. Wilkinson, unpublished data, 1974).

Mg concentrations of tall fescue. At available soil K levels of about 160 kg/ha or less; Mg concentrations in tall fescue increased rapidly to above 0.3% Mg (S. R. Wilkinson, unpublished data, Fig. 4-5). However, K in tall fescue did not reach 2% until soil levels exceeded about 140 kg K/ha. These results suggest that if K concentrations are to be maintained at 2% or higher, that relatively large inputs of Mg will be required to raise Mg levels to above the 0.2 Mg level.

Price and Moschler (1970) found that dolomitic limestone applications up to 54 tons/ha lowered Mg and increased Ca concentrations in tall fescue. Gross (1973) ranked tall fescue as a high Mg accumulator among various cool-season species studied in the greenhouse, and as an intermediate Mg accumulator in field studies. Legumes were ranked as high Mg accumulators. He reported that Mg content of tall fescue on a Hagerstown soil (a fine, mixed, mesic Typic Hapludalf) from Pennsylvania and Gilpin soil (a fine-loamy, mixed, mesic Typic Hapludult) from West Virginia fertilized with 112 kg Mg/ha/year increased 11 and 31%, respectively. Different soils are almost certain to vary in Mg fertilization response, particularly if they differ in K and Mg status.

DeGroot (1970) observed that the omission of K fertilization on one farm in the Netherlands prevented grass tetany. Pasture fertilization practices need to be reassessed from the perspective of grass tetany. Certainly, producers should avoid K applications before or during the grass tetany season because of the ability of grasses like tall fescue to absorb K in excess of their requirement for maximum yield, and the limited evidence that high K enhances the probability of grass tetany (for further discussion of grass tetany see Chapter 13).

Table 4-12—Response of tall fescue to microelements and sewage sludge treatment (average of eight sampling dates).†

Treatment description†	Yield	K	Cd	Cr	Cu	Pb	Zn
	metric tons/ha	%	μg/g dry matter				
Check	4.8 b*	1.2 c	2 c	5 c	10 c	12 c	25 b
N-P-K	12.0 a	2.1 a	2 c	5 c	10 c	12 c	25 b
N-P-K + ME	10.8 a	1.9 b	45 b	32 c	122 b	341 b	130 b
Sewage sludge	6.2 b	1.2 c	28 b	150 b	117 b	410 b	1,419 a
Sewage sludge + ME	5.8 b	1.3 c	107 a	196 a	306 a	847 a	1,622 a

* Duncan multiple range test at 5% level of significance, means of the column that do not have a letter in common differ significantly.   † Data from Boswell (1975). N-P-K applied was 224–94–202 kg/ha. Sewage sludge (5.6 metric tons/ha) supplied 378 kg/ha of N, 20 kg/ha of K, 3 kg Cd/ha, 30 kg Cr/ha, 11 kg Cu/ha, 46 kg Pb/ha, and 198 kg Zn/ha. Microelement application (ME) supplied 20 kg Cd/ha, 15 kg Cr/ha, 25 kg Cu/ha, 25 kg Pb/ha, and 50 kg Zn/ha from soluble sources. Microelements all supplied together (not separately).

C. Cycling of Calcium and Magnesium

When an 8,000-kg/ha yield of tall fescue containing 0.2% Mg and 0.4% Ca is harvested, 16 kg Mg/ha and 32 kg Ca/ha are removed. The removal of a 500-kg live-weight gain/ha removes approximately 5.9 kg Ca and 0.8 kg Mg. These losses are minor in a pasture system. Calcium and Mg are excreted primarily in the feces, and their turnover rates in a pasture system depend on the effectiveness of excreta return in time and space, as well as biological activity in incorporating the solid excreta into the soil. Cycling of Ca and Mg are reviewed by Wilkinson and Lowrey (1973).

MICRONUTRIENT REQUIREMENTS

There are no known documented tall fescue yield responses to micronutrients. However, in three separate experiments a complete micronutrient treatment was added to a N fertilization rate of 448 kg N/ha/year or higher without a yield response (Georgeville soil, a clayey, kaolinitic, thermic Typic Hapludult, South Carolina; Cecil soil, a clayey, kaolinitic, thermic Typic Hapludult, Georgia; and Taloka silt loam, Oklahoma). There have been several studies where micronutrient composition of tall fescue has been reported. Price and Moschler (1970) determined the effect of applying up to 54 metric tons/ha dolomitic limestone on the Mn, Fe, Cu, and Zn content of tall fescue, growing on Groseclose silt loam (a clayey, mixed, mesic Typic Hapludult). In the 4th year from liming, tall fescue samples (15 to 25 cm high) decreased in Mn, Fe, Cu, and Zn content over the pH range from 5.7 (without lime) to 7.4 (with 54 tons/ha of dolomitic lime). The greatest decrease was in Mn content. Boswell (1975) examined the effect of relatively heavy doses of microelements and sewage sludge on the growth of tall fescue and its chemical composition. Part of his results are given in Table 4-12. There were no visible signs of toxicity over the 2-year period of the study. However, forage yields were best without microelement addition. Tall fescue fertilized with sewage sludge was K deficient, whereas that receiving N-P-K fertilizer contained nearly adequate levels of K as judged by yield level. These results indicated that tall fescue is very tolerant to several heavy metals. However, since these heavy metals may accumulate to potentially toxic levels in or on the foliage, caution may be necessary in feeding such herbage.

Bingham et al. (1976) reported that Cd concentrations of 37 μg Cd/g in tall fescue clipped at early bloom stage were associated with a 25% yield depression. Tall fescue pastures fertilized with broiler litter contained an average of 0.042 ppm Se, while that receiving 224 kg N/ha as NH_4NO_3 contained an average of 0.053 ppm Se. Blood serum Se of cows grazing these pastures averaged 0.054 and 0.068 ppm, respectively (Stuedemann et al., 1975).

Waddington and Zimmerman (1972) determined the micronutrient content of tall fescue grown at water tables of 5 or 13 cm and did not observe any visual deficiency or toxicity symptoms, or abnormal concentrations of any of the micronutrients in the top growth of tall fescue over a 14-week period. Jensen and Lesperance (1971) reported that tall fescue was a low accumulator of Mo in field studies. In a greenhouse study, they reported that Mo accumulation was 3.5 ppm for each kg/ha Mo added. This accumulation was linear up to about 1,000 ppm. No toxicity symptoms were reported.

Fleming et al. (1974) found Al toxicity in tall fescue in nutrient solution and suggested that poor tall fescue growth on some very acid coal-mine spoils may be due to Al toxicity. Oertli and Kohl (1961) reported B toxicity symptoms in Alta tall fescue 8 days after tall fescue was transferred to nutrient solutions containing 10 ppm B. Toxicity symptoms were chlorosis, followed by leaf tip burn, and finally, necrosis. Necrotic leaf tissue contained from 1,510 to 8,200 ppm B, while green tissue contained 50 to 760 ppm B, which illustrates tall fescue's tolerance to abnormally adverse conditions.

SILICA ACCUMULATION

Jones and Handreck (1967) have shown that grasses accumulate much more silica than do legumes. Johnson et al. (1967) showed that contamination was a factor in their studies of silica distribution in *Festuca scabrella* grown at several locations. In their study, plant silica amounted to about one-half of the silica present. Van Soest and Jones (1968) reported evidence that silica decreases the digestibility of cell-wall carbohydrate, and indicated that it may rival the importance of lignin in digestibility of some species, like reed canarygrass (*Phalaris arundinacea* L.), and possibly tall fescue. Buckner et al. (1967) who found up to 3.9% SiO_2 in Kenwell tall fescue reported that an annual ryegrass-tall fescue amphiploid hybrid was lower in silica than was Kentucky 31 or Kenwell. Silica was positively associated with crude fiber and negatively related to protein, sugar, moisture, and digestibility. Morton and Jutras (1974) reported that SiO_2 concentrations of ungrazed tall fescue ranged from 0.3 to 2.65% and were generally highest during spring. Thereafter, concentrations remained relatively constant. Actively growing forages maintained in a vegetative state by grazing did not exhibit appreciable monthly differences in silica concentrations. Increases in silica were noted during winter dormancy of tall fescue.

Clippings from Kentucky 31 tall fescue growing in solution cultures containing 0, 50, and 100 ppm added SiO_2, contained 0.23, 2.35, and 4.12% SiO_2, respectively (J. R. Street. 1974. The influence of silica concentration on the chemical composition and decomposition of turfgrass, tissue and water absorption rates among three turfgrass species. Ph.D. Thesis. Ohio State Univ., Columbus). Silica concentrations for leaf and root tissue were highest in 'Pennstar' Kentucky bluegrass (*Poa pratensis* L.) > tall fescue > bermudagrass (*Cynodon dactylon* L.). Silica concentrations of various plant parts were in the order stubble > leaves > roots. Street also reported that increasing levels of N fertilization in the field produced systematic yield increases which were accompanied by decreases in silica concentration of leaf tissue.

Silicon exists in soil solution as monosilicic acid at soil pH values below 9, and as silicate ions at pH's above 9 (Jones and Handreck, 1967). Changes in soil pH from 5.4 to 7.2 decreased silica in soil solution from 70 to 23 ppm. Ferric and aluminum oxides absorb monosilicic acid, with aluminum oxides generally more effective than ferric oxide. Therefore, soils high in sesquioxides are likely to be low in soluble silica.

Accumulation of silica by tall fescue may be expected on soils low in ferric and aluminum oxides but not on highly weathered soils low in sesquioxides. Liming acid soils may decrease soluble silica in the soil solution (over pH range 5.0 to 6.8). Fertilization may also decrease tall fescue silica content by improving water-use efficiency. The increased tall fescue growth, particularly if harvested frequently, will likely contain a lower silica concentration.

Higher plants absorb silicon in the monosilicic acid form, and absorption is proportional to the amount in soil solution (Lewin and Reimann, 1969). Silica accumulation may also be proportional to the amount of water lost through evapotranspiration (suggesting passive uptake). However, Lewin and Reimann (1969) reviewed evidence that energy from respiratory reactions are also involved in silicon uptake by plant roots with silica concentrations usually increasing with plant age. Lewin and Reimann (1969) and Jones and Handreck (1967) reviewed the role of silicon in plant nutrition and the silica forms in plants.

LITERATURE CITED

1. Allaway, W. H. 1971. Feed and food quality in relation to fertilizer use. p. 533-555. *In* R. A. Olson, T. J. Army, J. J. Hanway, and V. J. Kilmer (eds.) Fertilizer technology and use. Soil Sci. Soc. Am., Inc., Madison, Wis.
2. ———. 1975. The effect of soils and fertilizer on human and animal nutrition. Agric.

Inf. Bull. No. 378. ARS and SCS, USDA, Washington, D.C. 52 p.
3. ─────, and J. F. Thompson. 1966. Sulfur in nutrition of plants and animals. Soil Sci. 101:240-247.
4. Barrow, N. J., and L. J. Lambourne. 1962. Partition of excreted nitrogen, sulfur, and phosphorus between feces and urine of sheep being fed—pasture. Aust. J. Agric. Res. 13:461-471.
5. Beeson, Kenneth C. 1945. The occurrence of mineral nutritional diseases of plants and animals in the United States. Soil Sci. 60:9-13.
6. Bertramson, B. R., M. Fried, and S. L. Tisdale. 1950. Sulfur studies of Indiana soils and crops. Soil Sci. 70:27-41.
7. Bingham, F. T., A. L. Page, R. J. Mahler, and T. J. Ganje. 1976. Yield and cadmium accumulation of forage species in relation to cadmium content of sludge-amended soil. J. Environ. Qual. 5:57-60.
8. Black, C. A. 1957. Soil-plant relationships. John Wiley & Sons, Inc. 332 p.
9. Blaser, R. E., and E. Lamar Kimbrough. 1968. Potassium nutrition of forage crops with perennials. p. 423-445. *In* V. J. Kilmer, S. E. Younts, and N. C. Brady (eds.) The role of potassium in agriculture. Am. Soc. Agron., Madison, Wis.
10. Boswell, Fred C. 1975. Municipal sewage sludge and selected element application to soil: Effect on soil and fescue. J. Environ. Qual. 4:267-272.
11. Bray, R. H. 1954. A nutrient mobility concept of soil-plant relationships. Soil Sci. 78:9-22.
12. Buckner, R. C., J. R. Todd, P. B. Burns, and R. F. Barnes. 1967. Chemical composition, palatability, and digestibility of ryegrass-tall fescue hybrids, 'Kenwell' and Kentucky 31 tall fescue varieties. Agron. J. 59:345-349.
13. Chamblee, D. S., and W. B. Gilbert. 1965. Effect of submersion in water on tall fescue, orchardgrass, and ladino clover. Agron. J. 57:502-504.
14. Clement, C. R., and M. J. Hopper. 1968. The supply of potassium to high yielding cut grass. Natl. Agric. Advis. Serv. Q. Rev. 79:101-109.
15. Cooper, C. S., M. G. Klages, and J. Schulz-Schaeffer. 1962. Performances of six grass species under different irrigation and nitrogen treatments. Agron. J. 54:283-288.
16. DeGroot, T. H. 1970. Some experience with the use of high levels of nitrogen on grasslands in the Netherlands with special reference to animal health. 7th Int. Grassl. Congr. Proc., Surfers' Paradise, Queensland, Australia. Plenary Paper A 107.
17. Denmead, O. T., J. R. Simpson, and J. R. freney. 1974. Ammonia flux into the atmosphere from a grazed pasture. Science 185:609-610.
18. DeWit, C. R., W. Dijkshoorn, and J. C. Noggle. 1963. Ionic balance and growth of plants. Versl. Landbouwkd. Onderz. 61:15.
19. Duell, R. W. 1965. Nitrogen-potassium utilization by three pasture grasses. Agron. J. 57:445-448.
20. ─────, and J. R. Trout. 1972. Quantitative removal of major nutrients by three pasture grasses. Agron. J. 64:739-743.
21. Duncan, C. C., M. Schupp, and C. M. McKell. 1969. Nitrogen concentration of grasses in relation to temperature. J. Range. Manage. 22:430-449.
22. Elkins, Charles B., Ronald L. Haaland, Carl S. Hoveland, and W. A. Griffey. 1977. Grass tetany potential of tall fescue as affected by soil oxygen. Agron. J. 70:309-311.
23. Embleton, T. W. 1966. Magnesium. p. 225-263. *In* H. D. Chapman (ed.) Diagnostic criteria for plants and soils. Div. Agric. Sci., Univ. of California, Berkeley.
24. Fleming, A. L., J. W. Schwartz, and C. D. Foy. 1974. Chemical factors controlling the adaptation of weeping lovegrass and tall fescue to acid mine spoils. Agron. J. 66:715-718.
25. Frame, J. 1973. The yield response of tall fescue/white clover sward to nitrogen rate and harvesting frequency. J. Br. Grassl. Sco. 28:139-148.
26. Fischer, R. A. 1968. Stomatal opening: Role of potassium uptake by guard cells. Science 160:784-785.
27. Fontenot, J. P. 1973. Magnesium in ruminant animals and grass tetany. p. 131-151. *In* J. B. Jones, M. C. Blount, and S. R. Wilkinson (eds.) Magnesium in the environment, soils, crops, animals, and man. Taylor County Printing Co., Reynolds, Ga.
28. Fuller, William N., William C. Elder, Billy B. Tucker, and Wilfred E. McMurphy. 1971. Tall fescue in Oklahoma—A review. Oklahoma State Univ. Agric. Exp. Stn. Prog. Rep. P-650. 22 p.
29. Gentry, C. E., R. A. Chapman, L. Henson, and R. C. Buckner. 1969. Factors affecting

alkaloid content of tall fescue (*Festuca arundinacea* Schreb.) Agron. J. 61:313-316.
30. George, J. R., C. L. Rhykerd, G. O. Mott, R. F. Barnes, and C. H. Noller. 1972. Effect of nitrogen fertilization of *Festuca arundinacea* Schreb. on nitrate nitrogen and protein content and the performance of grazing steers. Agron. J. 64:24-26.
31. Giddens, J., W. E. Adams, and R. N. Dawson. 1971. Nitrogen accumulation in fescuegrass sod. Agron. J. 63:451-454.
32. Grimes, R. C. 1967. The growth of lambs grazing tall fescue receiving high and low levels of nitrogen fertilizer. J. Agric. Sci. 69:33-41.
33. Gross, C. F. 1973. Managing magnesium deficient soils to prevent grass tetany. p. 88-92. *In* Plants, animals and man. Proc. 28th Annu. Meet. Soil Cons. Soc. Am., Hot Springs, Ark. Soil Cons. Soc. Am., Ankeny, Iowa.
34. Grunes, D. L., P. R. Stout, and J. R. Brownell. 1970. Grass tetany of ruminants. Adv. Agron. 22:331-374.
35. Hallock, D. L., R. H. Brown, and R. E. Blaser. 1965. Relative yields and composition of Kentucky 31 fescue and Coastal bermudagrass at four nitrogen levels. Agron. J. 57: 539-542.
36. ————, D. D. Wolf, and R. E. Blaser. 1973. Reaction of tall fescue to frequent summer nitrogen application. Agron. J. 65:811-812.
37. Hojjati, S. M., T. H. Taylor, and W. C. Templeton, Jr. 1972. Nitrate accumulation in rye, tall fescue, and bermudagrass as affected by nitrogen fertilization. Agron. J. 64: 624-627.
38. Hutchinson, G. L., R. J. Millington, and D. B. Peters. 1972. Atmospheric ammonia: Absorption by plant leaves. Science 175:771-772.
39. Jackson, W. A., and R. J. Volk. 1968. Role of potassium in photosynthesis and respiration. p. 109-140. *In* V. J. Kilmer, S. E. Younts, and N. C. Brady (eds.) The role of potassium in agriculture. Am. Soc. Agron., Madison, Wis.
40. Jensen, E. H., and A. L. Lesperance. 1971. Molybdenum accumulation by forage plants. Agron. J. 63:201-204.
41. Johnston, A., L. M. Berzeau, and S. Smolick. 1967. Variation in silica content of range grasses. Can. J. Plant Sci. 47:65-71.
42. Jones, L. H. P., and K. A. Handreck. 1967. Silica in soils, plants, and animals. Adv. Agron. 19:107-149.
43. Joy, Peter, Esko Lakanen, and Mikko Sillanpaa. 1973. Effects of heavy nitrogen dressings upon release of potassium from soils cropped with ley grasses. Ann. Agric. Fenn. 12:172-184.
44. Kaempffe, G. C., and O. R. Lunt. 1967. Availability of various fractions of urea-formaldehyde. J. Agric. Feed Chem. 15:967-997.
45. Kilmer, V. J., S. E. Younts, and N. C. Brady (eds.) 1968. The role of potassium in agriculture. Am. Soc. Agron., Madison, Wis. 509 p.
46. Leamer, R. W. 1963. The effect of fertilization on yield on an irrigated mountain meadow. J. Range Manage. 16:204-208.
47. Lechtenburg [sic], V. L., D. A. Holt, and H. W. Youngberg. 1972. Diurnal variation in nonstructural carbohydrates of *Festuca arundinacea* Schreb. with and without N fertilizer. Agron. J. 64:302-305.
48. Lewin, Joyce, and B. E. F. Reiman. 1969. Silicon and plant growth. Ann. Rev. Plant Physiol. 20:289-304.
49. Liebhardt, W. C. 1968. Effect of potassium on carbohydrate metabolism and translocation. p. 147-156. *In* V. J. Kilmer, S. E. Younts, and N. C. Brady (eds.) The role of potassium in agriculture. Am. Soc. Agron., Madison, Wis.
50. Lotero, J., W. W. Woodhouse, Jr., and R. G. Peterson. 1966. Local effect on fertility of urine voided by grazing cattle. Agron. J. 58:262-265.
51. Lunt, O. R., R. L. Branson, and S. B. Clark. 1966. Response of five grass species to phosphorus on six soils. p. 419-423. *In* Int. Grassl. Congr., Proc. 9th, Sao Paulo, Brazil. Brazil (State) Dep. in Producao Animal, Sao Paulo, Brazil.
52. Martin, W. E., and J. E. Matocha. 1973. Plant analysis as an aid in the fertilization of forage crops. p. 393-425. *In* L. M. Walsh and J. D. Beaton (eds.) Soil testing and plant analysis. Rev. Ed. Soil Sci. Soc. Am., Inc., Madison, Wis.
53. May, P. F., A. R. Till, and M. J. Cumming. 1973. Systems analysis of the effect of application methods on the entry of sulfur into pastures grazed by sheep. J. Appl. Ecol. 10: 607-626.
54. Mays, D. A., and G. L. Terman. 1969. Sulfur-coated urea and uncoated soluble nitrogen

fertilizers for fescue forage. Agron. J. 61:489-492.
55. ─────, and G. W. Bengston. 1974. Fertilizer effects on forage crops on strip-mined land in Northeast Alabama. TVA, Natl. Fert. Dev. Ctr. Bull. Y-74.
56. McKee, W. H., R. H. Brown, and R. E. Blaser. 1967. Effect of clipping and nitrogen on yields and stands of tall fescue. Crop Sci. 7:567-570.
57. Metson, A. J. 1973. Sulfur in forage crops. The Sulphur Inst. Tech. Bull. No. 20.
58. Meyer, B. S., and D. B. Anderson. 1952. Plant physiology. D. Van Norstrand Co., New York. 784 p.
59. Miaki, T. 1969. Studies on chemical composition and feeding value of forage crops. Effect of nitrogen fertilization on chemical composition and feeding value of Kentucky 31 tall fescue hay. J. Jpn. Soc. Grassl. Sci. 15:163-169.
60. Minderhoud, J. W., P. F. J. van Burg, B. Deinum, J. G. P. Dirven, and M. L. 't Hart. 1974. Effects of high levels of nitrogen and adequate utilization on grassland productivity and cattle performance with special reference to permanent pastures in the temperate regions. *In* V. G. Iglovikov, A. P. Movisissyants, F. F. Sarganova, and A. A. Stepanenko (eds.) Plenary paper presented at Grassl. Congr. Proc. 12th, Moscow, USSR. Four Continents Book Corp., New York.
61. Morton, B. C., and M. W. Jutras. 1974. Silica concentrations in grazed and ungrazed forage species. Agron. J. 66:10-12.
62. Mott, G. O. 1974. Nutrient recycling in pastures. p. 323-339. *In* D. A. Mays (ed.) Forage fertilization. Am. Soc. Agron. Crop Sci. Soc. Am. Soil Sci. Soc. Am., Madison, Wis.
63. ─────, C. J. Kaiser, R. C. Peterson, R. Peterson, and C. L. Rhykerd. 1971. Supplemental feeding of steers on *Festuca arundinacea* Schreb. pastures fertilized at three levels of nitrogen. Agron. J. 63:715-754.
64. National Research Council. Committee on Nitrate Accumulation. 1972. Accumulation of nitrate. NAS, NRC Publ. Washington, D.C. 106 p.
65. Neller, J. R., and C. E. Hutton. 1957. Comparison of surface and subsurface placement of superphosphate on growth and uptake of phosphorus by sodded grasses. Agron. J. 49:347-351.
66. Oertli, J. J., and H. C. Kohl. 1961. Some considerations about the tolerance of various plant species to excessive supplies of boron. Soil Sci. 92:243-247.
67. Peters, E. J., and S. A. Lowance. 1974. Fertility and management treatments to control broomsedge in pastures. Weed Sci. 22:201-205.
68. Pfander, W. H., and E. Rubic. 1972. Composition and supplements for fall grown fescue. J. Anim. Sci. 35(1):233.
69. Plancquaert, P. 1973. Study of the effect of N fertilizer on the yield of grasses. Publication, Institute Technque des Cereales et des Fourrages No. 3-2-01-24. Paris, France. 24 p.
70. Powell, A. J., R. E. Blaser, and R. E. Schmidt. 1967. Physiological and color aspects of turfgrasses with fall and winter nitrogen. Agron. J. 59:303-307.
71. Price, N. E., and W. N. Moschler. 1970. Residual lime effect in soils on certain mineral elements in barley, fescue, and oats. J. Agric. Food Chem. 18:5-8.
72. Rampton, H. H. 1950. Alta fescue for high yielding pastures. What's New Crops Soils 7:18-19, 25.
73. Reddy, G. E., and J. Giddens. 1975. Nitrogen fixation algae in fescue-grass soil crusts. Soil Sci. Soc. Am. Proc. 39:654-656.
74. Reid, R. L., and G. A. Jung. 1965. Influence of fertilizer treatment on the intake, digestibility and palatability of tall fescue hay. J. Anim. Sci. 24:615-625.
75. ─────, and ─────. 1974. Effect of elements other than nitrogen on the nutritive value of forages. p. 395-435. *In* D. A. Mays (ed.) Forage fertilization. Am. Soc. Agron., Madison, Wis.
76. ─────, A. J. Post, and G. A. Jung. 1970. Mineral composition of forages. W. Va. Agric. Exp. Stn. Bull. 589T. 35 p.
77. Rinne, S. L., M. Sillampaa, E. Huokuna, and S. L. Hiivola. 1974. Effects of heavy nitrogen fertilization on potassium, calcium, magnesium, and phosphorus contents in ley grasses. Ann. Agric. Fenn. 13:96-108.
78. Robbins, J. E., S. R. Wilkinson, and D. Burdick. 1973. Loline alkaloids of tall fescue seed and forage. p. 98-107. *In* Fescue Toxicity Conf. Proc., Univ. of Kentucky, Lexington. 31 May-1 June 1973. Univ. of Missouri, Columbia.
79. Rogers, J. A., and G. E. Davis. 1973. The growth and chemical composition of four grass species in relation to soil moisture and aeration factors. J. Ecol. 61:455-472.
80. Smith, Paul F. 1974. Soil absorption of atmospheric nitrogen in Florida. Commun. Soil Sci. Plant Anal. 5:341-353.

81. Spurway, C. H. 1941. Soil reaction (pH) preferences of plants. Spec. Bull. 306 Michigan Agric. Exp. Stn., East Lansing, Mich. 36 p.
82. Stewart, B. A. 1970. Volatilization and nitrification from urine under simulated feedlot conditions. Environ. Sci. Tech. 4:579-582.
83. Stuedemann, J. A., S. R. Wilkinson, D. J. Williams, H. Ciordia, J. V. Ernst, W. A. Jackson, and J. B. Jones, Jr. 1975. Long-term broiler litter fertilization of tall fescue pastures and health and performance of beef cows. p. 264-268. *In* Managing livestock wastes. 3rd Int. Sympos. on Livestock Wastes. Am. Soc. Agric. Eng. Publ. Proc. 275. Am. Soc. Agric. Eng., St. Joseph, Mich.
84. Teel, M. R. 1966. Nitrogen-potassium relationships and their influence on some biochemical intermediates and quality of crude protein in forages. p. 465-480. *In* Int. Potash Inst. Congr. Proc. 8th, Brussels, Belgium. Int. Potash Inst., Berne.
85. Templeton, W. C., and T. H. Taylor. 1966. Yield response of tall fescue-white clover swards to fertilization with nitrogen, phosphorous, and potassium. Agron. J. 58:319-332.
86. ————, ————, and J. R. Todd. 1965. Comparative ecological and agronomic behavior of orchardgrass and tall fescue. Kentucky Agric. Exp. Stn. Bull. 699. 18 p.
87. Terman, G. L., J. C. Noggle, and C. M. Hunt. 1976. Nitrate-N and total N concentration relationships in several plant species. Agron. J. 68:556-560.
88. Tesar, Milo B., and L. N. Shepherd. 1963. Evaluation of forage species on organic soil. Agron. J. 55:131-134.
89. U.S. Salinity Laboratory Staff. 1954. Plant response and crop selection for saline and alkali soils. p. 55-68. *In* L. A. Richards (ed.) USDA Handb. No. 60. U.S. Govt. Print. Off., Washington, D.C.
90. Van Burg, P. F. J. 1962. Internal nitrogen balance, production of dry matter and aging of herbage and grass. Versl. Landbouwkd. Onderz. 68:131.
91. Van Soest, P. J., and L. H. P. Jones. 1968. Effect of silica in forages upon digestibility. J. Dairy Sci. 51:1644-1648.
92. Vogel, W. G., and N. A. Berg. 1968. Grasses and legumes for cover on acid strip-mine spoils. J. Soil Water Conserv. 23:89-91.
93. Waddington, D. V., and T. L. Zimmerman. 1972. Growth and chemical composition of eight grasses under high water table conditions. Commun. Soil Sci. Plant Anal. 3:329-337.
94. Whitehead, D. C. 1964. Soil and plant nutrition aspects of the sulphur cycle. Soils Fert. 27:1-8.
95. ————. 1970. The role of nitrogen in grassland productivity. Commonw. Agric. Bur. Pasture Field Crops. Bull. 48. Hurley, Berkshire, England. 202 p.
96. ————. 1972. Chemical composition. p. 98-132. *In* C. R. W. Spedding and E. C. Diekmakus (eds.) Grasses and legumes in British agriculture. Commonw. Agric. Bur. Pasture Field Crops. Bull. 49. CAB, Farnham Royal, Bucks, England.
97. Wilkinson, S. R., J. A. Stuedemann, J. B. Jones, Jr., W. A. Jackson, and J. W. Dobson, Jr. 1973. Environmental factors affecting magnesium concentrations and tetanigenicity of pastures. p. 153-174. *In* J. B. Jones, Jr., M. C. Blount, and S. R. Wilkinson (eds.) Magnesium in the environment, soils, crops, animals, and man. Taylor County Publishing Co., Reynolds, Ga.
98. ————, and R. S. Lowrey. 1973. Cycling of mineral nutrients in pasture ecosystems. p. p. 248-309. *In* G. W. Butler and R. W. Bailey (eds.) Chemistry and biochemistry of herbage. Vol. 2. Academic Press, London and New York.
99. Wolfe, E. C., and A. Lazenby. 1973. Grass-white clover relationships during pasture development. 1. Effect of superphosphate. Aust. J. Exp. Agric. Anim. Husb. 13:567-574.
100. Zagallo, H. C., and W. B. Bollen. 1962. studies on the rhizosphere of tall fescue. Ecology 43:54-62.

Chapter 5   **Physiology of Growth and Development**

D. D. WOLF
Agronomy Department
Polytechnical Institute and State University
Blacksburg, Virginia

R. H. BROWN
Agronomy Department
University of Georgia
Athens, Georgia

R. E. BLASER
Agronomy Department
Polytechnical Institute and State University
Blacksburg, Virginia

INTRODUCTION

Growth and development of tall fescue (*Festuca arundinacea* Schreb.) involve many interacting processes that vary with changes in edaphic and environmental conditions (Youngner, 1969). Management of defoliation for pasture and hay production, or other uses such as turf, roadside stabilization, and erosion control can impose additional complex responses. The major consideration of all physiological studies should be aimed at improving the performance of the plants for the objectives intended. In this chapter, physiological responses during several developmental events such as germination, root growth, and herbage growth will be related to basic growth factors. Yield limiting processes will also be considered.

PHYSIOLOGY OF DEVELOPMENT

A. Seed Dormancy and Germination

Some physiological and biochemical development may sometimes be required after morphological ripening before seeds become viable (Boyce et al., 1976; Danielson and Toole, 1976). Viability is usually best maintained by drying to about 7% moisture and then storing at 15 to 20% relative humidity (R.H.) with temperatures not exceeding 32 C. Generally, as R.H. increases storage temperatures must be decreased to retain viability (Kulik and Justice, 1967). The sum of R.H. and temperature (Degrees F) should be usually less than 100.

Rate and percentage of germination are influenced by temperature,

---

Copyright © 1979 ASA-CSSA-SSSA, 677 South Segoe Road, Madison, WI 53711 USA. *Tall Fescue.*

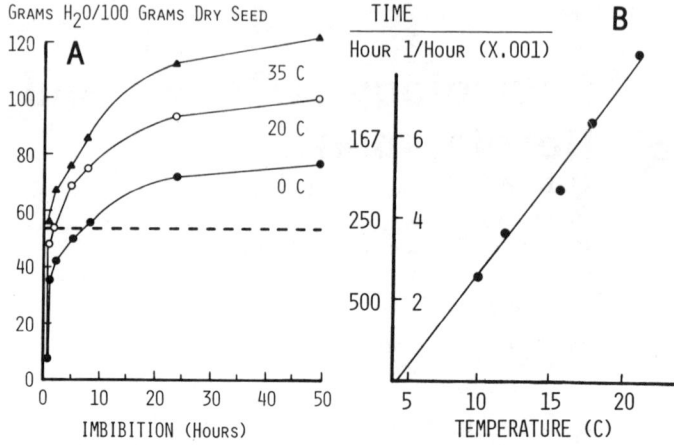

Fig. 5-1—Imbibition and germination rate of tall fescue. Left: Water content (g/100 g DM) of seeds at time intervals with favorable moisture. Dashed line indicates water content needed to sustain germination. Right: Time (hours) for coleoptile development to 10 mm at several temperatures. Threshold temperature is 4.5 C. Unpublished data by authors.

moisture, oxygen, and pathogens. Imbibition of water by seeds occurs rapidly and activates metabolic systems which initiate germination within a few hours under favorable conditions. About 56 g of water/100 g dry seed will sustain germination (Fig. 5-1). Subsequent enzymatic reactions, primarily regulated by temperature, govern germination. Water imbibition by tall fescue seed increased as temperatures increased from 0 to 35 C and a reciprocal plot of "time for coleoptile to reach 10 mm" shows a threshold germination temperature of 4.5 C (Fig. 5-1). Boyce et al. (1976) found that within a range of 13 to 33 C the optimum temperature for tall fescue germination was 20 to 25 C. Danielson and Toole (1976), using a range of 8 to 31 C, found 12 to 18 C to be optimum for germination. No explanation of the difference between these results is possible.

Seeds are rather tolerant of moisture fluctuations during germination. Elslik and Vogel (1959), after maintaining seed in soil with moisture potentials ranging between −0.3 and −15 bars for 2 weeks, found that germination was not depressed when the soil was hydrated. The seeds were resistant to desiccation injury during embryo dormancy before coleoptile emergence. After coleoptile emergence substantial tissue water reduction of the seedling may kill elongating roots. If some root primordia retain meristematic activity, seedlings will recover after the drought stress.

B. Root Development

Root growth in natural undisturbed environments is not well understood. Periodic determinations of root mass show only the balance between growth and decay with an increase in root mass occurring when growth exceeds rate of decomposition (Weinmann, 1948). Differentiating between living and dead roots and locating active root surfaces is difficult.

Several general principles are of value in managing herbage removal to encourage an aggressive root system. Regrowth of leaves after partial defoliation is usually faster than root development since metabolites required for growth are generally used preferentially in meristematic regions closest to the origin. Canopy and root growth are interdependent; thus, factors which increase the vigor of tops may reduce root growth. With intensive defoliation, rapidly growing roots may be deprived of sufficient energy and die (Crider, 1955). Since new roots originate from primordia near tiller bases just below the soil surface, roots may accumulate predominately in the upper soil profile. Oswalt et al. (1959) found that old grass roots died and began to decompose within 48 hours after defoliation. New roots did not penetrate to a 15-cm depth for 24 days. Wolf (1964), using plaster-of-paris moisture blocks at 30- and 120-cm depths to measure moisture depletion as an index of root activity, found that five cuttings resulted in more root activity in the upper 30 cm than three cuttings per season. Infrequent defoliation encouraged root activity into the 120-cm depth of the soil profile to provide more available water than with frequent cutting.

Root development generally occurs at lower temperatures than for top growth and continues during the autumn as soil temperatures are higher than air temperatures (Weinman, 1948). Adequate nutrition, especially N, during the cool autumn encourages root development (Oswalt, 1959). Photosynthesis is reduced less by cool temperatures than cell division and expansion, hence excess metabolites are available for root, tiller, and rhizome growth in warm soil environs (Treharne and Nelson, 1975).

C. Above Ground Development

Seasonal Growth

Active canopy growth begins in April followed by seed head development by late May and anthesis starts in early June at Blacksburg, Va. (37° N Lat, 80° W Long, and 600 m elevation). Flowering and stem elongation occurs only during the spring and all subsequent regrowth is leafy. The development of flowering tillers, as for other temperate grasses, depends on two conditions: a short day-long night photoperiod with low temperatures for induction of bud primordia during late autumn and winter, followed by a long day-short night photoperiod with cool temperatures during spring. About 45% of the annual dry matter production of well fertilized tall fescue occurs during the first 6 weeks of spring (Fig. 5-2). The rate of dry matter production of uncut canopies in spring reaches about 150 kg/ha/day for several weeks with a very sharp decline after anthesis.

Seasonal yields of tall fescue with adequate nutrition and soil moisture have exceeded 14,000 kg/ha when managed for hay (Table 5-1). Yields are reduced 35 to 50% when cut frequently to simulate grazing. Stands can re-

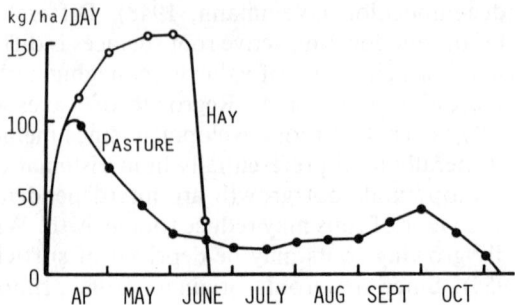

Fig. 5-2—Dry matter growth rates of tall fescue when managed for hay or pasture in spring. Growth after late June hay cutting is similar for hay and pasture management. Blacksburg, Va. Unpublished data by authors.

main vigorous and last indefinitely in the mid-Atlantic region when managed for hay, continuous or rotational grazing, or combinations of these practices. Dry matter production for pasture or hay management during summer with adequate soil moisture may exceed 50 kg/ha/day which is equal or higher than for other commonly used cool-season grasses. During 5 years near Blacksburg, Va., where summer rainfall averaged about 10 cm/month, irrigation increased yields during dry periods but did not increase seasonal production. In fact, yield of nonirrigated grass following relief of drought stress often exceeded yield of grass that had previously been irrigated.

High yields of tall fescue are partly attributed to its wide temperature adaptation, producing more dry matter than other grasses during late summer and autumn. Although growth is slow during late October and November, even with adequate N fertilization, net dry matter accumulation continues and total nonstructural carbohydrates (TNC) increase. Also, much of the foliage remains green during winter and deterioration of herbage is much less than for other cool-season species such as bluegrass (*Poa annua* L.), bromegrass (*Bromus* sp.), and orchardgrass (*Dactylis glomerata* L.) (Blaser et al., 1969; Taylor et al., 1976).

Tillering

Buds formed on rhizomes and in axes of leaves are major vegetative means of pereniation (Fig. 5-3). Tiller development depends largely on environment and management. Rate of tillering is often limited by and negatively correlated with rate of leaf appearance. Under stress, the buds either remain dormant or die before a tiller forms. Under optimum conditions, tiller numbers frequently increase exponentially since each developing tiller is a potential source of new tillers (Templeton et al., 1961).

Tillering is nil during summer even though buds are present. Yeh et al. (1976) associated low auxin levels at high temperatures with inactive sum-

GROWTH & DEVELOPMENT

Table 5-1—Growth parameters of tall fescue during three seasons of accumulating development. Nitrogen was applied at 112 kg/ha 10 days before each season and irrigation was used to limit soil moisture stress. 1971, Blacksburg, Va., 37°N lat, 80°W long, 600 M elevation. Unpublished data by authors.

Date observed		Yield	Leaf area index			Leaves	Crop growth rate	Two-week regrowth	Tillers at harvest				Nonstructural CHO	
Month	Day		Below 5 cm	Above 5 cm	Total				Population	Survival	Leaves		Herbage	Stubble
		kg/ha				%	kg/ha/day	kg/ha	No./dm²	%	No./tiller		%	
						Spring growth								
Apr.	19	240	2.2	0.5	2.7	100	--	870§	41[0]#	91	3.5		--	13.4
	27	680	2.0	1.5	3.5	100	53	1,060	43	92	4.1		15.2	12.0
May	4	1,670	1.1	3.4	4.5	92	141	1,590	51	82	4.7		15.7	13.9
	11†	2,450	1.2	3.5	4.7	67	113	1,950	54[0]	70	5.0		6.6	9.1
	18	3,970	0.5	4.8	5.3	56	214	580	49[16]	58	5.1		8.4	8.5
	25‡	5,990	0.2	7.3	7.5	56	290	380	50[45]	48	5.2		9.4	9.4
June	1	7,370	0.0	7.9	7.9	40	197	0	39[54]	42	5.2		--	8.9
	8	7,820	0.0	5.3	5.3	31	64	0	36[54]	39	5.2		6.0	6.8
						Summer growth (Cut to 5 cm on 5 July)									
July	5	0	0.6	0.0	0.6	--	--	1,560	21(10)¶	80	2.5		--	11.0
	16	320	1.1	0.8	1.9	100	20	740	29	85	3.0		--	8.9
	27	1,560	0.5	3.9	4.4	100	112	610	23	82	3.6		9.8	12.5
Aug.	9	2,250	0.6	5.6	6.2	100	53	580	22(33)	87	4.1		7.7	21.4
	20	3,120	0.8	7.8	8.6	94	79	550	19(48)	83	4.6		6.0	19.1
	31	4,190	0.5	8.5	9.0	93	97	0	19	92	5.2		9.6	28.5
						Fall growth (Cut to 5 cm 1 Sept.)									
Sept.	1	0	--	0.0	--	--	--	--	--(52)	--	--		--	--
Oct.	5	1,950	--	3.9	--	100	70	--	19(59)	97	--		19.3	27.1
Nov.	8	3,860	--	7.2	--	100	38	--	22(59)	95	--		22.2	35.1
Dec.	2	3,500	--	6.9	--	100	--	--	--	--	--		--	--

† Denotes early boot stage of development. ‡ Denotes full heading. Anthesis occurs about 4 to 8 June. § Growth and percent survival during 2 weeks after cutting to 5 cm at each date. ¶ Data in brackets are percent of tillers bearing seed heads. # Numbers in parentheses are lateral bud densities (Lopez et al., 1967) from geographic latitude similar to Blacksburg, Va.

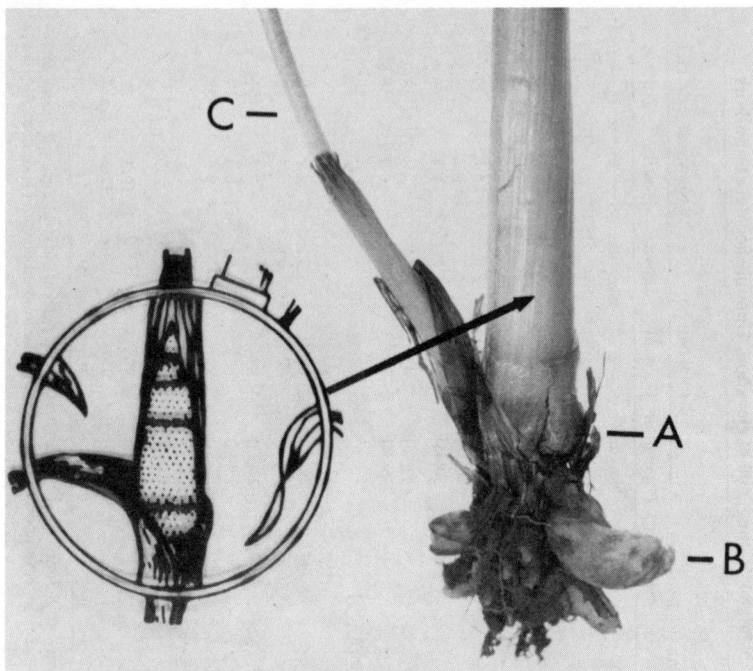

Fig. 5-3—A tiller (culm) showing the morphological form. A) a new bud; B) elongated bud; C) a newly developed (daughter) tiller all located on the stem base. Insert shows nature of the growing point located within the parent tiller.

mer tillering. During culm elongation apical dominance is an important factor influencing tiller development.

Longevity of tillers under various management regimes is generally unknown. Flowering terminates the life of a tiller. Also tillers that have been cut below the growing point die. Tillers emerging during autumn become inducted and during spring will flower and die. Many non-flowering spring tillers will probably remain active until the following spring when the inducted tillers will flower. The number of tillers that flower during spring should depend largely on tiller populations during the preceding late fall.

Tiller population is cyclic with dramatic increases occurring in early spring after the mean weekly temperature reaches about 4.5 C (Templeton et al., 1961). Increases in tiller populations continue until the jointing stage of development. The flush of tillering during spring is attributed to large bud populations developed during the preceeding autumn (Table 5-1). Developing canopies during spring become very dense as many tillers grow to a reproductive stage; hence, because of high leaf area and light competition, tiller populations decrease as small, short, nonflowering tillers die. Regrowth and tiller survival are often poor after cutting high yielding canopies and may be due to the drastic change in microenvironment. Harvesting hay in the early head stage usually reduces tiller losses and weed encroachment as compared with cutting after anthesis.

GROWTH & DEVELOPMENT

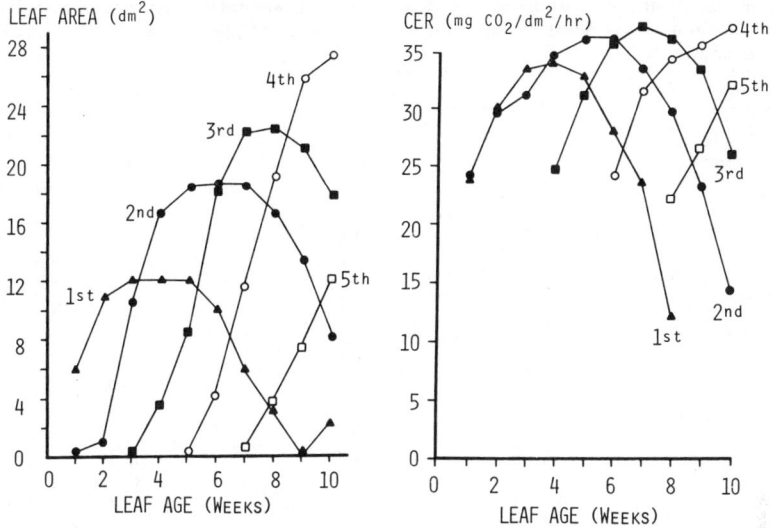

Fig. 5-4—Left, leaf area duration per tiller; Right, photosynthesis of successive tall fescue leaves during late spring growth (Alburquerque, 1967).

Adequate tillering and maintenance of stands are not usually problems with continuous rotational grazing or turf management. Partial defoliation after accumulating leaf area indexes (LAI) of about three to four (about 18-cm canopies) results in dense sods and high tiller populations. With favorable environments, grazing pressure should be adjusted to maintain canopies ranging from 3 to 15 cm in height with continuous grazing. Practices that favor increased tillering tend to reduce dry matter production per tiller. In general, as canopy accumulation increases, tiller populations decline. Some reduction in yield must be accepted to maintain vigorous, persistent, high quality pasturage from season to season.

GROWTH PRINCIPLES

A. Plant Characteristics

Usually only two leaves per tiller elongate at a given time (Alburquerque, 1967). Once the blade tip emerges from the whorl, cell division in the leaf ceases. Further blade enlargement depends on increases in cell size and is largely dependent on the environment. After the collar of a leaf is visible, blade elongation stops but sheath elongation may continue for a short period. During much of the growing season only about three leaves per tiller are active at any one time (Fig. 5-4). Individual leaves are photosynthetically active for about 6 weeks during spring and summer (Jewiss and Woledge, 1967) but may be less when canopies grow and accumulate rapidly. The activity of a given leaf is usually of longest duration during autumn. If a large

Table 5-2—Distribution of dry matter (percent of total) and total nonstructural carbohydrate (TNC) and protein levels among plant parts of tall fescue tillers at early anthesis. Adapted from Smith (1973).

Plant part	Distribution	TNC	Protein
		%	
Inflorescence	18.3	6.2	14.5
Blade 1†	1.7	3.9	20.3
2	3.5	4.4	15.9
3	4.8	6.3	12.0
	10.0	5.2‡	14.7‡
Sheath 1	6.7	7.4	8.9
2	5.7	11.0	6.2
3	5.2	13.7	4.4
4	3.6	8.0	4.2
	21.2	10.0	6.3
Internode 1	9.3	9.1	6.6
2	10.8	11.4	4.2
3	11.5	12.5	3.1
4	11.0	13.7	2.8
	42.6	11.8	4.1
Herbage	92.1	9.6	7.8
Stubble	7.9	15.8	2.9
Shoot (Total)	100.0	10.1	7.4

† Numbered from top to bottom of shoot.
‡ Weighted average for all blades, sheaths, or internodes.

portion of a blade is removed, elongation of that leaf soon terminates. New leaves emerge more quickly if some leaf area remains on a tiller than when totally defoliated. There is evidence that nondefoliated tillers stimulate regrowth of adjacent defoliated tillers (Matches, 1967).

Canopies of vegetative tillers are 100% leaf tissue. The growing points of vegetative tillers located in the tiller base, near or slightly below the soil surface, appear as inverted cones that are formed from nodes, compact internodes, and rudimentary leaves (Fig. 5-3).

Dry matter of tillers near anthesis is comprised of about 60% stem while only 10% of the weight is composed of light intercepting photosynthetic blades (Table 5-2). There is a high leaf area in the lowest strata of the canopy during early vegetative development. More than 50% of the leaf blades appear below a 10-cm height, when LAI values are less than about three (Fig. 5-5). As canopies grow the elongation of internodes elevates some of the bottom leaves, but many of them die causing a decreased LAI in the lowest strata and a dispersion of leaves in the upper canopy strata.

B. Nonstructural Carbohydrates

Metabolizable nonstructural carbohydrates occur in variable amounts in different plant tissues (Table 5-2). Concentrations of total nonstructural carbohydrates (TNC) are highest in the lowest internodes and sheaths at

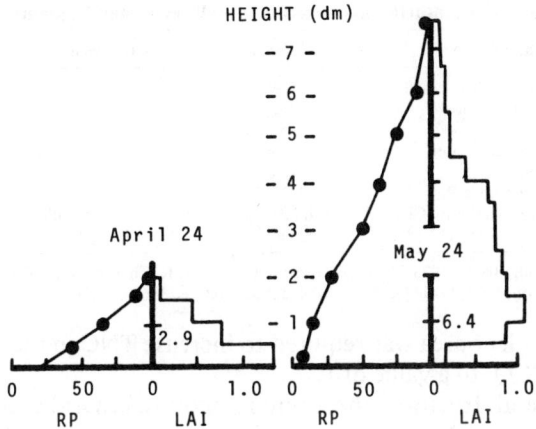

Fig. 5-5—Leaf area index (LAI) distribution and percent radiation penetration (R.P.) at different strata of tall fescue canopies when in leafy (LAI of 2.9) and full heading (LAI of 6.4) stages of development.[1]

base of the tiller. Growth of leaves after partial defoliation may come predominately from TNC in the subtending sheaths. A completely defoliated tiller with negligible photosynthetic capacity would die without a source of quickly metabolizable energy as from TNC degradation. After defoliation, tall fescue plants obtain energy from stored TNC for rapid development of leaves. Thus, TNC concentrations are cyclic (Table 5-1), depending on net energy balances of photosynthetic and growth rates in different environments. Low TNC values indicate that vigorous growth has occurred and tissue would have slow regrowth potential if defoliated. TNC accumulates when environmental factors such as low N, cool temperature, and soil moisture stress retard cell growth more than photosynthesis. TNC accumulations increase dry matter digestibility and forage quality especially during late fall growth (Brown et al., 1963).

TNC provides the major source of energy to sustain the biochemical processes of tall fescue plants. Respiration of stubble after total defoliation (measured by oxygen utilization) increased with increased TNC[1]. As regrowth occurred, TNC and respiration declined. After defoliation, growth (mm leaf elongation/day) was highly correlated (r = 0.96) with tillers initially high in TNC; however for tillers low in TNC, the correlation of TNC with growth was low (r = 0.52). This indicates that a source other than TNC provided some energy. The respiration quotient (R.Q.) for tillers with low TNC was less than 1.0, indicating that substantial energy was supplied by other substances, possibly lipids and protein, when TNC concentrations were too low for normal metabolism. After defoliation only a small

---

[1] Alburquerque, A. E. 1967. Leaf area index, light penetration and carbohydrate reserves during growth of Kentucky 31 tall fescue. M.S. thesis. Virginia Polytech. Inst. State Univ., Blacksburg.

Table 5-3—Similar photosynthetic characteristics of tall fescue and C_3 plants vs. C_4 species.

Photosynthesis characteristic†	Tall fescue	C_3 species	C_4 species
Maximum CER (mg CO_2/dm^2/hour)	35–40	35–40	60–80
First products of CO_2 fixation	3-phosphoglyceric acid	3-phosphoglyceric acid	Malic, aspartic and oxaloaceltic acids
CO_2 compensation concentration	35 ppm	35 ppm	10 ppm
Inhibition of CER by 21% O_2	30%	30%	5%
Optimum temperature for CER	20–25 C	20–30 C	30–40 C
Irradiance required for maximum CER	⅓ to ½ of full sunlight	¼ to ½ of full sunlight	Near full sunlight

† Data obtained from Black, 1973; Charles-Edwards et al., 1971; Chen et al., 1970; Chen et al., 1971; Treharne and Nelson, 1975; Woledge, 1971; Woledge and Jewiss, 1969.

amount of new leaf area was required to increase TNC in the tillers enough to restore the R.Q. to a value of 1.

Glucose and fructose, the primary monosaccharides in tall fescue, occur in nearly equal proportions and concentrations and do not change during most of the growing season; however, these sugars increase during cool temperatures in late fall. Sucrose, a disaccharide, may account for 30% or more of the TNC, especially when TNC values are low (Ojima and Isawa, 1968). Fructosan is the primary polysaccharide occurring in vegetative tissues and is the primary constituent in TNC measurement (Smith, 1968). Fructosans appear mainly as levans with 2-6 fructofuranose repeating units. Polymerization of these linear units occurs in varying degrees, with short chain lengths of 30 to 40 units predominating. These nonstructural carbohydrates accumulate to very high levels during cool temperatures and may contribute to the high digestibility and acceptance during the fall and winter months.

C. Photosynthesis and Photorespiration

Grasses of temperate climatic origin, such as tall fescue, assimilate CO_2 by the Calvin (C_3) cycle having 3-phosphoglyceric acid as the primary end product and exhibiting characteristics of photorespiration. Grasses of tropical origin assimilate CO_2 by carboxylation of phosphoenolpyruvate (C_4 cycle) to form the 4-carbon dicaboxylic acids of malate, asparatate, and oxaloacetate (Black, 1973) which is typical of nonphotorespiring plants. Ribulose-1,5-diphosphate carboxylase (RuDP), the primary carboxylating enzyme of the C_3 cycle, shows high levels of activity in tall fescue, as it does in other plants fixing CO_2 by the C_3 cycle. Phosphoenolpyruvate carboxylase, the primary carboxylase in C_4 species, occurs in small quantities in tall fescue (Chen et al., 1971; Nelson et al., 1975). The characteristics of CO_2 fixation processes of tall fescue and other C_3 species were similar and differed sharply with those of C_4 plants (Table 5-3). The C_4 plants require nearly full sunlight for maximum CO_2 exchange rate (CER); whereas C_3 plants such as tall fescue become light saturated in 25 to 50% full sunlight.

Another important characteristic of CO_2 fixation by the C_3 cycle is the sharp 30% reduction in CER when changed from 2 to 21% O_2 atmosphere as compared to little reduction with C_4 species. The inhibition of CO_2 fixation by O_2 is apparently due to competition by O_2 for the CO_2 fixation site on the RuDP carboxylase. Elimination of this oxygen response through plant breeding could increase productivity, especially during summer months.

The irradiance required for saturation decreases as leaves age and when subjected to low irradiance for long periods. Woledge (1971) found that CER decreased as leaves aged and that such decreases were more marked when maintained at 30.3 vs. 2.5 klux. She also showed that increasing irradiance on leaves after full leaf expansion increased their photosynthetic capacity. Blenkensop and Dale (1974) with barley (*Hordeum vulgare* L.), associated such increases in CER with an increase in total soluble protein. Alburquerque[1] found that tall fescue leaves on field grown plants were light saturated at about 60.6 to 80.8 klux when young but required only 30.3 klux when leaves were old. These saturation irradiances are higher than generally reported for tall fescue. Chen et al. (1970) observed saturation of tall fescue to occur at about 40.4 klux for plants grown in a greenhouse.

Sharp differences occur in irradiance within dense canopies because less than 5% of the total light may reach basal leaves (Fig. 5-5). Such low irradiance may be of little value for energy fixation in a canopy, but may encourage survival of the basal shaded leaves to augment yield and quality of forage. Woledge (1971) demonstrated that leaves grown in low irradiance (18 w/m^2 from 320 to 730 nm light) or transferred from high (72 w/m^2) to low irradiance resulted in reductions in CER, dark respiration, and specific leaf weight (SLW). Decreases in respiration and SLW of bottom leaves may show that photosynthate from upper leaves are not transported to severely shaded leaves. Thus, mature leaves with low CER gradually die.

Photosynthetic capacity of tall fescue changes with leaf age as it does for other herbaceous plants. Woledge and Jewiss (1969) found that CER of leaves produced under warm conditions (20/15 C day/night temperatures) decreased more rapidly than those grown under a 10/5 C day/night regime. The fourth leaf on tall fescue lived 30 days at the warm temperature and 60 days at the cooler temperature. The fourth leaf lived for 80 days after full expansion on plants transferred from the high to low temperature regime. The long photosynthetic life of leaves at low temperatures may be important in compensating for a slow leaf initiation rate and for maintaining live leaves during the autumn and winter grazing season.

Maximum CER values in field grown leaves occurred 28 to 35 days after emergence and then decreased. Full leaf expansion coincided with maximum CER[1]. By the 3rd and 4th week after full leaf expansion, CER dropped to 20% of the maximum (Fig. 5-4). Declines in CER with age varied with leaf position and previous cutting practices. The first leaves that appeared after cutting had lower CER and aged more rapidly than subsequent leaves.

D. Environmental Factors

Temperature

Specific optimum temperatures do not usually occur but may be defined as temperatures where growth parameters are within 20% of maximum (Robson, 1972). The optimum temperature range of lamina area, lamina length and sheath length extension, number of days between successive leaves, and duration of leaf growth is 20 to 30 C. Optimum temperatures for relative growth rate were between 22 and 27 C. Optimum temperature was between 10 and 20 C for SLW, percentage of TNC, number of tillers, and root-shoot weight ratio. Temperature variations of 10 to 30 C did not influence percentage N, lamina width, or number of leaves. Increases in root growth, tiller initiation, and TNC occur at low temperatures, indicating relatively higher rates of CER than vegetative development. Growth extension from cell elongation increases dramatically when temperatures switch from cool to warm, as Williams and Biddiscombe (1965) demonstrated with surges of growth after temperature increases.

Temperatures during leaf development have a significant influence on stomata densities. Stomata density per unit leaf area was higher on leaves from plants grown at 25 C than 10 C (Treharne and Nelson, 1975). Reduction of stomata density on adaxial and absence on abaxial surfaces, as a result of leaf development at 10 C, may indicate that stomata mother cells did not differentiate to form mature guard cells. Apparently, a stomata density of more than $50/cm^2$ leaf area would not reduce stomata resistance and thus not influence CER. However, a reduction to 21 stomata/cm^2 (as on the abaxial surface) increased the resistance and reduced CER. An optimum density of stomata has not been established nor has the relative efficiency of stomates on upper vs. lower leaf surfaces been documented.

Canopy Structure

The microenvironment inside a canopy differs markedly from that observed at standard height for weather observations. Leaf orientation and distribution influences radiation penetration into canopies and subsequent CER in canopy layers. CER of tall, natural canopies never become light saturated on sunny days due to mutual leaf shading. Adaptive responses to long term radiation changes include morphological and biochemical modifications. High radiation generally increases TNC, SLW, respiration, and yield, while leaf length and cell size decrease. CER per unit land area increases as LAI increases, but progressive declines in irradiance per leaf develop which decreases CER and respiration per unit leaf area. With increased age or prolonged limited radiation leaves die without energy from translocated assimilates. Management practices should be adopted that encourage nearly complete radiation interception yet maintain a low enough

LAI so that bottom leaves are not below light compensation for long periods. Well-grazed tall fescue should utilize old leaves and encourage predominately young leaves that are relatively efficient radiation interceptors.

YIELD LIMITATIONS

A. Source and Sink

Growth depends on the supply of assimilates (source), storage capacity (sinks), and capability of translocating assimilates from source to sink. The source of assimilates in tall fescue is predominately from photosynthesis in leaf blades; thus, favorable management practices are needed to attain maximum environmental, edaphic, and genetic potentials. Major sinks for assimilates in tall fescue are division and expansion of cells during root and herbage growth, and accumulation of nonstructural organic compounds. Growth of new leaves and daughter tillers are dominant sink locations. An accumulation of TNC or other assimilates may have a distinct influence on rates of photosynthesis, in that maximum CER capacity cannot be attained without cell division and expansion. Diurnal accumulations of TNC in leaf tissue decreases CER because transport mechanisms may not supply assimilates to growth regions rapidly enough. If CER is a consequence of growth, rather than the cause, then increasing CER or selecting for high CER rates may not result in yield increases. As noted earlier, large primary tillers with small axillary tillers had the highest CER because the many growing points caused high demands for metabolites. These, therefore, served as excellent sinks. Data from other species indicate increases in CER of leaf portions remaining after partial defoliation. If this applies to tall fescue, then compensatory CER responses occur to maintain the source supply with partial defoliation.

Evidence of temporary sinks or readily available stored TNC is supported by the fact that the growth rates do not decline with short periods of cloud cover, even though CER may be reduced considerably. For example, growth rate was rather constant during 10-hour dark periods (Williams and Biddiscombe, 1965). The momentum of growth can apparently be maintained during several consecutive cloudy days. Subsequently, the low levels of metabolites cause higher than normal CER when bright sunshine again occurs.

The growth limitations caused by inefficient translocation between source and sink are difficult to quantify. Increased tall fescue yields might be attained by improving the mechanisms of translocation and developing more active sinks. Progress for increasing the source (CER) through genetics appears limited (Asay et al., 1974) because CER and yields are not always closely related (Nelson et al., 1975). Selection of strains or management practices that would stimulate leaf appearance rates, larger leaves, and rapid tillering could increase yields.

Table 5-4—Total nonstructural carbohydrates in stem bases of floral tall fescue tillers 0, 4, and 8 days after spring hay harvest. Roots were trimmed from tillers and buds were removed from one-half the tillers. Tillers were held in plastic bags at two temperatures. Unpublished data by authors.

Day	Axillary buds	Temperature (C)		Average
		22	35	
0		7.0	7.0	7.0
4	Present	5.3	3.0	4.6
	Removed	5.4	3.2	4.3
8	Present	1.8†	3.3	2.6
	Removed	4.7	3.4	4.0

† Indicates presence of well developed axillary tillers from buds.

B. LAI and TNC

The cyclic concentrations of TNC in plants are not necessarily correlated with growth rate (Brown and Blaser, 1963; Jung et al., 1976). TNC's are the major components of a dynamic energy balance system and are interrelated with growth, leaf area, and CER. Any conditions that stimulate growth and an energy need above that supplied by CER result in a net loss of TNC regardless of LAI values. Reductions in TNC occur during the night, with low radiation, after liberal N fertilization, and with shifts from low to high temperatures. Conversely, TNC accumulation occurs when respiration for maintenance and growth is lower than CER as with cool temperature, low N availability, and limited soil moisture.

The relationship between TNC and residual leaf area is important when making defoliation management decisions. The relative contributions of TNC and LAI to regrowth will change with season, plant morphology, and environmental conditions. Building materials for leaf and root development after defoliation are obtained from TNC in plant tissues and new metabolites. TNC's are very important for growth after periods of environmental stress or when defoliation is severe, as after cutting tall canopies for hay. In an experiment where roots and axillary buds were removed from tillers, only 33% TNC was lost by respiration, but when buds were left on tillers and began to develop there was an additional 50% decrease in TNC (Table 5-4). At high temperatures, where new tillers did not develop, there was an additional decrease in TNC and much TNC loss could be attributed to maintenance of respiration.

Prediction of growth responses on the basis of TNC concentration has not been consistent. Alburquerque[1] showed that regrowth may be influenced by the TNC status of the plants at the time of defoliation. He associated fast regrowth with tall fescue plants that were accumulating TNC at the time of defoliation and slower regrowth (irrespective of the TNC level) if TNC was decreasing at the time of cutting (Fig. 5-6). Apparently, TNC concentrations begin to increase after supporting vigorous root systems, new tillers, and leaf regrowth when CER exceeded respiration. Thus,

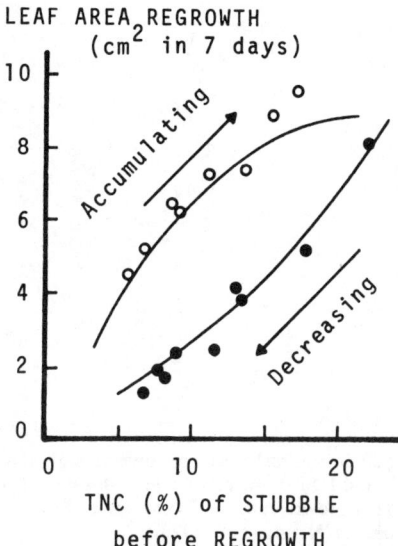

Fig. 5-6—Relationships of leaf regrowth influenced by initial TNC in stubble when TNC's were accumulating or decreasing (net utilization) at time of defoliation (Alburquerque, 1967).

the status of morphological and physiological mechanisms caused rapid regrowth after defoliation. In practice, defoliation of tall fescue during rapid growth has been observed to give rapid regrowth. When morphological changes and physiological processes caused declining TNC at the date of defoliation, high TNC is apparently needed to stimulate regrowth. LAI values of about 3.5 were needed during spring and summer growth to initiate increases in TNC in stem bases and rapid regrowth.[1] With LAI values around 1 or less, TNC decline was rapid and growth was very slow after defoliation.

The ratio of relative growth rate to relative leaf area growth rate (RGR/RLaGR), as obtained from classical growth analyses data, has biological significance in describing the partitioning of assimilates. A negative ratio indicates a loss in total plant weight, hence leaf area expansion occurs at the expense of some other plant fraction such as energy reserves in stubble and roots. A ratio between zero and unity indicates that leaf area expansion (the dominant component of growth as an energy sink), occurs at a faster rate than total plant weight. When the ratio exceeds unity, then rates of total plant growth exceed leaf area expansion rates, suggesting that sinks other than leaves are making major demands for energy supplies. These ratios identify three phases of leaf regrowth: leaf expansion may be exploitative, competitive, or noncompetitive with respect to energy needs. At unity, a state of balance exists between growth rates of leaves and total weight and is called the photosynthetic entity (Whitehead and Myerscough, 1962). The RGR/RLaGR ratio applied to regrowth data of tall fescue with varying TNC and leaf area after defoliation shows that growth increased

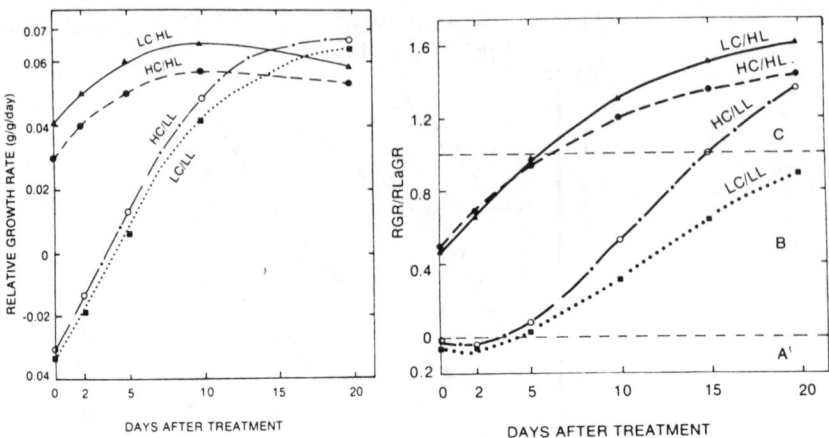

Fig. 5-7—Growth during 20 days as influenced by low and high TNC (LC and HC) and by low and high leaf area (LL and HL). Left, relative growth rate; Right, ratio of relative growth rate and relative leaf area growth rate. A) Exploitative; B) Competitive; and C) Noncompetitive leaf regrowth phases (Nelson et al., 1975).

initially at the expense of total plant weight when leaf area was low (Fig. 5-7). High TNC did not prevent this exploitative phase. Plants with high initial leaf areas began regrowth in the competitive phase which was soon nullified. It may be concluded that both leaf area and TNC contribute energy to regrowth and that the severity of defoliation or environmental stress determines the degree of dependence of energy from TNC. Because both LAI and TNC augment regrowth, management practices during the growing season should leave an adequate LAI after cutting for hay or grazing.

C. Improving Yields and Quality

Yield and quality increases can be attained by better management of given cultivars using existing knowledge. Several principles are important:
1. Encourage rapid leaf development to intercept radiation;
2. Provide adequate soil amendments;
3. Manage so that continuous rapid growth is maintained;
4. Maintain legume associations.

Allowing excessive canopy accumulation except when needed for winter grazing is not desirable. Close grazing, especially during the autumn and winter, encourages dense swards of new tillers and aids in establishment of late winter clover plantings. Drastic defoliation during periods when regrowth may be slow (e.g., a hot, dry summer) would be undesirable since soil temperature may increase and allow enchrochment by weedy species.

Improving forage quality may be realized through selection of genotypes with high carbohydrates in the herbage, rapid tillering, and increased leaf appearance rates. Presently, the improvement of quality by increased nonstructural carbohydrate concentration and concurrent yield increases seems impossible since rapid growth decreases TNC levels. For example, N fertilization increases yield and protein but percentage of utilizable energy for ruminants is not increased because of reduced TNC. Management practices and genotypes that will exploit an extended season with greater production in very early spring, late fall, and during summer would be a desirable objective for tall fescue.

LITERATURE CITED

1. Alburquerque, A. E. 1967. Leaf area and age and carbohydrate reserves in the regrowth of tall fescue tillers. Ph.D. Thesis. Univ. Microfilms. Ann Arbor, Mich. (Diss. Abstr. 68-055-49).
2. Asay, K. H., C. J. Nelson, and G. L. Horst. 1974. Genetic variability for net photosynthesis in tall fescue. Crop Sci. 14:571-574.
3. Black, C. C., Jr. 1973. Photosynthetic carbon fixation in relation to net CO_2 uptake. Annu. Rev. Plant Physiol. 24:253-286.
4. Blaser, R. E., H. T. Brant, R. C. Hommes, Jr., R. L. Bowman, J. P. Fontenot, and C. T. Pollen. 1969. Managing forages for animal production. Virginia Polytech. Inst. State Univ. Res. Bull. 45.
5. Blenkensop, P. G., and J. E. Dale. 1974. The effects of shade treatment and light intensity on ribulose-1,5-diphosphate carboxylase activity and fraction I protein level in the first leaf of barley. J. Exp. Bot. 25:899-912.
6. Boyce, K. G., D. F. Cole, and D. O. Chilcote. 1976. Effect of temperature and dormancy on germination of tall fescue. Crop Sci. 16:15-19.
7. Brown, R. H., R. E. Blaser, and J. P. Fontenot. 1963. Digestibility of fall grown Kentucky 31 fescue. Agron. J. 55:321-324.
8. Charles-Edwards, D. A., J. Charles-Edwards, and J. P. Cooper. 1971. The influence of temperature on photosynthesis and transpiration in ten temperate grass varieties. J. Exp. Bot. 22:650-662.
9. Chen, T. M., R. H. Brown, and C. C. Black. 1970. CO_2 compensation concentration, rate of photosynthesis and carbonic anhydrase activity of plants. Weed Sci. 18:399-403.
10. ————, ————, and ————. 1971. Photosynthetic $^{14}CO_2$ fixation products and activities and enzymes related to photosynthesis in bermudagrass and other plants. Plant Physiol. 47:199-203.
11. Crider, R. J. 1955. Root growth stopage resulting from defoliation of grass. USDA Tech. Bull. 1102.
12. Danielson, H. R., and V. K. Toole. 1976. Action of temperature and light on the control of seed germination in Alta tall fescue. Crop Sci. 16:317-320.
13. Elslick, R. F., and W. Vogel. 1959. Effect of soil moisture tension on ultimate emergence of grass and legume seeds. Proc. Assoc. Off. Seed Analysts N. Am. 49:151-155.
14. Jewiss, O. R., and J. Woledge. 1967. The effect of age and the rate of apparent photosynthesis in leaves of tall fescue (*Festuca arundinacea* Schreb). Ann. Bot. 31:661-671.
15. Jung, G. A., P. E. Kocher, C. F. Grass, C. C. Berg, and O. O. Bennett. 1976. Nonstructural carbohydrate in the spring herbage of temperate grasses. Crop Sci. 16:353-359.
16. Kulik, M. M., and O. L. Justice. 1967. Some influence of storage fungi temperatures and relative humidity on the germinability of grass seeds. J. Stored Prod. Res. 3:335-343.
17. Lopez, R. R., A. G. Matches, and J. D. Baldridge. 1967. Vegetative development of organic reserve of tall fescue under conditions of accumulated growth. Crop Sci. 7:409-412.
18. Matches, A. G. 1967. Influence of intact tillers and height of stubble on growth response of tall fescue (*Festuca arundinacea* Schreb). Crop Sci. 14:6-28.
19. Nelson, C. J., K. H. Asay, and G. L. Horst. 1975. Relationship of leaf photosynthesis to forage yield of tall fescue. Crop Sci. 15:476-478.
20. ————, K. J. Treharne, and E. J. Lloyd. 1975. Genetic variation in enzyme activity of

tall fescue leaf blades. Crop Sci. 15:771-774.
21. Ojima, K., and T. Isawa. 1968. The variation of carbohydrates in various species of grasses and legumes. Can. J. Bot. 46:1507-1511.
22. Oswalt, D. L., A. R. Bertrand, and M. R. Teal. 1959. Influence of nitrogen fertilization and clippings on grass roots. Soil Sci. Soc. Am. Proc. 23:228-230.
23. Robson, M. J. 1972. The effect of temperature of growth on S-170 tall fescue. Part I. Constant temperature. J. Appl. Ecol. 9:643-653.
24. Rogers, J. A., and G. E. Davis. 1973. The growth and chemical composition of four grass species in relation to soil moisture and aeration factors. J. Ecol. 61:455-472.
25. Smith, Dale. 1968. Classification of several North American grasses as starch or fructosan accumulators in relation to taxonomy. J. Br. Grassl. Soc. 23:306-309.
26. ―――. 1973. Distribution of dry matter and chemical constituents among the plant parts of six temperate—origin forage grasses at early anthesis. Univ. of Wisconsin, Madison. Res. Rep. R 2552.
27. Templeton, W. C. 1961. Some effects of temperature and light on growth and flowering on tall fescue. 2. Floral development. Crop Sci. 1:283-286.
28. ―――, G. O. Mott, R. J. Bula. 1961. Some effects of temperature and light on growth and flowering of tall fescue. 1. Vegetative development. Crop Sci. 1:216-219.
29. Treharne, K. J., and C. J. Nelson. 1975. Effect of growth temperature on photosynthetic and photorespiratory activity in tall fescue. p. 175-198. In R. Marcelle (ed.) Environmental and biological control of photosynthesis. Dr. W. Junk-Publisher, The Hague.
30. Weinmann, H. 1948. Underground development and reserve of grasses, a review. J. Br. Grassl. Soc. 3:115-140.
31. Whitehead, F. H., and P. J. Meyerscough. 1962. Growth analysis of plants. The ratio of mean relative growth rate to mean relative growth rate of leaf area increase. New Phytol. 61:314-321.
32. Williams, C. N., and E. F. Biddiscombe. 1965. Extension growth of grass tillers in the field. Aust. J. Agric. Res. 16:14-22.
33. Woledge, J. 1971. The effect of light intensity during growth on the subsequent rate of photosynthesis of leaves of tall fescue (*Festuca arundinacea* Schreb). Ann. Bot. 35:311-322.
34. ―――, and O. R. Jewiss. 1969. The effects of temperature during growth on the subsequent rate of photosynthesis in leaves of tall fescue (*Festuca arundinacea* Schreb). Ann. Bot. 33:97-117.
35. Wolf, D. D. 1964. Soil moisture extraction trends of several legume grass mixtures as effected by cutting frequency and nitrogen fertilization. Agron. J. 56:467-469.
36. Yeh, J. R., A. G. Matches, and R. L. Larson. 1976. Endogenous growth regulators and summer tillering of tall fescue. Crop Sci. 16:409-413.
37. Youngner, V. B. 1969. Physiology of growth and development. In A. A. Hanson and F. W. Juska (eds.) Turfgrass science. Agronomy 14:187-216. Am. Soc. Agron., Madison, Wis.

Chapter 6 **Cytogenetics and Genetics**

CLYDE C. BERG
SEA-USDA, U.S. Regional Pasture Research Laboratory
University Park, Pennsylvania

G. T. WEBSTER
Agronomy Department
University of Kentucky
Lexington, Kentucky

PREM P. JAUHAR
SEA-USDA
Agronomy Department
University of Kentucky
Lexington, Kentucky

Festuca is a diverse, widely adapted genus of grass. It includes species that range from diploid (2n = 14) to decaploid (2n = 70). Higher ploidy levels can be produced artificially. Fescues of agricultural importance include the diploid meadow fescue (*F. pratensis* Huds.) (2n = 14), two hexaploid (2n = 6x = 42) species, tall fescue (*F. arundinacea* Schreb.), and red fescue (*F. rubra* L.). Meadow fescue and tall fescue belong to the broad-leaved section Bovinae, whereas red fescue is classed under the section Ovinae.

Cytotypes from tetraploid to decaploid levels have been grouped under the species *F. arundinacea.* The most widely grown ecotypes and cultivars of tall fescue are bivalent forming allohexaploids. Despite the great agricultural importance of *Festuca,* its cytology and genetics have been studied by a rather limited number of scientists.

CYTOLOGY

A. Chromosome Number and Pairing

Evans (1926) determined the chromosome number as n = 21 from the pollen mother cells (PMC's) of tall fescue, and later Peto (1933) reported that 21 bivalents were usually formed at metaphase I. Crowder (1953a) studied the meiotic behavior of 247 plants from various sources. Of the 10,150 cells examined, he found 85% with 21 bivalents, 11% with 20 bivalents plus two univalents, and 4% with mostly bivalents plus univalents and multivalents. On a chromosomal basis, 99% of the chromosomes occurred as bivalents, 0.7% as univalents, and 0.3% as multivalents. Malik

Copyright © 1979 ASA-CSSA-SSSA, 677 South Segoe Road, Madison, WI 53711 USA. *Tall Fescue.*

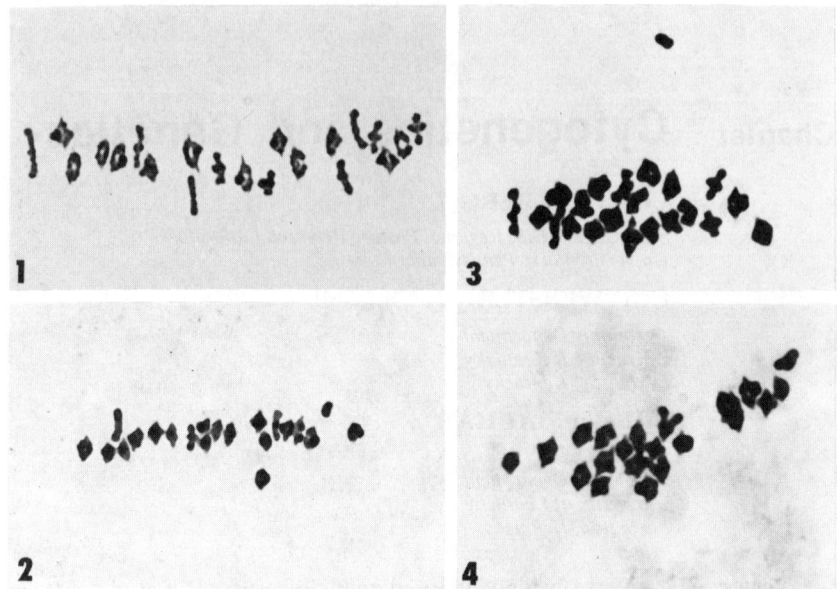

Fig. 6-1 to 6-4—Chromosome pairing at meatphase I in euploid and aneuploid plants of hexaploid tall fescue. Fig. 6-1—A typical metaphase cell with 21_{II} in euploid ($2n=6x=42$) tall fescue. Fig. 6-2—$20_{II} + 1_{I}$ in a monosomic ($2n-1=41$). Fig. 6-3—$21_{II} + 1_{I}$ in a trisomic ($2n+1=43$) plant from the British cultivar S170. Fig. 6-4—$1_{III} + 20_{II}$ in the trisomic.

and Thomas (1966c) and Jauhar (1975a, b) also reported that almost all chromosomes associated as bivalents in geographically diverse populations of tall fescue. Figure 6-1 shows a typical metaphase cell with 21 bivalents. Other reports of high frequencies of bivalent formation in tall fescue include Malik and Tripathi (1970), Chandrasekharan and Thomas (1971b), and Evans et al. (1973). Six plants studied by Myers and Hill (1947) had a higher degree of meiotic irregularity (primarily quadrivalents). In a population of 79 S_1 plants derived from a hexaploid clone of tall fescue, Carnahan and Hill (1962) observed on the average 178% as many anaphase I bridges and fragments and 235% as many quartets with micronuclei as were found in the parent clone. The increased meiotic irregularity was attributed to a reduced homeostasis in the inbred population, segregation of recessives deleterious to meiosis, and possible increased pairing of nonhomologous chromosomes.

B. Karyotype

Karyotype analysis is difficult in *Festuca* because the chromosomes are small and form a continuum from longest to shortest with little difference in arm length. Malik and Thomas (1966b) studied karyotypes of several species of *Lolium* and *Festuca,* including one indigenous Welsh cultivar 'S170' and six exotic populations of *F. arundinacea* (2n = 42). Within these populations of tall fescue they observed a gradual transition from long to short chromosomes with median to submedian centromeres. Idiograms developed by Raicu et al. (1974) for a Romanian population showed some structural differences between the Romanian population and those studied by Malik and Thomas (1966b). These variations in size and arm ratio of chromosomes and in number and location of secondary constrictions between populations have been interpreted to indicate that chromosomal differentiation has taken place within tall fescue. Malik and Thomas (1966b) contend that their karyotype analysis does not support the hypothesis that *Lolium* contributed a genome in the evolution of tall fescue. Other evidence presented later in this chapter, however, does support the hypothesis that one genome in tall fescue came from *Lolium*.

C. Genomic Constitution

Tall fescue is a typical allohexaploid with three closely related genomes designated as AABBCC (Jauhar, 1975b). The A genome was contributed by diploid meadow fescue (Nilsson, 1940; Malik and Thomas, 1966b, 1967) and probably occurs throughout the polyploid fescues (Chandrasekharan and Thomas, 1971a). Although the karyotypes developed by Malik and Thomas (1966b) do not support the proposal of Carnahan and Hill (1961) that *Lolium* donated a genome to tall fescue, the crossability of *Lolium* to tall fescue and chiasmate chromosome pairing in the hybrids (2n = 28) strongly suggest a close relationship. A similar relationship exists between *Lolium* and red fescue (Jauhar, 1975b). Perhaps the donor of B genome is extinct or it was a progenitor of one or more of the modern species of *Lolium*.

Malik and Thomas (1967), Malik and Tripathi (1970), and Malik and Mary (1971) considered the B and C (designated by them as B and B') genomes to be very similar. Therefore, they described tall fescue as an autoallohexaploid. Chandrasekharan and Thomas (1971b) also suggested that these two genomes were very closely related. In fact, as in other polyploids like wheat (*Triticum aestivum* L.), the three genomes of tall fescue are closely related and their meiotic integrity is maintained by genetic control of chromosome pairing (Jauhar, 1975a, b, 1977b; Clarke et al., 1976).

It has been proposed that *F. pratensis* and *F. arundinacea* var. *glaucescens* are the progenitors of hexaploid tall fescue (Malik and Thomas, 1967; Chandrasekharan and Thomas, 1971a). This proposal, however, awaits confirmation.

The donor of C genome is also not yet known; perhaps it is another *Festuca* species (Jauhar, 1975b, e). The identification of the donor of C genome is of particular importance because it is believed that the regulatory gene(s) which confer diploid-like meiotic behavior to tall fescue are located in this genome (Jauhar, 1975d, 1977b). In the bivalent-forming tetraploid species of *Festuca,* which can probably be genomically designated as AAXX (Jauhar, 1975e; Borrill, 1976), the identification of the donor of XX (probably CC) genome could provide an answer to this problem.

The karyotypes of *Festuca* are apparently rather flexible and vary even among ecotypes of tall fescue (Malik and Thomas, 1966b). If that is indeed true then tall fescue has changed enough that it might be very difficult to ascertain precisely the donors of all the three genomes.

D. Genetic Control of Diploid-Like Chromosome Pairing

Although tall fescue has three closely related genomes, it regularly forms 21 bivalents (Fig. 6-1). The basis of precise bivalent formation has been investigated recently. Chromosome pairing was analyzed in several euploid (2n = 42) hybrids and a monosomic (2n − 1 = 41) progeny (Bn 90-64-7) from a cross between the Algerian (Bn 273) and Israeli (Bn 488) ecotypes, and in another three monosomics isolated from crosses involving geographically diverse ecotypes of tall fescue (Jauhar, 1975a, b, c, d, e). Whereas all the euploid progeny from the cross Bn 273 × Bn 488 and its reciprocal regularly formed 21 bivalents, the sister monosomic (Bn 90-64-7) formed various multivalents in addition to bivalents. The other three monosomics of a different genotypic background also showed normal bivalent pairing. Jauhar concluded, therefore, that the missing chromosome in the monosomic Bn 90-64-7 was critical for normal bivalent pairing, and that the multivalents resulted from homoeologous pairing presumably caused by the removal of the "genetic regulator" on the missing chromosome. It was further inferred that the diploidizing gene(s) must at least be disomic in dosage to be effective in suppressing homoeologous pairing. Hence it would have no influence upon pairing in the tall fescue polyhaploids as well as in the haploid complements of tall fescue in its hybrids with other species, e.g., *L. multiflorum* × *F. arundinacea,* which show extensive homoeologous pairing. However, in the amphiploids derived from these hybrids, which have the full complement of tall fescue, and hence the regulator in the double dose, predominantly bivalent pairing and fertility are largely restored (Jauhar, 1975b). The synaptic behavior of several other hybrids and amphiploids involving polyploid fescues reported in the literature can be satisfactorily explained on the basis of Jauhar's regulation theory.

Clarke et al. (1976), studying chromosome pairing in several hybrids between hexaploid tall fescue and the tetraploid *F. pratensis* var. *apennina,* provided evidence to support this regulation theory. Jauhar (1977b) has given similarities and differences between the regulatory mechanisms controlling diploid-like pairing in wheat, oats (*Avena sativa* L.), and tall fescue.

Malik and Thomas (1966a) and Malik and Tripathi (1970) have previously suggested some type of genetic control of chromosome pairing in tall fescue. But they considered tall fescue as an autoallohexaploid. A diploidizing gene system will, however, probably not work on four allegedly homologous sets of chromosomes of tall fescue as suggested by these workers.

E. Other Ploidy Levels

Tetraploid (2n = 28), octoploid (2n = 56), and decaploid (2n = 70) ecotypes have been reported and described (Borrill et al., 1971, 1976). These include *F. arundinacea* var. *glaucescens* Boiss. (2n = 28), *F. arundinacea* var. *atlantigena* St. Yves forma *pseudo-mairei* Lit. and Maire (2n = 56), and *F. arundinacea* var. *letourneuxiana* St. Yves and *cirtensis* St. Yves (2n = 70). Malik and Thomas (1966a) examined five decaploid (2n = 70) plants and found typically 91.4% bivalents, 1.6% univalents, and 7.0% multivalents. Laggards appeared in 5% of the cells at anaphase I, and fewer than 3% of the quartets had micronuclei.

F. Polyhaploids

Polyhaploids of tall fescue have been reported by Malik (1967). However, information on chromosome pairing is available only on one polyhaploid (2n = 3x = 21) studied by Malik and Tripathi (1970). They reported an average of $4.0_{II} + 13.0_{I}$ per cell. About 40% of the cells had five to seven bivalents. These results were interpreted to indicate close affinity between two of the genomes of tall fescue which, according to the authors, "casts serious doubts on its being alloploid." Malik and Tripathi (1970), therefore, considered tall fescue essentially an autoallohexaploid. Basing such phylogenetic conclusions on the basis of chromosome pairing in polyhaploids derived from natural polyploids is, however, dangerous because most of these polyploids might have developed a genetic control on chromosome pairing (Jauhar, 1975b, e, g). The homoeologous pairing observed in the polyhaploid can be satisfactorily explained on the basis of the hemizygous ineffectiveness of gene(s) which regulate chromosome pairing in the hexaploid tall fescue (Jauhar, 1975b, 1977b).

G. Isolation of Aneuploids

Monosomics (2n = 6x − 1 = 41) have been isolated from tall fescue at the Welsh Plant Breeding Station (Jauhar, 1975a, b). Chromosome pairing in three monosomics usually formed $20_{II} + 1_{I}$ (Fig. 6-2). The monosomic Bn 90-64-7 formed many multivalents, presumably because pairing was disturbed by the lack of a critical chromosome (Jauhar, 1975b).

Some primary trisomics (2n = 6x + 1 = 43) have also been reported by Jauhar (1978). In a trisomic isolated from the British cultivar S170 of tall fescue, a mean of $0.36_{III} + 20.61_{II} + 0.68_{I}$ were observed; 59% of the cells showed $21_{II} + 1_{I}$ (Fig. 6-3) and 36% showed $1_{III} + 20_{II}$ (Fig. 6-4).

The establishment of a complete series of monosomics, and possibly nullisomics, could help in the elucidation of the regulatory mechanism controlling chromosome pairing in tall fescue.

H. B Chromosomes

Several investigators have reported on the occurrence of B chromosomes in tall fescue or its interspecific hybrids and derivatives (Peto, 1933; Myers and Hill, 1947; Crowder, 1953a, b; Malik and Thomas, 1966b; Chandrasekharan and Thomas, 1971b). Jauhar (1978) has recorded from one to 10 B chromosomes in several populations of tall fescue from Switzerland.

B chromosomes are known to affect a wide range of morphological and physiological characters. They have also been implicated in the suppression of homoeologous chromosome pairing (Hovin and Hill, 1966, p. 706; Evans and Macefield, 1972, 1973). In an F_1 hybrid of *L. perenne* × *F. arundinacea,* Bowman and Thomas (1973) found that the presence of a B chromosome was associated with a significant suppression of homoeologous pairing and hence in the reduction of chiasmata per cell. A pair of B chromosomes also brought about a reduction of multivalents in the derived amphiploids. Bowman and Thomas (1973) concluded that B chromosomes could help produce stable amphiploids if the "genetic information" from the B chromosomes could be incorporated into the normal chromosome complement. However, Jauhar (1975e, f, 1976, 1977a) has questioned the breeding significance of B chromosomes. Working on different *Lolium-Festuca* hybrids he concluded that B chromosomes are unlikely to be of much use in plant breeding because it may be very difficult, if not impossible, to transfer the "information" from B chromosomes on to A chromosomes.

INTRASPECIFIC HYBRIDS

A. Hexaploids

The nature of reproductive isolation mechanisms between the geographically diverse populations of tall fescue was investigated by Jauhar (1975a, b, c, d, e). He studied meiosis in 45 different hybrid families and some of their reciprocals from a 10 × 10 diallel set of crosses involving ecotypes from central Russia, Finland, United Kingdom, northern France, central France, Spain, Portugal, Tunisia, Algeria, and Israel. Hybrids between ecotypes of northern Europe had normal chromosome pairing and formed bivalents, but varying numbers of univalents and multivalents were observed in hybrids between the Mediterranean and the north European ecotypes.

Crosses involving the Algerian ecotype (Bn 273) and all other lines except the Israeli ecotype had extremely irregular meiosis resulting from extensive multivalent formation. In certain hybrid families both multivalents and univalents were observed at metaphase, the univalents resulting predominantly from desynapsis of the initially paired chromosomes (Jauhar,

1975b). These results were explained on the basis of genetic regulation of chromosome pairing. The Algerian genotype interferes with the regulatory mechanism in its hybrids, or the genetic interaction between the Algerian and other ecotypes leads to the inactivation of the genetic control resulting in high frequency of multivalents and, hence, in complete sterility. This is an unprecedented method of the creation of reproduction isolation barriers between ecotypes of the same species.

Evans et al. (1973) also found that hybrids between American and North African hexaploid tall fescue genotypes had highly irregular meiosis and were male-sterile. This study included nine hybrids between American selections and populations from France, the Netherlands, Israel, and North Africa. All parental genotypes showed normal pairing at first metaphase.

Malik and Thomas (1966c) studied the cytological behavior of hybrids obtained by pollinating 5 plants of each of 10 introductions with the Welsh cultivar S170. Hybrids involving populations from Israel, Spain, Morocco, and Portugal had high levels of bivalent pairing and good pollen fertility. Hybrids of one population from Israel and one from Morocco had fewer bivalents, some univalents and multivalents and less fertile pollen. Hybrids involving two populations from Algeria and two from Tunisia had very irregular meiosis with many univalents and multivalents. These plants produced very little stainable pollen. Malik (1970) studied pachytene of one cytologically irregular hybrid, Bn 271 (from Tunisia) × S170 (from Wales). Structural changes (deletions, translocations, and inversions) as well as unpaired and loosely paired segments were observed and were presumed to be major causes of sterility.

B. Tetraploids, Octoploids, and Decaploids

Chandrasekharan and Thomas (1971b) obtained a hexaploid hybrid from *F. arundinacea* var. *glaucescens* (2n = 28) × *F. arundinacea* var. *atlantigena* forma *pseudo-mairei* (2n = 56). Over 90% of the chromosomes paired with the majority associated as bivalents. Univalents and multivalents were infrequent and most cells had from one to three anaphase I laggards. Anthers, however, did not dehisce and the pollen was sterile. Crosses between *F. arundinacea* (2n = 42 + 2B) × *F. arundinacea* var. *atlantigena* f. *pseudo-mairei* (2n = 56 + 2B) resulted in four hybrids. The average cell had 35 chromosomes paired as bivalents, five chromosomes in multivalent configurations, and nine unpaired chromosomes. Pollen stainability varied and some had dehiscent anthers. Fifteen F_2 plants were established.

A high frequency of bivalent pairing occurred in the three hybrids from *F. arundinacea* var. *letourneuxiana* and *cirtensis* (2n = 70) × *F. arundinacea* var. *atlantigena* f. *pseudo-mairei* (2n = 56). From 78.7 to 94.2% of the pollen stained; anthers dehisced normally; 5.6 to 12.5% of the florets set seed on selfing. A completely male-sterile hybrid with a high frequency of bivalents resulted from the cross *F. arundinacea* var. *cirtensis* (2n = 70) × *F. arundinacea* var. *glaucescens* (2n = 28).

Tall fescue as such has a wide geographical range. However, a large proportion of the introduced material studied by Malik and Thomas (1966c) came from populations that were small and isolated. These populations probably formed small restricted breeding units that would encourage random fixation of chromosomal rearrangements. Tall fescue's wide geographical distribution evidences its high adaptive potentialities, which may be due in part to polyploidy. Chromosomal deviations have a far better opportunity to survive in polyploids than in diploids. Reorganizations could enhance the selective value of an organism in a particular environment. There is a good chance for altered chromosomal arrangements to be maintained in a population because tall fescue reproduces vegetatively, is a hexaploid, and is a perennial. Malik and Thomas (1966b) point out that extensive polyploidy, wide geographical distribution, and occurrence of subspecies, varieties, and subvarieties within the several species of *Festuca* suggest that the genus is phylogenetically very old. However, they (Malik and Thomas, 1966c) also suggest that early stages of speciation are occurring because of the effective isolating mechanisms that exist among populations. According to Jauhar (1975b) the reproductive isolation, which stems from the breakdown in the hybrids of the regulatory mechanism controlling chromosome pairing, is clearly of evolutionary significance in: (a) permitting introgression of characters between related taxa; (b) the creation of effective sterility barriers for further speciation.

INTERSPECIFIC HYBRIDS

Numerous attempts have been made to intercross different species of *Festuca*. Some produced hybrids but many were unsuccessful. Often embryos aborted or seed failed to germinate. Better success was frequently obtained when a plant with lower chromosome number was used as the female parent. The success or failure of an attempted cross could be attributed in some instances to the genotypes used. Quite often small numbers of plants were used. Some of the earlier hybrids were spontaneous, but most were the result of controlled pollinations with or without emasculation.

A. Tall Fescue × Meadow Fescue

Hybrids between *F. pratensis* Huds. (2n = 14) and *F. arundinacea* have been recorded by several investigators (Jenkin, 1933, 1955; Nilsson, 1940; Myers and Hill, 1947; Crowder, 1953c, 1956; Hertzsch, 1961; Sulinowski, 1966a, 1972a; Sulinowski et al., 1968; Malik, 1967; Malik and Thomas, 1967; Chandrasekharan and Thomas, 1971a). All reported male-sterile hybrids with 2n = 28 had indehiscent anthers. Nilsson (1940) observed seven bivalents and 14 univalents as normally occurring in the F_1 hybrids, with occasional trivalents and quadrivalents. All the progeny of the F_1's had chromosome numbers higher than 2n = 28.

Crowder (1953c, 1956) found all stages of meiosis to be irregular in hybrids from reciprocal crosses. Multivalents and univalents were prevalent at metaphase I. However, 14 bivalents occurred occasionally, indicating that the meadow fescue genome apparently paired with those from tall fescue and that the tall fescue chromosomes paired among themselves.

In addition to the expected tetraploid hybrids ($2n = 28$), Malik and Thomas (1967) obtained three pentaploids ($2n = 35$) from *F. pratensis* × *F. arundinacea*. In the pentaploids 52% of the chromosomes were bivalents (9.1 chromosomes/cell), 34% were multivalents (10.0/cell), and 14% were unpaired (4.8/cell). These interspecific pentaploids may have resulted from an unreduced *F. pratensis* gamete being fertilized by 21-chromosome pollen from *F. arundinacea*. In the tetraploid hybrids the mean number of bivalents per cell was 9.4 (range = 5 to 14). Sulinowski et al. (1968) also obtained hybrids with $2n = 35$. Ten F_1's were $2n = 35$ and 141 were $2n = 28$. Five of the pentaploid hybrids had indehiscent anthers, whereas the other five had partial release of pollen. Under open pollination the plants that produced some normal-appearing pollen produced some germinable seed.

Sulinowski (1972a) obtained two 28-chromosome F_1 hybrids by crossing *F. pratensis* ($2n = 14$) × *F. arundinacea* ($2n = 42$). These hybrids showed the typical meiotic aberrations of interspecific hybrids; lagging chromosomes and bridges at anaphase and micronuclei in tetrads. Colchicine treatments of both F_1's yielded partially fertile plants in the C_1 generation, with a preponderance of $2n = 42$ and three with $2n = 56$. Some of the sterile C_1's had chromosome complements of $2n = 35$, 42, and 77. Partially fertile hexaploid C_2's displayed considerable aberrations of meiotic divisions in PMC's. The partially fertile offspring of octoploid forms were hexaploid and octoploid. The sterile C_2 progeny included plants with chromosome complements of 35, 39, 42, and 70.

B. Tall Fescue × Giant Fescue

Most reports on giant fescue (*F. gigantea* (L.) Vill.) show $2n = 42$, although Sulinowski (1972a) reported Polish forms with $2n = 28$. F_1 plants produced by Jenkin (1933) from *F. arundinacea* × *F. gigantea* and reciprocal cross were male-sterile and failed to set seed when pollinated by either parent. The cytological study of these plants by Peto (1933) showed $2n = 42$ and an average of 13.9 univalents per cell (range = 10 to 19). Nilsson (1935) found meiosis of F_1 hybrids of tall fescue × giant fescue to be similar to that shown by Peto. Two plants were established from seed produced by surrounding the F_1 plant with parent plants. One resembled a hybrid between tall fescue and meadow fescue and was completely sterile ($2n = 28$). The other derivatives had 84 chromosomes and were meiotically stable with viable pollen.

Sulinowski (1972b) treated a $2n = 42$ sterile hybrid from *F. gigantea* × *F. arundinacea* with colchicine. Treated plants with dehiscent anthers and stainable pollen were found to be $2n = 56$, 70, and 84. Those with negligible anther dehiscence had 2n = 56, 58, 61, and 63 and forms with indehiscent anthers included 2n = 50, 53, and 54 chromosomes. About 65% of the C_2 plants from self-pollination of two 2n = 70 C_1 plants reproduced the maternal cytotype (2n = 70) with other forms carrying 2n = 49, 53, 56, 59, and 63 chromosomes. About 50% of the offspring from self-pollination of the 2n = 56 C_1 plant carried the maternal cytotype, with the remaining C_2 plants carrying 2n = 47, 49, 51, and 53 chromosomes.

Hybrids and hybrid derivatives from the reciprocal cross *F. arundinacea* × *F. gigantea* were morphologically and cytologically similar (Buckner et al., 1976). The F_1 hybrids were sterile and meiotically irregular. They showed extensive homoeologous pairing as evidenced by the formation of hexavalents, pentavalents, quadrivalents, and trivalents (Jauhar et al., 1978b). Amphiploids were highly male-fertile, based on anther dehiscence (Buckner et al., 1976). First generation amphiploids varied from 2n = 80 to 84, with high frequencies of bivalent formation and with mean numbers of micronuclei per quartet ranging from 1.2 to 2.5. Twenty-five fourth generation amphiploids varied from 2n = 53 to 84 with eight having 2n = 84. During four generations, a total of 424 amphiploids ranged in seed set from 0.01 to 0.9 g per panicle with a mean of 0.32 g (tall fescue averaged 0.34 g/panicle).

Genetic information can probably be transferred from one species of *Festuca* to another. The frequency of success in obtaining hybrids perhaps depends upon the genotype of the parents. The high incidence of interspecific pairing in the hybrids suggests that genetic information could be exchanged. However, the frequent occurrence of male sterility and sometimes female sterility limits the success of transferring genetic information from one species to another. It appears that genetic information can be transferred among many of the species and varieties of *Festuca,* but the process will probably be time consuming.

C. Tall Fescue × Other Fescues

Chandrasekharan and Thomas (1971a) studied more than 100 hybrids from five genotypes of *F. arundinacea* var. *glaucescens* (2n = 28) crossed to six genotypes of *F. mairei* (2n = 28). Most of the chromosomes were associated as bivalents, but pollen stainability was low (5 to 16%) and the anthers were nondehiscent. Pollination of the hybrid with *F. mairei* indicated that the hybrid was female-sterile.

Chandrasekharan and Thomas (1971b) obtained four hybrids from crossing *F. arundinacea* var. *atlantigena* f. *pseudomairei* (2n = 56) × *F. mairei* (2n = 28). A large proportion of the PMC's had 21 bivalents, but they varied in chiasma frequency. They were partially fertile with dehiscent anthers and 34 to 51% sterile pollen. On selfing, 12% of the florets set seed, 35% of which germinated to give a population of 21 plants. Crosses between the hexaploid F_1's and the natural hexaploid *F. arundinacea* were unsuccessful.

Only one hybrid plant was obtained from *F. arundinacea* var. *cirtensis* (2n=70) × *F. mairei* (2n=28). Instead of the expected number of 49, this hybrid had 63 chromosomes, and most of them paired as bivalents. Anthers dehisced, 57.6% of the pollen stained, and there was a 9.3% seed set from selfing (Chandrasekharan and Thomas, 1971b).

F. arundinacea (2n=42) × *F. mairei* (2n=28) yielded five hybrids with 35 chromosomes plus one B chromosome (Chandrasekharan and Thomas, 1971b). Because of extreme chromosome stickiness, only two of the five were studied cytologically, and these hybrids were irregular meiotically. Pollen stainability was low and the anthers failed to dehisce.

INTERGENERIC HYBRIDS

A. Ryegrass × Tall Fescue

Hybrids and hybrid derivatives of crosses involving both annual (*L. multiflorum* Lam.) and perennial (*L. perenne* L.) ryegrass and tall fescue have been studied (Peto, 1933; Crowder, 1953c; Buckner, 1960; Buckner et al., 1961, 1963, 1965; Hill and Buckner, 1962; Hill and Carnahan, 1962; Hovin and Hill, 1965; Lewis, 1966; Malik and Thomas, 1966a; Sulinowski, 1966b; Webster and Buckner, 1971; Jauhar, 1975b; Jauhar et al., 1978a). Terrell (1966) summarized the results with regard to taxonomic relationships.

In general, the cytological observations of F_1 hybrids reported by different investigators were similar. With one exception, all showed 2n=28, presumably seven chromosomes from ryegrass and 21 from tall fescue. Two F_1 plants from *L. perenne* × *F. arundinacea* reported by Peto (1933) had 2n=30. He indicated that the two extra chromosomes might have resulted from fragmentation because they were smaller than the others. In all cases anthers failed to dehisce and little or no viable seed was produced.

Meiotic configurations varied greatly within and among plants. A high percentage of pairing was bivalent with a lower frequency for univalents and multivalents. An average of all meiotic configurations approached seven bivalents plus seven univalents, suggesting a phylogenetic relationship between the ryegrass genome and one from tall fescue. Occasionally 14 bivalents were observed, indicating not only pairing of chromosomes from tall fescue with those from ryegrass, but also inter-genomic pairing within the tall fescue complement.

From the backcross of one *L. perenne* × *F. arundinacea* hybrid to *L. perenne,* Peto (1933) observed two plants with 23 and one with 22 chromosomes. He found one plant from the F_1 × *F. arundinacea* with approximately 37 chromosomes. All four plants were meiotically irregular and pollen-sterile. Further backcrosses to tall fescue were all male-sterile with variable chromosome numbers. Progeny of second and third backcrosses to *L. perenne* had chromosome numbers ranging from 14 to 18, with meiosis more regular than earlier generations and, in general, viable pollen.

In more than 100 backcrosses of one to four generations involving both annual and perennial ryegrass × tall fescue F_1 hybrids backcrossed to annual ryegrass and to tall fescue, and 2n = 56 colchicine-induced amphipoloids and progenies, chromosome numbers varied from 28 to 56 (Hill and Buckner, 1962). About 4% had 28 chromosomes, 12% had 35, 81% had 42, 1% had 49, and 1% had 56. The occurrence of progenies with chromosome numbers higher than either parent implies that nonreduction or partial nonreduction occurred in both megaspores and microspores. Little relationship was found between meiotic regularity and fertility as measured by pollen stainability.

Malik and Thomas (1966a) obtained 42-chromosome hybrids from *L. multiflorum* (2n = 14) × *F. arundinacea* (2n = 70) crosses. The seven *Lolium* chromosomes were easily distinguishable by their large size and pairing was extremely variable. The most frequent association was 14 bivalents and 14 univalents. A high percentage of ring bivalents suggested close homologies between the two sets of *Festuca* genomes.

Fertility in hybrids between ryegrass and tall fescue has been restored by colchicine treatment. Buckner et al. (1961, 1963, 1965) obtained several fertile 56-chromosome hybrids of annual ryegrass and tall fescue. A high percentage of the progenies of amphiploids maintained the 56-chromosome level of ploidy through three generations of selection. However, a few 42- and 49-chromosome derivatives were found. Progenies of crosses between 28-chromosome hybrids and 56-chromosome amphiploids would normally be expected to have 42 chromosomes. However, in addition to two plants with 42 chromosomes, 14 were found to have 56, indicating a high degree of nonreduction in the egg cells of the F_1 hybrids. Subsequent interpollination of 56-chromosome amphiploids selected for fertility and seed set through six generations resulted in meiotically unstable plants representing nearly the entire range of 42 to 56 chromosomes. Within one line, meiotically stable plants (2n = 42) were recovered from crosses between amphiploids and F_1 hybrids of annual ryegrass and tall fescue (Webster and Buckner, 1971).

Lewis (1966) obtained fertile amphiploids between *L. multiflorum* and *F. arundinacea* by doubling either the chromosome numbers of the hybrids or of the parents before crossing. Cytological analysis indicated a degree of inter- and intra-genomic pairing among the chromosomes of the sterile F_1 hybrids and largely bivalent pairing in the amphiploids. The C_1 generation showed a drastic rate of chromosome loss from the C_0 generation, with nearly 50% of the plants having fewer than 54 chromosomes. The degree of aneuploidy was appreciable in the C_1 generation but more restricted in the C_2. Jauhar (1975b) has described chromosome pairing in several euploid (2n = 56) and aneuploid (2n = 52, 53, 54) amphiploids. It was observed that the aneuploids generally had more homoeologous pairing resulting in higher multivalent formation than the euploids which showed predominantly bivalent pairing. The excessive homoeologous pairing in the aneuploids was interpreted as being due to the loss of some chromosome(s) carrying the diploidizing gene(s).

Jauhar et al. (P. P. Jauhar, R. C. Buckner, and L. P. Bush. 1977. Chromosomal status and meiotic stability of KENHY—an intergeneric hybrid derivative. p. 59. Agron. Abstr. Am. Soc. Agron., Madison, Wis.) studied the cytology of the cultivar 'Kenhy' ($2n=42$), an intergeneric hybrid derivative. It has most of the chromosomes, and hence characters, of tall fescue. However, some of the tall fescue chromosomes have probably been substituted by the ryegrass chromosomes. Moreover, Kenhy exhibits chromosomal polymorphism due to the addition of a presumably alien centric fragment and also due to segmental interchange of *Lolium-Festuca* chromosomes which could have occurred as a result of rigorous selection for superior forage quality of the hybrid derivatives. Since the regulatory mechanism controlling chromosome pairing (Jauhar, 1975b) seems intact in Kenhy it is remarkably stable meiotically. Thus, Kenhy is probably the first "alien substitution line" ever released for general cultivation.

B. Perennial Ryegrass × Meadow Fescue × Tall Fescue Trispecies Hybrids

Hill and Carnahan (1962) developed trispecific hybrids involving perennial ryegrass, meadow fescue, and tall fescue. Triploid F_1 hybrids from *Lolium perenne* ($2n=14$) × *Festuca pratensis* ($2n=28$) crossed with tall fescue ($2n=42$) were tetraploid ($2n=28$) and pentaploid ($2n=35$). Autoallohexaploids ($2n=42$) crossed with tall fescue produced sterile trispecific hybrids ($2n=42$). Various derivatives of the triploid F_1 hybrids, including colchicine-induced autoallohexaploids and open-pollinated progenies, gave chromosome numbers of 14, 21, 28, 35, 42, and 49.

In a follow-up study, Hovin and Hill (1965) found that all 33 BC_1 plants, 180 open-pollination progenies, 160 polycross, and 47 single-cross progenies had $2n=42$ chromosomes, which demonstrated numerical stability at the hexaploid level of the BC_1 generation.

GENETICS

Studies on character inheritance in tall fescue are very meager. Crowder (1953b) described the ciliated auricles of tall fescue as a readily identifiable character that distinguished this species from meadow fescue. Various investigators have noted dominance or partial dominance of this character in hybrids with species having nonciliated auricles.

Hovin and Hill (1965) placed a value of 25% heritability for seed yield through parent-offspring regression in their trispecific hybrids. Burton and DeVane (1953) calculated broad-sense heritability for several traits from 49 tall fescue clones. The expected gain from selection (S) for forage yields, forage yield ratings, greenness, and disease resistance indicated that an advance of 33 to 72% over the population mean could be made by selecting the top 5% of the population. The S value for seed yield was 161% greater than the mean.

Buckner et al. (1973) obtained broad-sense heritability estimates ranging from 57 to 80% over a 3-year period for the alkaloid perloline in ryegrass-tall fescue hybrids that had previously been selected for improved forage quality. Cornelius et al. (1974) studied perloline content among nine clones derived from hybrids of tall fescue and annual ryegrass. A modified partial diallel analysis showed that concentration of perloline was controlled by a few major genes with a high degree of dominance for low perloline. However, some genes with smaller effects may exist. Asay et al. (1974) noted significant differences in net C exchange among 25 genetically diverse clonal lines and polycross progenies. On dates when differences among parental lines were significant, broad-sense heritability values ranged from 57 to 83% and narrow-sense values ranged from 22 to 44%. In those instances, expected genetic advance with selection of the upper 5% of the population ranged from 0.77 to 3.13 mg CO_2 dm^{-2} $hour^{-1}$. Genetic variation was from four to six times greater among the parental lines than their polycross progenies. Asay et al. (1974) concluded that genetic progress could be made through selection for net C exchange in tall fescue but that the environment had to be considered during selection.

Wilkins et al. (1974) found that a *L. multiflorum* × *F. arundinacea* amphiploid hybrid was resistant to several diseases that occurred on one or the other of the parent species. Backcrossing to *L. multiflorum* resulted in decreasing resistance to diseases that normally occur on this ryegrass. A desired combination of traits obtained through hybridization may be lost by backcrossing.

The limited number of genetic studies that have been made with tall fescue prevents us from reaching definitive conclusions about the genetics of many traits in tall fescue. However, as tall fescue is a hexaploid, has a wide geographical distribution, and rather high heritability estimates for most of the traits that have been studied, it seems reasonable to predict that selection can be successful for almost any trait for which selection can be applied. Additional genetic variability surely must exist in closely related species and genera (*Lolium*). In some situations it may take considerable time and effort to exchange genetic information among specific genotypes.

LITERATURE CITED

1. Asay, K. H., C. J. Nelson, and G. L. Horst. 1974. Genetic variability for net photosynthesis in tall fescue. Crop Sci. 14:571-575.
2. Borrill, M. 1972. Studies in *Festuca*. III. The contribution of *F. scariosa* to the evolution of polyploids in sections Bovinae and Scariosae. New Phytol. 71:523-532.
3. ―――. 1976. Temperate grasses. *Lolium, Festuca, Dactylis, Phleum, Bromus* (Gramineae). p. 137-142. In N. W. Simmonds (ed.) Evolution of crop plants. Longman, London and New York.
4. ―――, B. F. Tyler, and M. Lloyd-Jones. 1971. Studies in *Festuca*. I. A chromosome atlas of *Bovinae* and *Scariosae*. Cytologia 36:1-14.
5. ―――, ―――, and W. G. Morgan. 1976. Studies in *Festuca*. 7. Chromosome atlas (Part 2). An appraisal of chromosome race distribution and ecology, including *F. pratensis* var. *appennina* (De Not) Hack.-tetraploid. Cytologia 41:219-236.
6. Bowman, J. G., and Hugh Thomas. 1973. B chromosomes and chromosome pairing in

Lolium perenne × *Festuca arundinacea* hybrid. Nature New Biol. (London) 245:80–81.
7. Buckner, R. C. 1960. Cross-compatibility of annual and perennial ryegrass with tall fescue. Agron. J. 52:409–410.
8. ―――, L. P. Bush, and P. B. Burrus, II. 1973. Variability and heritability of perloline in *Festuca* sp., *Lolium* sp., and *Lolium-Festuca* hybrids. Crop Sci. 13:666–669.
9. ―――, Helen D. Hill, and P. B. Burrus, II. 1961. Some characteristics of perennial and annual ryegrass × tall fescue hybrids and of the amphiploid progenies of annual ryegrass × tall fescue. Crop Sci. 1:75–80.
10. ―――, ―――, A. W. Hovin, and P. B. Burrus, II. 1963. Cytogenetic and morphological characteristics of progenies of crosses of annual ryegrass × tall fescue hybrids and their amphiploid derivatives. Crop Sci. 3:453–454.
11. ―――, ―――, ―――, and ―――. 1965. Fertility of annual ryegrass × tall fescue amphiploids and their derivatives. Crop Sci. 5:395–397.
12. ―――, G. T. Webster, P. B. Burrus, II, and L. P. Bush. 1976. Cytological, morphological and agronomic characteristics of tall × giant fescue hybrids and their amphiploid progenies. Crop Sci. 16:811–816.
13. Burton, Glenn W., and E. H. DeVane. 1953. Estimating heritability in tall fescue (*Festuca arundinacea*) from replicated clonal material. Agron. J. 45:478–481.
14. Carnahan, H. L., and Helen D. Hill. 1961. Cytology and genetics of forage grasses. Bot. Rev. 27:1–162.
15. ―――, and ―――. 1962. Increased meiotic irregularities in inbred progenies of a *Festuca arundinacea* clone with apparent inversion heterozygosity. Crop Sci. 2:445–446.
16. Chandrasekharan, P., and Hugh Thomas. 1971a. Studies in *Festuca*. 5. Cytogenetic relationships between species of *Bovinae* and *Scariosae*. Z. Pflanzenzüchtg. 65:345–354.
17. ―――, and ―――. 1971b. Studies in *Festuca*. 6. Chromosome relationships between *Bovinae* and *Scariosae*. Z. Pflanzenzüchtg. 66:76–86.
18. Clarke, J., P. Chandrasekharan, and Hugh Thomas. 1976. Studies in *Festuca*. 9. Cytological studies of *Festuca pratensis* var. *apennina* (DeNot) Hack. (2n=28). Z. Pflanzenzüchtg. 77:205–214.
19. Cornelius, P. L., R. C. Buckner, L. P. Bush, P. B. Burrus, II, and J. Byars. 1974. Inheritance of perloline and content in annual ryegrass × tall fescue hybrids. Crop Sci. 14:896–898.
20. Crowder, Loy V. 1953a. A survey of meiotic chromosome behavior in tall fescue grass. Am. J. Bot. 40.348–354.
21. ―――. 1953b. A simple method for distinguishing tall and meadow fescue. Agron. J. 45:453–454.
22. ―――. 1953c. Interspecific and intergeneric hybrids of *Festuca* and *Lolium*. J. Hered. 44:195–203.
23. ―――. 1956. Morphological and cytological studies in tall fescue (*Festuca arundinacea* Schreb.) and meadow fescue (*F. elatior* L.). Bot. Gaz. 117:214–223.
24. Evans, Gwilym. 1926. Chromosome complements in grasses. Nature (London) 118:841.
25. Evans, G. M., K. H. Asay, and R. G. Jenkins. 1973. Meiotic irregularities in hybrids between diverse genotypes of tall fescue (*Festuca arundinacea* Schreb.). Crop Sci. 13:376–379.
26. ―――, and A. J. Macefield. 1972. Suppression of homoeologous pairing by B chromosomes in a *Lolium* species hybrid. Nature New Biol. (London) 236:110–111.
27. ―――, and ―――. 1973. The effect of B chromosomes on homoeologous pairing in species hybrids. I. *Lolium temulentum* × *L. perenne*. Chromosoma (Berlin) 41:63–73.
28. Hertzsch, W. 1961. Gattungskreuzungen zwischen den Gattungen *Festuca* und *Lolium*. C. Die F_1-Bastarde, ihr Verhalten und ihr Aussehen. Z. Pflanzenzüchtg. 45:345–360.
29. Hill, Helen D., and R. C. Buckner. 1962. Fertility of *Lolium-Festuca* hybrids as related to chromosome number and meiosis. Crop Sci. 2:484–486.
30. ―――, and H. L. Carnahan. 1962. *Lolium perenne* L. × induced tetraploid *Festuca elatior* L. and hybrids with *F. arundinacea* Schreb. Crop Sci. 2:245–248.
31. Hovin, A. W., and Helen D. Hill. 1965. Intergeneric *Lolium-Festuca* trispecific derivatives with increased fertility. Crop Sci. 5:257–260.
32. ―――, and ―――. 1966. B chromosomes, their origin and relation to meiosis in interspecific *Lolium* hybrids. Am. J. Bot. 53:702–708.
33. Jauhar, Prem P. 1975a. Genetic control of diploid-like meiosis in hexaploid tall fescue. Nature (London) 254:595–597.
34. ―――. 1975b. Genetic regulation of diploid-like chromosome pairing in the hexaploid

species, *Festuca arundinacea* Schreb. and *F. rubra* L. (Gramineae). Chromosoma (Berlin) 52:363-382.

35. _____. 1975c. Genetic control of chromosome pairing in polyploid fescues and its breeding implications. p. 167-170. *In* B. Nüesch (ed.) Proc. EUCARPIA Conf. on Polyploidy in Fodder Plants. Zurich, 1975. Eucarpia, Zurich-Reckenholz.

36. _____. 1975d. Grass cytogenetics. Annu. Rep. of the Welsh Plant Breed. Stn. for 1974. p. 71-73.

37. _____. 1975e. Genetic control of chromosome pairing in polyploid fescues: its phylogenetic and breeding implications. Annu. Rep. Welsh Plant Breed. Stn. for 1974. p. 114-127.

38. _____. 1975f. Chromosome relationships between *Lolium* and *Festuca* (Gramineae). Chromosoma (Berlin) 52:103-121.

39. _____. 1975g. Polyploidy, genetic control of chromosome pairing and evolution in the *Festuca-Lolium* complex. Heredity (London) 35:430.

40. _____. 1976. Chromosome pairing in some triploid and trispecific hybrids in *Lolium-Festuca* and its phylogenetic implications. p. 165-177. *In* P. L. Pearson and K. R. Lewis (eds.) Chromosomes today Vol. 5. John Wiley and Sons, New York.

41. _____. 1977a. B-chromosomes in relation to forage breeding. *In* R. W. Downes (ed.) Proc. 3rd Int. Congr. of SABRAO, Canberra, section 14(b):34-39. Canberra, Australia, 11-18 Feb. 1977.

42. _____. 1977b. Genetic regulation of diploid-like chromosome pairing in *Avena*. Theor. Appl. Genet. 49:287-295.

43. _____. 1978. Primary trisomy in tall fescue. J. Hered. 69:217-223.

44. _____, Robert C. Buckner, and Lowell P. Bush. 1978b. Chromosome pairing in interspecific hybrids and amphiploids between tall fescue and giant fescue. Proc. XIV Int. Congr. Genetics, Moscow. 21-30 Aug. 1978. USSR Acad. Sci., Moscow. (In press).

45. _____, _____, _____, E. M. Thacker, and P. B. Burrus, II. 1978a. Synthesis and meiotic stability of low-perloline derivatives produced by intergeneric hybridization between ryegrass and tall fescue. Genetics 88:s47-s48.

46. Jenkin, T. J. 1933. Interspecific and intergeneric hybrids in herbage grasses. Initial crosses. J. Genet. 28:205-264.

47. _____. 1955. Interspecific and intergeneric hybrids in herbage grasses. IX. *Festuca arundinacea* with one other *Festuca* species. J. Genet. 53:81-93.

48. Lewis, E. J. 1966. The production and manipulation of new breeding material in *Lolium-Festuca*. p. 688-693. *In* A. G. G. Hill (ed.) Int. Grassl. Congr., Proc. 10th Helsinki, Finalnd, 7-16 July 1966.

49. Malik, C. P. 1967. Hybridization of *Festuca* species. Can. J. Bot. 45:1025-1029.

50. _____. 1970. Pachytene studies in the intraspecific hybrid of *Festuca arundinacea* Schreb. (2n=42). Genet. Pol. 11:379-383.

51. _____, and T. H. Mary. 1971. Probable genotype of the B˙B˙ genome in *Festuca arundinacea* Schreb. (2n=42). Z. Biol. (Munich) 116:501-506.

52. _____, and P. T. Thomas. 1966a. Meiosis in the intergeneric hybrid beween *Lolium multiflorum* (2n=14) × *Festuca arundinacea* (2n=70) and its amphiploid (2n=84). Z. Pflanzenzüchtg. 55:81-94.

53. _____, and _____. 1966b. Karyotypic studies in some *Lolium* and *Festuca* species. Caryologia 19:167-196.

54. _____, and _____. 1966c. Chromosomal polymorphism in *Festuca arundinacea*. Chromosoma (Berlin) 18:1-18.

55. _____, and _____. 1967. Cytological relationships and genome structure of some *Festuca* species. Caryologia 20:1-39.

56. _____, and R. C. Tripathi. 1970. Mode of chromosome pairing in the polyhaploid tall fescue (*Festuca arundinacea* Schreb. 2n=42). Z. Biol. (Munich) 116:332-339.

57. Myers, W. M., and Helen D. Hill. 1947. Distribution and nature of polyploidy in *Festuca elatior* L. Bull. Torrey Bot. Club 74:99-111.

58. Nilsson, Fredrik. 1935. Amphiploidy in the hybrid *Festuca arundinacea* × *F. gigantea*. Hereditas 20:181-198.

59. _____. 1940. The hybrid *Festuca arundinacea* × *F. pratensis* and some of its derivatives. Bot. Not. 1940:33-50.

60. Peto, F. H. 1933. The cytology of certain intergeneric hybrids between *Festuca* and *Lolium*. J. Genet. 28:113-156.

61. Raicu, P., Rodica Chirila, and E. Kellner. 1974. Chromosomal complement of some Romanian populations of *Lolium* and *Festuca*. Rev. Roum. Biol. 19:205-209.
62. Sulinowski, S. 1966a. Preliminary studies on interspecific and intergeneric hybrids in grasses of the *Festuca* and *Lolium* genera. Genet. Pol. 7:13-25.
63. ―――. 1966b. Intergeneric hybrid *Lolium multiflorum* Lam. (2n=14) ×*Festuca arundinacea* Schreb. (2n³42). Part I. F_1 characteristics. Backcrossing results obtaining of alloploids. Genet. Pol. 7:81-98.
64. ―――. 1972a. Induced alloploids in grasses of *Festuca* and *Lolium* genera. Part I. *Festuca pratensis* Huds. (2n=14) × *F. arundinacea* Schreb. (2n=42) hybrid derivatives. Genet. Pol. 13(1):91-106.
65. ―――. 1972b. Induced alloploids in grasses of *Festuca* and *Lolium* genera. III. *Festuca gigantea* (L.) Vill. (2n=42) × *F. arundinacea* Schreb. (2n≈42) hybrid derivatives. Genet. Pol. 13(4):91-107.
66. ―――, K. Sekowska, and U. Luty. 1968. Results of interspecific crosses *Festuca pratensis* Huds. (2n=14) × *Festuca arundinacea* Schreb. (2n=42) performed without applying emasculation. Genet. Pol. 9:77-85.
67. Terrell, Edward E. 1966. Taxonomic implications of genetics in ryegrass (*Lolium*). Bot. Rev. 32:138-164.
68. Webster, G. T., and R. C. Buckner. 1971. Cytology and agronomic performance of *Lolium-Festuca* hybrid derivatives. Crop Sci. 11:109-112.
69. Wilkins, P. W., A. J. H. Carr, and E. J. Lewis. 1974. Resistance of *Lolium multiflorum/Festuca arundinacea* hybrids to some diseases of their parent species. Euphytica 23:315-320.

Chapter 7

Breeding and Cultivars

K. H. ASAY
SEA-USDA
Utah State University
Logan, Utah

ROD V. FRAKES
Farm Crops Department
Oregon State University
Corvallis, Oregon

ROBERT C. BUCKNER
SEA-USDA
University of Kentucky
Lexington, Kentucky

Tall fescue (*Festuca arundinacea* Schreb.) which was introduced from Europe in the mid-1800's, is an allohexaplod ($2n = 6x = 42$) with a genomic constitution AABBCC (Jauhar, 1975a, b; Clarke et al., 1976). The species received little attention prior to 1930, partially because of its early confusion with meadow fescue (*Festuca elatior* L.). The species was regarded as *F. elatior* var. *arundinacea* (Schreb.) Winn. until 1950 when it was given the name *F. arundinacea* Schreb. (Cowan, 1956).

BREEDING PROCEDURES

Initial breeding efforts were based on isolation of selected introductions or naturalized populations. Two cultivars, 'Kentucky 31' and 'Alta', both of which arose largely through natural selection, were instrumental in the early expansion of the species and they still form the bulk of the tall fescue acreage in the United States. The development of these two cultivars is discussed in Chapter 1.

After the release of Kentucky 31 and Alta, tall fescue rapidly became established as a major pasture grass. With the increased popularity came the demand for new cultivars with improved forage quality and adaptation to more specific environmental and management regimes. As with most cross-fertilized forage species, breeding procedures were adapted from those used in other crops, primarily corn (*Zea mays* L.). Breeding approaches varied according to the objectives of the program, resources available, and personal preferences of the breeder. Modification of mass selection and recurrent selection schemes as outlined by Hanson and Carnahan (1956) have been commonly used. Many breeders have used one to two cycles of pheno-

Copyright © 1979 ASA-CSSA-SSSA, 677 South Segoe Road, Madison, WI 53711 USA. *Tall Fescue.*

typic selection in spaced-plant nurseries followed by progeny testing under sward conditions to facilitate genetic improvement in quantitatively inherited traits.

Johnson (1952) reviewed the progeny testing procedures used in forage breeding. He concluded that polycross (PC), open pollination (OP), and topcross (TC) progeny tests were all useful in screening non-inbred plants for general combining ability. He recognized the need for single crosses (SC) to evaluate breeding lines for specific combining ability. Echeverri (1964) found that selfed (S_1), SC, OP, and PC progeny tests were equally effective in detecting differences in general combining ability for seed yield and associated characters in tall fescue clonal lines. Thomas and Frakes (1967) later reported that results from clonal evaluation were similar to those obtained with S_1 and SC progeny tests for seed yield and seed weight. They concluded that OP, PC, and F_2 progeny tests were generally least useful in detecting genetic differences for seed characters in tall fescue. Their data also showed that clonal evaluation would be quite useful in screening large numbers of clones for synthetic cultivars and would seldom result in the loss of superior germplasm. Frakes and Matheson (1973) found that SC, OP, PC, and F_2 progeny tests all identified the same genotypes with high and low potential for forage yield.

In the tall fescue breeding program at the University of Missouri, clonal selections from space-planted nurseries were established by sprigging in replicated sod plots (Asay, 1971). Plots transplanted in September were sufficiently established by the next fall for determination of forage yield and quality, disease resistance, and other characters under sward conditions. The experimental error in these plantings was comparable to that from seeded progeny trials and genetic differences were much more evident (Asay et al., 1974). In addition, the procedure saves at least 2 years per cycle compared to schemes involving PC progeny tests. Some of the genetic differences are conditioned by nonadditive effects and would not be fully recovered in progenies from selected clones. However, experimental synthetics developed from parental clones evaluated in this manner were equivalent to those in which PC progeny tests were used, and the synthetics produced significantly more forage yield than check cultivars KY 31 and 'Fawn' (Sleper and Mitchell, 1975). A possible modification would be to include clonal lines in sprigged plots along with their OP progenies in seeded plots to determine the genetic potential of genotypes previously selected from source nurseries.

GENETIC VARIATION

The nature and extent of genetic variability for quantitative traits in tall fescue has received comparatively little attention. Most studies show that forage yield, seed yield, quality components, and related characters can be substantially improved through mass or recurrent selection (Burton and Devane, 1953; Thomas, 1967; Buckner et al., 1973; Frakes and Matheson, 1973; Asay et al., 1974) (Nelson, C. J., K. H. Asay, and D. A. Sleper. 1975.

A characterization of yield of tall fescue Am. Soc. Agron. Abstr., p. 83; Frakes, R. W., and J. R. Cowan. 1957. Dry matter and protein heritability estimates in tall fescue, *Festuca arundinacea* Schreb. Am. Soc. Agron. Abstr., p. 53). In most cases, conclusions were based on data from replicated spaced plantings. Application of broad-sense heritabilities in their computations assumes that superiority, conditioned by both additive and nonadditive genetic effects, would be realized in the derived populations. This would require clonal propagation of selected lines or use of breeding procedures to fix desired genetics combinations in F_1 hybrids.

Neither alternative should be considered out of the realm of possibility. Vegetative propagation has been used extensively in the tall fescue breeding program at the University of Missouri (Asay, 1971). Cowan (1956) obtained 200 vegetative propagules from each of 40 2-year-old clones in Oregon. The propagules were established in the greenhouse in August and transplanted to the field in October. Each plant yielded 40 more sprigs the next fall, thus giving about 8,000 plants from each original clone during the study. Cowan (1956) also reported that chopped rhizomes could be used with some success to establish tall fescue and that plants removed from the field in February yielded many more tillers than those lifted in September or December.

The feasibility of developing and using F_1 hybrid cultivars of tall fescue has not been adequately studied. Obviously, the gain in productivity of the hybrids must offset the increased cost of producing the seed. Echeverri (1964) found little heterosis in crosses involving parental clones from an early maturity group for plant height, culms per plant, and seed yield. Similarly, Matheson (1965) reported little heterosis within intermediate- and late-maturity groups of tall fescue for plant width, plant height, leaf length, plant density, and forage yield. Thomas (1967) could not detect significant heterosis for seed yield and its components within early- and intermediate-maturity tall fescue groups.

In other species, including alfalfa (*Medicago sativa* L.) and corn, the expression of hybrid vigor has been most pronounced in crosses involving genetically diverse parentage (Wilsie, 1958; and Moll et al., 1962). Moutray and Frakes (1973) compared the heterotic responses of tall fescue hybrids from within and between maturity groups. They found greater heterosis for plant height, panicle number, seed yield, and autumn vigor in crosses involving parents with contrasting anthesis dates. To allow mutual pollination between maturity groups, flowering of early clones was delayed by transferring them from the greenhouse to a growth chamber at the head stage and holding them at 3 C. When panicles of the late clones had completely emerged from the boot, the early clones were returned to the greenhouse and subjected again to a 16-hour day length and a temperature of 20 to 22 C. Development of hybrids from genetically diverse parentage merits additional study; however, a system that synchronizes flowering in large scale plantings would have to be devised. The magnitude of heterosis among distantly related parents of similar maturity should also be investigated. It may also be possible to alter the floral induction or floral initiation requirements of diverse genotypes through selection (Bean, 1969, 1970).

INBREEDING

Although tall fescue is predominantly self-sterile, self-fertile types are present in the species (Cowan, 1956). Intensive work at the Kentucky Agricultural Experiment Station indicates that selection within inbred lines can be successfully used to develop improved cultivars of tall fescue (Buckner, 1960a). The synthetic cultivar 'Kenwell' was derived from S_3 inbred lines that had been screened for disease resistance and grazing preference during each generation of inbreeding. Bennett et al. (1958) inbred selectively grazed clones of tall fescue for a number of generations. They found polycross progenies and synthetic strains developed from selected inbred lines to be significantly more palatable than Kentucky 31. Buckner and Fergus (1960) concluded that lines were relatively homozygous for characters under consideration after three generations of inbreeding.

Inbreeding has reduced the number of panicles per plant, seed per panicle, seed yield per plant, culm height, and forage yield (Buckner et al., 1969). However, Buckner (1960a) found that crossing inbreds to produce synthetic strains restored vigor sufficiently to warrant the use of the procedure in a breeding program. In addition, some lines are lost during the inbreeding cycle, but Buckner and Fergus (1960) indicated that enough lines remained after four generations of inbreeding to make meaningful progress. Insufficient seed is often produced in pollination bags to provide an adequate population for the next cycle of selection or for progeny testing. Buckner (1960b) proposed the use of polyethylene tubes to enclose the entire plant. His studies indicated that substantially more seed was produced with the tubes and more plants could be selfed with less labor than with conventional means.

Carnahan and Hill (1962) reported that inbred plants of tall fescue had more meiotic irregularities than that found in noninbred germplasm. They attributed the problem to reduced hemeostasis of inbreds, segregation of recessive polygenes that were deleterious to normal meiosis, and possible increased pairing of homeologous chromosomes.

GERMPLASM RESOURCES

Tall fescue is closely related to a complex group of *Festuca* and *Lolium* species with chromosome numbers ranging from $2n = 14$ to $2n = 70$. Workers at the Welsh Plant Breeding Station (Breese and Davies, 1973; and Borrill et al., 1971) have studied the characteristics, chromosome number, and specific relationships of *Festuca* species collected from throughout Europe and the Mediterranean region. They found the hexaploid ($2n = 42$) *F. arundinacea* Schreb. to be the most widely distributed of the broadleaved fescues in these areas, except in Morocco where it is apparently replaced by a decaploid ($2n = 70$) fescue.

Hexaploid races have undergone considerable genetic diversification and morphological differences associated with area of adaptation are apparent. The hexaploids of Northwest Iberia possess a complex of characters associated with rhizomaty. Ecotypes from Tunisia have an unusually lax type of leaf and are not as harsh in general appearance as is normally the case (Borrill et al., 1971).

Introductions from North Africa exhibit superior fall and winter growth in temperate regions of Europe and the United States (Lewis, 1963; Frame, 1972; Asay et al., 1974). However, these ecotypes are unusually dormant and nonproductive during the summer and lack winterhardiness. Chatterjee (1966) found that North African types produced more leaf area and had a higher net assimilation rate (NAR) under low temperatures than 'S-170', primarily because of their inherently greater physiological efficiency under winter conditions. Robson and Jewiss (1968) found temperature to be the major environmental factor controlling seasonal differences in growth between North African and European germplasm. Nelson and Asay (C. J. Nelson, and K. H. Asay. 1969. Temperature responses of tall fescue genotypes. p. 28. Agron. Abstr. Am. Soc. Agron., Madison, Wis.) reported that North African lines had a comparatively high NAR under temperature regimes above optimum for tall fescue, but assimilates were apparently used for root growth and energy reserves. North African ecotypes are also characterized by greater tillering capacity and finer leaves than cultivars adapted to more temperate regions. Apparently, North African fescues have been genetically isolated from European and North American populations long enough to allow extensive karyotypic differentiation. Hybrids between the two groups are plagued with meiotic irregularities and nearly complete sterility (Evans et al., 1973; Lewis, 1963).

Although additional and more thorough plant exploration is needed, tall fescue breeders have a comparatively large supply of genetic variability available. European gene banks and breeding programs such as that at the Welsh Plant Breeding Station, maintain sizable collections (Davies et al., 1972). Germplasm stocks are also maintained by active public and commercial breeding programs in the United States. The USDA has assigned the Western Regional Plant Introduction Station the primary responsibility for collecting, evaluating, maintaining, and distributing tall fescue introductions. More than 235 seed accessions from over 25 foreign countries are included in their expanding inventories. In addition, a wealth of genetic variability exists in naturalized strains, old established pastures, released cultivars, experimental strains and breeding lines, and derivatives from interspecific and intergeneric hybridization.

FORAGE QUALITY

The agronomic advantages of tall fescue are well documented. The species produces abundantly over a wide range of environments, tolerates adverse soil and climatic conditions, and often thrives on sites where other grasses fail to persist. However, the quality of tall fescue forage has been criticized. Consequently, breeding for improved quality has been a major objective in tall fescue breeding programs, but progress has been hindered by a poor understanding of the relationship between plant factors and animal performance. For example, Phillips et al. (1954) and later Sullivan et al. (1956) found tall fescue to be high in nutritive value compared to seven other forage grasses, based on content of protein, N-free extract, fructosan, soluble ash, lignin, fiber, and cellulose. The species has also compared favorably with other temperate grasses in terms of digestibility (Miles et al., 1964; Baker et al., 1965; Bryan et al., 1970). However, animal performance on tall fescue is variable and often lower than that on other species (Gross et al., 1966; Jacobson et al., 1957; Mott et al., 1971; Peterson et al., 1959).

Toxicity problems in cattle grazing tall fescue have been encountered in several instances. The fescue foot syndrome was first noted by Cunningham (1949) in Australia. Research has yet to clearly identify the toxic principle(s). Subclinical levels of such entities may be largely responsible for erratic animal performance. Bush and Buckner (1973) reviewed the various forms of fescue toxicity, including poor animal performance, fat necrosis, and fescue foot.

A. Palatability

The USDA-ARS has cooperated with the Kentucky Agricultural Experiment Station in a breeding program to develop tall fescue cultivars that are more acceptable to livestock, particularly during the summer. Two methods of measuring relative palatability of tall fescue breeding lines were tested (Buckner and Burrus, 1962). One method consisted of sampling plots before and after grazing to measure dry matter consumption. The other was an observational technique in which visual ratings of grazing intensity were made after the grazing period. Significant differences were detected among breeding lines with both methods and the two sets of data were closely correlated. The researchers concluded that the observational technique was simpler and more suitable for screening large populations for palatability. Peterson et al. (1958) also successfully used an observational technique to detect consistent differences among tall fescue genotypes for palatability to sheep. The number of sheep feeding on a plot was recorded at 5-min intervals over a period of 7 to 10 hours. Relative palatability was derived from the number of times that a plot was grazed during the observation period. Estimates of dry matter consumption before and after grazing gave unreliable estimates of palatability, primarily because of large sampling errors.

Using an observational technique, Buckner and co-workers found sufficient heritable variation for palatability in tall fescue to make meaningful genetic advance (Buckner, 1960a, 1973; Buckner and Fergus, 1960; Buckner et al., 1969). Inbred lines, developed and selected from grazing trials in space-plant nurseries, demonstrated improved palatability in sod plots. Polycross progenies and synthetic strains subsequently derived from selected inbreds were significantly better grazed than naturalized and commercial check cultivars. Third generation inbred lines from this screening program formed the parentage of the synthetic cultivar Kenwell. However, grazing trials failed to show that genetic advances in palatability produced significant improvement in animal performance. Apparently, preferential grazing did not solve the problem of erratic performance of cattle (Buckner, 1973).

Craigmiles et al. (1964) obtained a positive and significant correlation in palatability between tall fescue clones in a space-planted nursery and their polycross progenies in sod plots. Calves tended to select clones with broad, thick leaves over those with narrow leaves. The broad-leaf types were also more succulent, but less winterhardy than the narrow-leaf plants.

Although ample genetic variability is apparently available to facilitate selection for high or low levels of silica, total sugars, or crude protein, these entities are apparently not related to palatability of tall fescue to cattle (Buckner et al., 1969). Van Soest (1964, 1965) reported a poor correlation between chemical components such as lignin, acid detergent fiber, and in vitro digestibility and voluntary intake of tall fescue. He concluded that a taste factor or toxic substance in the extract of the tall fescue forage may have been influential. Buckner and Burrus (1962) found little relationship between vigor of inbred lines and palatability to cattle.

B. Perloline

Gentry et al. (1969) categorized tall fescue as an alkaloid bearing plant after finding as many as 11 different types of alkaloids in a strain of Kentucky 31. Perloline, a close relative of perlolidine, is the predominant alkaloid in the species. Workers at Kentucky (Bush et al., 1970, 1972; Buckner, 1973) found that perloline inhibited in vitro ruminal cellulose digestion, production of fatty acids, and growth of ruminal cellulytic bacteria. They postulated that such inhibition of microbial activity would decrease the rate of digestion in the rumen which in turn would curb the rate of ingesta passage and voluntary intake. Later, Boling et al. (1975) found that perloline decreased the digestibility of crude protein and cellulose when added to the diets of lambs.

The environmental and genetic effects of perloline content in tall fescue were studied at the University of Kentucky (Buckner, 1973). Perloline concentrations in the plant varied during the growing season and were positively correlated with levels of N in the soil. Perloline levels increased from March to August and declined thereafter. These trends were negatively related to the seasonal changes in the performance of animals feeding on tall fescue. For example, Boling et al. (1973) reported that dry matter disappearance of tall fescue forage was at a low level in August and increased during the fall.

Apparently perloline content is under genetic control in tall fescue and sufficient genetic variability exists to facilitate selection. Gentry et al. (1969) reported that Kenwell, Kentucky 31, and Alta had high, medium, and low levels of perloline, respectively. Although sizable genotype × environment interactions were encountered, Buckner et al. (1973) obtained broad-sense heritability values for perloline content ranging from 0.57 to 0.80. Positive and highly significant parent-progeny correlations gave additional evidence that perloline levels in tall fescue and fescue-ryegrass (*Lolium multiflorum* Lam.) derivities can be altered through selection. Cornelius et al. (1974), using a partial diallel series involving nine derivatives of ryegrass × tall fescue, showed that genes for low perloline were dominant over genes for high perloline content. Their results were confirmed by Watson (1976).

C. Laboratory Methods

Barnes (1973) discussed the laboratory methods used to evaluate the quality of forage samples. He categorized these methods into four groups: 1) chemical, including solubility indices; 2) physical; 3) in vivo, with the rumen or cecum as the living incubator; 4) in vitro rumen fermentation. He concluded that for practical feeding situations the most valuable information required would be content of total N, Ca, P, K, and Mg, along with some measure of available energy. He also indicated that balances or excesses of certain minerals or the presence of toxic compounds may be an important consideration in some forages, including tall fescue.

The introduction and refinement of the two-stage in vitro rumen fermentation technique (Tilley and Terry, 1963; Barnes, 1966) has been a major contribution to most grass breeding programs where quality of large numbers of relatively small forage samples must be determined. The process, which involves a 48-hour incubation period with rumen fluid and a buffer solution followed by 24-hour incubation with acid pepsin, is relatively simple and results are highly correlated with in vivo digestibility values. Cellulolytic enzymes have been used with some success as digesting agents to reduce errors due to variability in the potency of rumen inoculum (Jones, 1974).

Van Soest (1963, 1964, 1965) proposed the use of detergents to separate the relatively insoluble cell wall constituents (neutral detergent fiber or NDF) from the highly digestible cell contents. Treatment of NDF with progressively stronger detergents removes the hemicellulose, yields acid detergent fiber (ADF), and ultimately separates the cellulose from lignin to yield acid detergent lignin (ADL).

Various combinations of the Tilley-Terry in vitro and Van Soest procedures have been proposed. Van Soest et al. (1966) replaced the acid-pepsin stage of the Tilley-Terry method with a neutral detergent. They found that this modification required less time and predicted dry matter digestibility more accurately than the Tilley-Terry method.

Less complex solubility methods have been devised to qualify determination of small forage samples. Donefer et al. (1966) measured the dry matter disappearance (DMD) of forage in a solution of 0.2% pepsin dissolved in 0.75 N HCL. Their solubility indices of 35 grasses and 14 legumes were highly correlated with digestible energy intake potential (nutritive value index) derived from in vivo studies. These procedures, which are relatively simple and inexpensive, have been modified by Kendall et al. (1970) for use in the tall fescue breeding program at Kentucky.

Infrared reflectance spectroscopy has recently been proposed as a means for rapidly predicting the quality of relatively small forage samples (Norris and Barnes, 1976; Norris et al., 1976, and Shenk et al., 1976). Using up to nine wavelengths in analyses and prediction equations, these workers accurately estimated crude protein, neutral detergent fiber, digestibility, and other quality parameters of several forages including tall fescue. Their results suggest that infrared reflectance may permit breeders to evaluate larger populations over more harvests, locations, years, and generations than previously possible.

Flexibility of leaves has been used as an index of forage quality in tall fescue at the Station d'Amelioration des Plantes Fourrageres des Lusignan in France (Gillet and Judas-Hecart, 1965). Their method consists of rating plants from one to five (very stiff to very flexible) after passing hands through the forage. Although flexibility was not closely related to chemical indicators of forage quality, the character was positively correlated with palatability to sheep and was apparently related to the size of the fibro vascular bundles in the leaves. These researchers concluded that flexibility did not explain all differences in palatability but that it should be considered for inclusion in selection indices for quality of tall fescue forage, especially during the summer. An experimental cultivar, recently developed based on selection for this character, will soon be added to the French list of cultivars.

Mineral deficiencies have been observed in temperate grasses, including tall fescue (Patil and Jones, 1970). Required levels are not clearly defined and concentrations of minerals actually available to the animal are difficult to determine without animal trials; however, selection for better mineral balance is apparently possible in grass species (Cooper, 1973). Sleper et al. (1977) detected significant genetic variation in Mg, Ca, K, and P content among tall fescue clones and single-cross progenies. Genetic effects were mostly additive and narrow-sense heritability estimates indicated that mineral balance could be effectively altered through selection. The K/(Mg+Ca) ratio was also heritable, suggesting that tall fescue could be bred for low grass tetany potential. Hill and Guss (1976) proposed the use of family selection in breeding for chemical traits, including mineral balance. This method, which entails bulking plants within a family, reduces the number of samples for analysis and should improve response to selection over that based on individual plants when heritabilities are intermediate to low.

D. Bioassays

The potential of laboratory animals to assay the quality of tall fescue forage has not been fully exploited. The house cricket (*Acheta domesticus*) has been proposed as a possible test animal (Stone, 1953). The insect is relatively free of disease, easy to maintain, and it has a well-developed gastrointestinal tract. Crickets also have a rapid growth rate and are sensitive to nutritional deficiencies. Stone (1963) detected large differences among tall fescue forage samples based on cricket performance. Later, Stone and Matches (1966) successfully used cricket feeding trials to detect changes in quality of tall fescue forage associated with plant maturity.

Asay et al. (1975) conducted feeding trials with crickets to determine their usefulness as test animals for evaluating the forage quality of tall fescue breeding lines. Certain genotypes consistently ranked high on the basis of cricket growth rate and survival, whereas other lines produced almost complete mortality in all trials. They concluded that the high rate of mortality was apparently due to the inhibitory substance in the cation fraction of the forage. This entity is probably not responsible for fescue foot, because an earlier work (Williams et al., 1973) has placed the toxic principle for this syndrome in the anion fraction. Furthermore, the anionic fraction, previously shown to be toxic to cattle, did not adversely affect cricket performance. However, alkaloids are in the cation fraction and may be related to cricket responses. With refinement and additional ruminant data, cricket trials and other bioassay procedures may prove useful in evaluating forage quality in the tall fescue breeding programs.

Final evaluation of new tall fescue cultivars must be expressed in terms of animal performance, because of the inability of laboratory procedures to predict animal responses in the species. Grazing and feeding trials have been established at the University of Missouri as part of their forage research program (Martz et al., 1976). Objectives are 1) to evaluate experimental strains from the tall fescue breeding program under grazing conditions and 2) to gain a better understanding of forage characteristics related to animal performance in order to develop more effective selection criteria. These trials consist of a series of replicated 0.405 ha (1-acre) pastures with movable electric fences used to provide strip grazing. Each grazing season is divided into three periods of 40 to 50 days each. Preliminary results indicate that genetic variability for quality factors, including average daily gain is available in tall fescue. Heifers consistently gained faster on 'Kenhy', a new cultivar of *Lolium-Festuca* parentage developed at Kentucky, and 'Missouri 96', a recently released cultivar of French parentage from the Missouri program, than the check cultivars Fawn, 'Kenmont', and Kentucky 31. Buckner and co-workers at Kentucky routinely include their advanced experimental strains in grazing trials at Lexington and other locations (Buckner, 1973).

E. Maturity Effects

Jones et al. (1973) emphasized the need to consider changes in quality associated with stage of maturity and date of harvest. Selection for digestibility or other quality traits on a given harvest date irrespective of growth stage would tend to favor plants that are less mature at sampling. The net result is likely to be a later maturing strain. Conversely, earlier lines are favored when selection is based on a stage of growth, because they are sampled on an earlier calendar date. Possible solutions are to 1) select within maturity groups; 2) base selection on means adjusted through co-variance analysis; 3) alter the maturity date. Late-maturing strains of tall fescue may offer advantages when grown in association with legumes or when used to extend the summer grazing season. Some loss of early spring growth may be encountered, however.

FORAGE YIELD

Increased productivity of dry matter, at least on a seasonal basis, has been a major objective of most tall fescue breeding programs. Results from past yield trials indicate that cultivars released since Kentucky 31 and Alta do not represent major breakthroughs in forage yield. Raymond (1969) stated that progress in forage crop breeding has been limited because objectives have not been as clearly defined as in other crops. A grass cultivar, may be grown with or without companion grasses or legumes, at high or low levels of fertility or management intensity, may be used for grazing or stored as hay or silage and fed to animals at all levels of production potential. Raymond concluded that new cultivars must be bred with narrow but clearly defined utilization objectives, specified in advance by the breeder.

Studies of genetic variability indicate that selection for improved forage yield under a given management system would be effective (Burton and Devane, 1953; Frakes and Matheson, 1973). The performance of newly developed experimental strains show that substantial genetic progress is forthcoming. In Missouri trials conducted over three harvests, 2 years, and at two locations, Kenhy and a Missouri experimental synthetic 'H-1' yielded 12 and 13% more forage respectively than Kentucky 31 (Sleper et al., 1976). Preliminary results from other trials indicate that other experimental synthetics from the Missouri breeding program show an even greater yield advantage over commonly used check cultivars. Experimental strains in commercial breeding programs such as those from North American Plant Breeders and Farmers Forage Research Cooperative have also shown potential advantages in forage yield (J. B. Moutray and S. J. Baluch, personal communication). Progress in forage yield is apparently being made in European breeding programs as well (Frame et al., 1974).

A. Fall Growth

Tall fescue has been acclaimed as a valuable source of pasture during the fall and winter, particularly in the lower Midwest and Southeast. Quality of the forage in terms of digestibility and animal performance increases substantially as the season progresses from midsummer to fall and often surpasses that of other temperature species such as orchardgrass (*Dactylis glomerata* L.) and reed canarygrass (*Phalaris arundinacea* L.) (Baker et al., 1965; Bryan et al., 1970; Wedin et al., 1966).

The exceptional fall growth potential of ecotypes indigenous to the Mediterranean region has attracted considerable attention. Breeding programs have been initiated to incorporate this germplasm into new cultivars that produce more and higher quality forage during the fall (Asay et al., 1974; Frame, 1972; M. Gillet and R. Haaland, personal communication). Procedures consist of 1) selecting within Mediterranean accessions for improved agronomic potential and cold resistance and 2) combining the fall growth potential of these introductions with the desired attributes of adapted cultivars. Two new cultivars, 'Maris Jebel', a hexaploid, and 'Maris Kasba', a decaploid, have been developed by the Plant Breeding Institute at Cambridge from material collected in North Africa (H. H. Rogers, personal communication). The Division of Plant Industry at Canberra has recently released the cultivar 'Melik' from their breeding program with Mediterranean germplasm (CSIRO, 1972).

Attempts to transfer genes between Mediterranean and temperate ecotypes have not been immediately successful. Although ample hybrid seed is usually produced in crosses, F_1 hybrids are plagued with meiotic irregularities and resultant sterility (Evans et al., 1973; Gillet, personal communication; Lewis, 1963). Hunt and Sleper (K. L. Hunt and D. A. Sleper. 1976. Hybridization of two geographic races of tall fescue. Am. Soc. Agron. Abstr., p. 53.) reported that intermediate clones, which are at least partially interfertile with the two extreme populations, may be used in "bridge crosses" to affect genetic transfer.

DISEASE RESISTANCE

Although some diseases had been reported in tall fescue (primarily *Helminthsporium* and *Rhizoctonia* spp.), Cowan (1956) considered the species to be relatively free of disease. However, as has been the case with many other crops, the expansion of acreages has been accompanied by the increased incidence of disease. As early as 1967, 'Goar' showed heavy infestations of crown rust (*Puccinia coronata* Corda. var. Coronata). In 1971, crown rust infestations were reported on Kentucky 31 throughout Alabama and to a lesser extent, in Missouri, Tennessee, Kentucky, and North Carolina (Berry, 1973).

Berry and Gudauskas (1972a) studied the factors associated with storage and infectivity of *P. coronata* uredospores and developed procedures to screen large populations of tall fescue for rust resistance. They screened 8,400 Goar seedlings and identified 173 (2%) apparently resistant plants. They obtained 472 (7%) apparently resistant plants from 7,100 seedlings of Kentucky 31. Selected lines are presently being progeny tested for rust resistance in the field.

Based on the susceptibility reported in Alabama trials and the increasing incidence of crown rust in the field, more intensive breeding efforts for disease resistance would be justified. Much of the tall fescue acreage in the United States is based on a single cultivar, Kentucky 31. Although this cultivar is highly heterogeneous, virulent races of *P coronata* and other disease organisms can develop rapidly and produce serious consequences. Berry and Gudauskas (1972b) isolated two races of *P. coronata* from Goar and Kentucky 31. The race collected from Goar was pathogenic almost exclusively on Goar, but the race isolated from Kentucky 31 parasitized both cultivars with equal facility.

Haaland (personal communication) reported losses in tall fescue stands in Alabama due to infestations of nematodes. These organisms damage the root system, making the plant more susceptible to drought and soil borne diseases. Breeding for nematode resistance has been initiated at Auburn University, and significant genetic variation has been observed in early trials.

SEED YIELD

The amount and nature of genetic variability indicate excellent opportunities for improving seed yield in tall fescue through breeding. In trials conducted at Oregon, seed yield among 1,500 clonal lines ranged from less than 0.1 to 1.6 g/panicle (Cowan, 1956). The cultivar, 'Fawn,' which was developed from this breeding program, represents a substantial improvement in seed yield. This cultivar produced 24, 30, and 22% more seed than Alta, Goar, and Kentucky 31, respectively, over a 5-year period in Oregon (R. V. Frakes, unpublished data).

Bean (1972) studied the heritability and genetic variation among replicated clones of S 170. Broad-sense heritability on a mean basis computed over 2 years was 0.55 for seed weight per plant. An experimental strain (BN 845) compounded from six selected clones yielded 20% more seed than the parental population in subsequent trials. The superiority of the selected population was due almost entirely to increased seed weight per inflorescence, which was a reflection of floret fertility and 1,000-seed weight. The number of inflorescences per plant remained unaltered in the selected line. Bean concluded that seed yield should be increased by improving the efficiency of the reproductive system (percentage of fertile florets and seed weight) rather than by increasing its size (number and size of inflorescences). He postulated that selection for more and larger inflorescences would result in more reproductive tillers and larger primordia, which may adversely affect aftermath growth and ultimately affect perenniality.

Bean's conclusion that selection on the basis of clonal evaluation would be an effective means of increasing seed yield, confirmed the findings of Thomas and Frakes (1967). They reported that clonal evaluation was as effective as any of the five methods of progeny testing used in their studies.

Cowan (1956) concluded from earlier studies that concurrent selection for seed and forage yield would be effective. Burton and DeVane (1953) reported that selection could improve seed yield, but their data suggested a negative relationship between seed and forage yield.

Stems and reproductive parts of forage grasses are usually of lower forage quality than the more leafy vegetative growth. Consequently, seed yield is often negatively related to forage quality (Carlson, 1974; Ross et al., 1970) and genetic improvement in one character may be at the expense of the other. Thus, the relative merits and interrelationships of the two characters must be considered in the refinement of selection criteria.

TURF

Tall fescue is productive over a wide range of environments; however, it is best adapted to the transition zone that separates the northern and southern regions of the United States. This area is south of the adaptive range of many temperate turf grasses yet too far north for the optimum growth and longevity of warm-season species.

Juska et al. (1969) tested the turf potential of Kentucky 31 for 8 years under different cutting heights and fertility levels. Although tall fescue tended to clump in mixtures with Kentucky bluegrass, they concluded that it was suitable for expansive turf areas where its relatively coarse texture was not objectionable. The species is rapidly gaining acceptance as a turf species in the transition zone, particularly in play areas, air fields, roadsides, and to a lesser extent, parks and lawns. Cultivars developed specifically for turf have been an objective of the breeding program at the University of Missouri (Asay et al., 1974). Selections from broadly based source nurseries and from areas subjected to close mowing, such as golf course fairways, have been screened for fine texture, short culms, basal leaves, tolerance to close mowing, greenness under environmental stress, and rhizomatous growth habit. Experimental synthetics have been developed for preliminary testing. The Weibullsholm Plant Breeding Institute in Sweden (Hans-Arne Jonson, personal communication) has also exploited the turf potential of tall fescue. The cultivar 'Backafall', which was released from this program, has more extensive rhizomes and a shorter growth habit than normal tall fescue.

The turf cultivar, 'Fortune,' was released in 1968 by the Oregon Agricultural Experiment Station. This cultivar, which was derived primarily from introductions from Portugal and Tunisia, is tolerant of close mowing and has fine leaves and dark green color (Hanson, 1972). Soon after its release, however, some serious limitations, including poor seedling vigor, susceptibility to disease, and lack of winterhardiness, became evident. The cultivar was subsequently withdrawn from the seed market for additional breeding.

CULTIVARS

Encouraged by increasing demands for improved tall fescue germplasm, breeders in the United States and Europe have added substantially to the list of available cultivars during the last decade.

A. Kentucky 31 and Alta

The interesting history and development of these cultivars is discussed in Chapter 1. Kentucky 31, which was derived from a naturalized strain from Menifee County, Kentucky, was released by the Kentucky Agricultural Experiment Station in 1943. It is characterized by good fall and winter growth, medium maturity, and greater resistance to leaf diseases than that in Alta, Goar, or Fawn. It is a heterogeneous and broadly adapted cultivar. Consequently, natural selection over a wide range of environments has produced several ecotypic variations.

Alta originated from ecotypes evaluated and selected by H. A. Schoth, USDA, in cooperation with the Oregon Agricultural Experiment Station. The cultivar was registered by the American Society of Agronomy in 1945 (Hollowell, 1945) and was the first forage crop to be so certified. Alta rapidly gained acceptance in the Pacific Northwest and later in California and Arizona as a major constituent in irrigated pastures. It is about 5 days earlier than Kentucky 31; however, considerable introgression has occurred between the two cultivars, particularly in uncertified seedlots.

B. Fawn

This synthetic cultivar was developed by R. V. Frakes and J. R. Cowan at the Oregon Agricultural Experiment Station. Eight parental clones were selected from an original nursery of 9,000 plants. Evaluation was based on chromogen content, crude protein, digestibility (in vivo and in vitro), forage yield, seed yield, and low self-fertility. Types that carried the leaves high on the culms were also selected. Final evaluations were based on progeny tests. The experimental strain was distributed for testing as Oregon Syn E and was released by the Oregon Station in 1964.

Fawn is included in the early maturity group and is particularly well adapted for seed and forage production in the Williamette Valley of Oregon. In Oregon yield trials, Fawn produced a 5-year average of 13, 16, and 18% more forage than Alta, Goar, and Kentucky 31, respectively. It produced 24, 30, and 22% more seed than the same three cultivars over the same 5-year period. The cultivar has been somewhat susceptible to crown rust in Missouri and to *Helminthsporium* spp., *Rhizoctonia solani*, and crown rust in Kentucky trials.

Fawn is grown primarily in the northwest and central U.S. Breeder seed is maintained by the Oregon Agricultural Experiment Station. certification requirements limit seed production to one generation of each of breeder, foundation, and certified seed classes (Hanson, 1972; R. V. Frakes, unpublished data).

C. Kenwell

This cultivar was developed by R. C. Buckner, USDA, in cooperation with the Kentucky Agricultural Experiment Station. The parental germplasm was obtained from naturalized strains in Kentucky, and selection was based primarily on palatability to cattle and resistance to foliar diseases. Selected clones were selfed each generation to produce populations for subsequent cycles of selection. The synthetic was compounded from 43 clones representing the three S_3 inbred lines 45-50, 42-33, and 61-48. Final selection of the parental clones was based on the performance of polycross progeny in animal preference trials (Buckner and Burrus, 1965; Hanson, 1972).

Kenwell was significantly more palatable than Kentucky 31 in Kentucky trials (Buckner and Burrus, 1965). Subsequent grazing trials have not demonstrated any advantage of Kenwell over Kentucky 31 in terms of animal performance, indicating that the two cultivars are essentially the same nutritionally (Buckner, 1973). Kenwell appears to be more leafy than other cultivars and maintains relatively good color during hot, dry periods. The cultivar is from 5 to 7 days later than Kentucky 31 and may be more compatible with legumes than more competitive earlier types (Buckner and Burrus, 1965). Kenwell has exhibited resistance to leaf diseases caused by *Helminthsporium* and *Rhizoctonia* species (Gentry et al., 1968; Flores et al., 1969).

The Kentucky Agricultural Experiment Station released the cultivar in 1965 and currently maintains breeder seed. Certified seed is produced on a limited generation basis, as specified by the Kentucky Seed Improvement Association.

D. Kenmont

The parental germplasm for Kenmont was derived from a naturalized strain from southeastern Kentucky. Initial collection and seed increase was made by E. N. Fergus and R. C. Buckner at the Kentucky Agricultural Experiment Station. The strain, distributed for testing as 59 G1-32, performed well at several locations in Montana and was released in 1963 by the Montana Agricultural Experiment Station (Hanson, 1972). In Montana, Kenmont produced significantly more forage than Kentucky 31 and Goar (Joppa and Roath, 1964). The cultivar has also performed well in Missouri and Ohio yield trials (Sleper et al., 1976; VanKeuren and Niehaus, 1970) and is often cited for its midseason productivity.

Kenmont had the best average palatability ratings to sheep, the highest crude protein content, and of the four cultivars tested in Montana was easiest to establish on saline soils (Joppa and Roath, 1964). Kenmont is similar to Kentucky 31 in general appearance and relative maturity. The cultivar is grown primarily in the Southeast and northern Great Plains. Breeder seed is maintained by the Montana Agricultural Experiment Station.

E. Goar

Final evaluation, screening, and release of Goar was done by L. G. Goar at the El Centro Experiment Station in the Imperial Valley of California. Professor W. Southworth of the University of Manitoba obtained the original seed introductions from D. Dagon, Budapest, Hungary. In early trials conducted in southern Manitoba, the accession displayed good yield potential and drought tolerance, although the forage was not considered to be of high quality. Seed was sent to P. B. Kennedy at the University of California, Davis, in 1925, who included the new line in grass nurseries at Davis and Berkeley. The accession, which was labeled T.O. 899, was received by L. G. Goar at the El Centro Experiment Station in 1941. There the accession was subjected to testing and selection for tall fescue types. Seed from selected lines was then sent to the SCS Plant Materials Center at Pleasanton, Calif., where it was assigned accession number P-13847. After additional testing, the cultivar was released and certified by the California Crop Improvement Association in 1946. Breeder seed is maintained by the Plant Center in Pleasanton, Calif. (Cowan, 1956; Hanson, 1972).

Goar is from 1 to 2 weeks earlier than Kentucky 31 and has exceptionally good seedling vigor. Compared to other tall fescue cultivars, Goar produces well under high temperatures and is adapted to heavily textured soils. It is one of the cultivars most susceptible to crown rust, (Berry and Gudauskas, 1972a). Goar is rapidly gaining acceptance in Alabama, where it has produced substantially more winter and early spring forage than Kentucky 31 (Hoveland, 1967; Hoveland et al., 1974).

F. Missouri 96

Missouri 96 was released in 1977 by the Missouri Agricultural Experiment Station. The cultivar, which was developed from French germplasm was extensively tested as 'I-96'. It is softer textured and has finer leaves than other tall fescue cultivars. In grazing trials with cattle at the Missouri Station, Missouri 96 produced from 35 to 40% more daily gain than Kentucky 31 (Martz et al., 1976; D. A. Sleper, personal communication). It is in the same maturity group and produced about the same amount of forage as Kentucky 31.

G. 'Asheville'

This strain was collected and increased by Paul Tabor and J. D. Powell at the Soil Conservation Service (SCS) Nursery, Americus, Ga. The original accession was obtained from a naturalized stand near Asheville, N.C., and was designated SC-20-764. Although Asheville does not yield more than Kentucky 31 or Alta, it is considered to be more stable to adverse environmental conditions than they are. It is particularly well adapted to wet soils during the winter in North Carolina. The strain has not been officially released, but seed was distributed by the SCS in 1952, and breeder seed is maintained at the Plant Materials Center, Americus, Ga. (Hanson, 1972).

H. S170

S170 is a 24-clone synthetic cultivar developed by T. J. Jenkin at the Welsh Plant Breeding Station. Sixteen of the parental clones were from indigenous pastures in Buckinghamshire, England and Pembrokeshire, Wales. The remaining parental germplasm was derived from collections made in France, South America, and the United States (Baker and Chard, 1964; G. M. Evans, personal communication). The cultivar is distributed throughout Europe and is now the most widely grown cultivar in Great Britain. It has been included in trials coordinated by the three Scottish Agricultural Colleges (Hunt et al., 1972; Frame et al., 1974). Based on forage yield, forage quality, and persistency, they gave S170 their top merit rating. It is included in the early maturity group and is valued primarily for its productivity in the early spring and autumn.

I. 'Grasslands 4710'

The cultivar Grasslands 4710 was recently developed by the Department of Scientific and Industrial Research, Palmerston North, New Zealand (Anderson, 1976). Selection criteria included forage yield, persistency, disease resistance, and low tensile strength of the leaves. It has a more lax growth habit, a lighter green color, and is later than S170. Based on cattle and sheep observations, Anderson concluded that G4710 is also more acceptable to livestock than S170.

J. 'Melik'

The CSIRO Division of Plant Industry, Canberra, Australia, developed the cultivar Melik from an Israeli introduction. It has shown promise as a long-lived perennial with good capacity for fall growth and tolerance to drought (CSIRO, 1973).

K. 'Hokuryo'

The cultivar Hokuryo is a 27-clone synthetic developed by the Hokkaido National Agricultural Experiment Station, Sapporo, Japan. It was released in 1972 by the Ministry of Agriculture and Forestry of Japan. The parental ecotypes were obtained from natural stands at the Hokkaido National Experiment Station. The origins of these ecotypes are not known. The cultivar matures about a week later than Kentucky 31 in Hokkaido and is described as high yielding with superior digestibility in the spring. It has good winterhardiness in Japan and is resistant to *Helminthosporium dictyoides.* Hokuryo is recommended particularly for pastures in northern Japan. Breeder seed is maintained at the Hokkaido Experiment Station, and certified seed is available (Kawabata et al., 1972).

L. 'Yamanami'

This cultivar is also a 27-clone synthetic developed at the Hokkaido National Agricultural Experiment Station. The parental clones were selected from introductions from the United States and Canada. Yamanami produces abundant forage and has good seasonal distribution of growth. It matures somewhat earlier and is a better seed producer than Kentucky 31. The cultivar is adapted throughout Japan and has some resistance to *Helminthosporium dictyoides.* Breeder seed is maintained at the Hokkaido Experiment Station, and certified seed is available (Kawabata et al., 1972).

M. Backafall

Backafall was developed primarily for turf in the 1950's at the Weibullsholm Plant Breeding Institute in Sweden. The original selections were made from pastures and shore slopes on the Island of Ven between Denmark and Sweden. The cultivar is more prostrate than other tall fescue and has undergone some selection pressure for rhizomatous growth habit. It is correctly described as *Festuca arundinacea* Schreb. *forma aspera.* Although Backafall was not developed for agricultural use, it has performed well in forage trials. Forage yields of Backafall were equivalent to S170 in Weibullsholm trials and equal to Kentucky 31 and Alta in trials conducted by North American Plant Breeders of Brookston, Ind. (Jonson, 1976, personal communications).

N. 'Rozelle'

Rozelle, formerly known as 'McGillsmith Early', was developed in Great Britain. It is the only cultivar to consistently outyield S170 in Scottish trials. It is an early maturing cultivar and along with S170 has been given a merit rating of one by West of Scotland Agricultural College indicative of their top recommendation (Frame et al., 1970, 1974; Hunt et al., 1972).

O. 'Demeter'

Demeter was derived from French introductions at the Armidale Experiment Station in New South Wales, Australia. According to Cowan (1956), the original introduction was classified as *Festuca mairei*. Demeter is particularly noted for its productivity during the fall and winter.

P. Other Cultivars

'Manade' is a French cultivar with comparatively soft leaves and a yield potential about equal to S170 (Frame, 1963). Other noteworthy cultivars include 'Ludion' and 'Ludine' from France; 'Festal' and 'Motall' from the Netherlands; 'Brudzynska' and 'Pulawska' from Poland; 'Garton's Own Leafy' from Britain; 'Ottawa Syn A' from Canada; and 'Zapadnaya' (Western) from the USSR.

INTERGENERIC AND INTERSPECIFIC HYBRIDIZATION

Intergeneric and interspecific hybridization of *Lolium* and other *Festuca* species with tall fescue has attracted the attention of several workers during recent years. The genetic and cytological behavior of these hybrids is discussed in Chapter 6. Hybridization of tall fescue with *Lolium* and *Festuca* species offer new sources of germplasm from which superior cultivars may originate. Hybridization may be used as a method of breeding when desirable characters lacking in a species are present in related species or genera. Chromosome pairing and reassortment of chromosomes and genes in hybrids and their amphiploids result in a recombination of specific characters and a new range of genetic variation.

Hybridization programs involving *Festuca* and *Lolium* are presently underway in the United States and several European countries. Breeders in Europe use the procedures to improve the winterhardiness, seedling vigor, and early spring growth of tall fescue (Lewis, 1966; Djikstra, 1974).

In a cooperative breeding program between the USDA and the University of Kentucky, palatability and other nutritious qualities of annual ryegrass, perennial ryegrass (*L. perenne* L.), and giant fescue (*Festuca gigantea* (L.) Vill.) are being combined with the excellent agronomic qualities of tall fescue (Buckner et al., 1961, 1967, 1973, 1976).

A. *Lolium* × *Festuca* Hybrids

The F_1 hybrids (2n = 28) from annual and perennial ryegrass (2n = 14) × tall fescue (2n = 42) are male-sterile. Fertility is achieved by backcrossing to either parent, by treatment with colchicine to produce amphiploids, and by backcrossing the F_1 hybrids to the amphiploids (Buckner et al., 1961; Buckner, 1973; Jenkin, 1933). Backcrossing to the tall fescue parent restores meiotic stability, the 2n = 42 chromosome number, and fertility in one to three generations (Hill and Buckner, 1962).

Colchicine treatment restores fertility more effectively in hybrids between annual ryegrass and tall fescue than in hybrids between perennial ryegrass and tall fescue (Buckner et al., 1961). Irregular meiosis is common and chromosome numbers range from 42 to 56 each amphiploid generation (Webster and Buckner, 1971; and Lewis, 1966). Unstable meiotic behavior of the amphiploids made it difficult for Kentucky workers to secure true breeding, fertile, 56-chromosome populations. Advanced generation populations derived from the 56-chromosome amphiploids tended to drift to the 42-chromosome level of the tall fescue parent. Meiotically stable and fertile 42-chromosome populations derived from the 56-chromosome amphiploids were morphologically similar to tall fescue, but were intermediate to annual ryegrass and tall fescue in forage quality characteristics. The fluorescent seedling character that occurs in annual ryegrass but not in tall fescue was found in the 42-chromosome progenies of the amphiploids, indicating that sufficient chromosomal pairing occurred between the two parental species to permit the transfer of genes from annual ryegrass to tall fescue (Buckner et al., 1975; Saiga et al., 1976).

Performance of Kenhy and Other Hybrids

The cultivar Kenhy was derived from annual ryegrass × tall fescue hybridization. Initially, 56-chromosome amphiploids of annual ryegrass × tall fescue were used as male parents in crosses with their corresponding 28-chromosome F_1 hybrids. A high percentage of progenies from these crosses had a somatic chromosome number of 56 (Buckner et al., 1963). These 56-chromosome progenies were male-fertile and showed a wide range in seed set. They were morphologically similar to the 56-chromosome male parent, and chromosome number varied from 42 to 56 in each generation. During the fourth generation, vigorous plants with good seed set were selected for further study. Eleven clones with a high moisture content during summer drought stress and with soft, lax leaves were established in an isolated polycross block. These clones were meiotically stable with 42 chromosomes and apparently carried some genes or chromosomes from ryegrass (Webster and Buckner, 1971).

Seed from the polycross block was blended to produce breeder seed of a synthetic cultivar that was eventually released as Kenhy after extensive evaluation.

The ryegrass × tall fescue backcrossed hybrids, amphiploids, and their derivatives were evaluated during 1971 and 1972 in replicated tests established in solid seedings at Lexington, Ky. The 56-chromosome amphiploid was considerably superior in forage quality to the backcross hybrids and the cultivars (Table 7-1). Although a portion of the superior forage quality of the 56-chromosome amphiploid progenies can be attributed to annual ryegrass, a part of this superiority may be a consequence of their chromosome number. Sullivan and Myers (1939) found that chromosomal and genetic reduplication were associated with an increase of sugar content in *L. perenne*. Sullivan (1944), in a further comparison of diploid and tetraploid *L. perenne*, stated that, in general, an increase of chromosome number was frequently associated with an increase in moisture and soluble constituents and with a decrease in structural constituents. Thus, an increase in chromosome number and the resultant increase in cell size and cell volume of the amphiploid progenies may be associated with their superior forage quality.

The forage quality of the perennial ryegrass × tall fescue backcross hybrid was not different from that of the standard check cultivar Kentucky 31, but the annual ryegrass × tall fescue backcross and Kenhy were significantly superior in quality to Kentucky 31. Perloline also occurred at significantly lower levels in the hybrids than in Kentucky 31 (Table 7-1).

B. Interspecific Hybrids

Giant fescue is a soft-leaved, high quality forage species normally found in moist, shady areas of Europe. It does not grow well in the transition zone of the United States where tall fescue is well adapted. Both tall and giant fescue belong to the Bovine (broad-leaf) section of the genus *Festuca*, and are bivalent forming hexaploids ($2n = 42$). However, Jenkin (1955) considered them morphologically so dissimilar that he could not regard them as extreme forms of the same species. In fact, giant fescue has often been known as *Bromus giganteus* L.

Hybridization between tall and giant fescue has been studied at Lexington, Ky. as a possible means of improving the forage quality of tall fescue. Four generations of colchicine-induced amphiploid progenies (ca. $2n = 84$) of tall fescue ($2n = 42$) × giant fescue ($2n = 42$) hybrids were male-fertile, and seed-set per panicle was equivalent to that of tall fescue (Buckner et al., 1976). Chromosome number of the fourth-generation progenies varied from $2n = 53$ to 84. Dodecaploids ($2n = 84$) occurred most frequently, indicating that rigid selection can rapidly secure stable populations at the

Table 7-1—Dry matter yields, palatability, nutritive value index (N.V.I.), and peroline content of ryegrass × tall fescue backcrossed hybrids, amphiploids, and their derivatives.

Entry[†]	Yields		Palatability[‡]		Perloline		N.V.I.[§]	
	1971	1972	1971	1972	1971	1972	1971	1972
	— metric tons/ha —				— $\mu g/g$ —			
(pt)t	8.7	4.9	6.5	7.0	59	125	81	77
(at)t	8.3	5.3	3.0	2.0	89	167	81	78
AT	--	--	1.0	1.0	151	--	89	86
Kenhy	10.0	5.2	2.0	3.0	114	190	83	75
Kentucky 31	8.6	4.9	7.5	7.5	185	259	81	73
Fawn	8.5	4.4	9.0	9.0	88	239	80	74
L.S.D. at 0.05	0.9	0.5	1.9	1.8	39	91	1	3
C.V. (%)	6.5	6.8	38	39	30	32	2	4

[†] Backcrossed progenies of (perennial and annual ryegrass × tall fescue) × tall fescue are designated (pt)t and (at)t, respectively, and amphiploids of annual ryegrass × tall fescue are designated (AT).
[‡] Entries scored 1 were most frequently grazed; and those scored 9 were most poorly grazed.
[§] Nutritive value index is a laboratory method used to estimate digestibility in ruminants.

Table 7-2—Dry matter and clean seed yields and nutritive value index (N.V.I.) of tall × giant fescue hybrids and tall fescue cultivars during 1974.

Entries	Hay	Aftermath	Total		N.V.I.[†]
			Forage	Seed	
	——————— kg/ha ———————				%
TG Hybrid[‡]	3,003	2,488	5,492	25	68.5
Kenhy	4,192	3,878	8,070	125	59.3
Ky 31	3,363	2,981	6,344	110	61.4
Fawn	2,936	3,587	6,523	116	56.3
Penngreen	2,869	3,318	6,187	56	55.1
L.S.D. at 0.05	425	425	628	60	2
C.V. (%)	7	8	6	35	4

[†] Nutritive value index is a laboratory method used to estimate digestibility in ruminants.
[‡] Amphiploid progenies of tall fescue × giant fescue.

$2n = 84$ chromosome level. Sulinowski (1972) studied two generations of colchicine-induced progenies of a giant × tall fescue F_1 hybrid. He reported that 10 to 14% of the self-pollinated progenies from two dodecaploids had the maternal chromosome number. Early studies with these hybrids suggested that it was easier to stabilize alloploids at the octoploid and decaploid level than at the dodecaploid level.

Fourth-generation amphiploid progenies were included in a cultivar evaluation trial at Lexington, Ky. When managed as hay, dry matter production of the progenies was not significantly different from Kentucky 31; however, they yielded significantly less than any of the tall fescue cultivars under simulated pasture management (Table 7-2). The ampliploid progenies failed to produce a dense stand of panicles, which helped explain the significantly lower clean seed yields than that of the tall fescue cultivars. Although relatively good stands of the amphiploids were obtained during the first year, the sod was noticeably thinner than that of the tall fescue cultivars a year later.

Nutritive Value Index (NVI) of hay and accumulated growth from the hybrids, which were permitted to grow from August until December, was significantly higher than that of the tall fescue cultivars (Table 7-2). The high digestibility of the hay from the hybrids may have resulted from the low ratio of stems to leaves in this material; however, this does not explain the superior values obtained for the forage accumulated until late autumn. Although a portion of the superior digestibility of the amphiploid progenies may be attributed to the quality of the giant fescue parent, a part of this superiority may be a consequence of their chromosome number, as was postulated for the *Lolium-Festuca* amphiploids. The cytological and agronomic performance of the colchicine-induced progenies suggests considerable potential for producing a highly nutritious forage crop at the dodecaploid level.

LITERATURE CITED

1. Anderson, L. B. 1976. Grasslands G4710 tall fescue *Festuca arundinacea* (Schreb.). N.Z. Grassl. Assoc. Proc., Palmerston, New Zealand. 36:198-199.
2. Asay, K. H. 1971. Breeding grasses for forage production. p. 1-4. *In* Western Grass Breeders Conf. Proc., Eugene, Oreg. 22-27 June 1971.
3. ─────, T. R. Minnick, G. B. Garner, and B. W. Harmon. 1975. Use of crickets in a bioassay of forage quality in tall fescue. Crop Sci. 15:585-588.
4. ─────, C. J. Nelson, and G. L. Horst. 1974. Genetic variability for net photosynthesis in tall fescue. Crop Sci. 14:571-574.
5. ─────, ─────, A. G. Matches, F. A. Martz, G. B. Garner, J. H. Dunn, and D. O. Randall. 1974. Grass breeding and genetics. Research in Agronomy. Missouri Agric. Exp. Stn. Misc. Publ. 74-2:55-59.
6. Baker, H. H., and J. R. A. Chard. 1964. Tall fescue (S-170) in the sward. Agriculture 71: 509-512.
7. ─────, ─────, and W. E. Hughes. 1965. A comparison of cocksfoot and tall fescue—dominant swards for out-of-season production. J. Br. Grassl. Soc. 20:84-90.
8. Barnes, R. F. 1966. The development and application of in vitro rumen fermentation techniques. p. 434-438. *In* G. G. Hill (ed.) Int. Grassl. Congr. Proc. 10th, Helsinki, Finland. 7-16 July 1966. Univ. of Helsinki, Finland.
9. ─────. 1973. Laboratory methods of evaluating herbage feeding value. p. 179-214. *In* G. W. Butler, and R. W. Bailey (eds.) Chemistry and biochemistry of herbage. Vol. 3. Academic Press, New York.
10. Bean, E. W. 1969. Environmental and genetic effects upon reproductive growth in tall fescue (*Festuca arundinacea* Schreb.). J. Agric. Sci. 72:341-350.
11. ─────. 1970. Short-day and low temperature control of floral induction in *Festuca*. Ann. Bot. 34:57-66.
12. ─────. 1972. Clonal evaluation for increased seed production in two species of forage grasses, *Festuca arundinacea* Schreb. and *Phleum pratense* L. Euphytica 21:377-383.
13. Bennett, H. W., N. L. Taylor, M. E. McCullough, and G. W. Burton. 1958. The use of animals to evaluate the palatability of breeding lines in forage crops improvement. p. 32-34. South Pasture and Forage Crop Improvement Conf. Proc. 5th., Starkville, Miss.
14. Berry, C. D. 1973. Breeding for rust resistance in tall fescue. p. 119-120. *In* South. Pasture and Forage Crop Improvement Conf. Proc. 30th. 29-31 May 1973. Univ. of Kentucky, Lexington.
15. ─────, and R. T. Gudauskas. 1972a. Susceptibility of tall fescue *Festuca arundinacea* Schreb., to crown rust. Crop Sci. 12:101-102.
16. ─────, and ─────. 1972b. Races of *Puccinia coronata* var. *coronata* on tall fescuegrass. Plant Dis. Rep. 56:614-615.
17. Boling, J. A., R. C. Buckner, and L. P. Bush. 1973. Physiology and animal nutrition. p. 91-97. *In* South. Pasture and Forage Crop Improvement Conf. Proc. 30th. 29-31 May

1973. Univ. of Kentucky, Lexington.
18. ———, L. P. Bush, R. C. Buckner, and L. C. Pendlum, P. B. Burrus, S. G. Yates, S. P. Rogovin, and H. L. Tookey. 1975. Nutrient digestibility and metabolism in lambs fed added perloline. J. Anim. Sci. 40:972-976.
19. Borrill, M., B. Tyler, and M. Lloyd Jones. 1971. Studies in *Festuca* I. A chromosome atlas of bovinae and scariosae. Cytologia 36:1-14.
20. Breese, E. L., and W. E. Davies. 1973. Herbage breeding. Annual Rep. Welsh Plant Breeding Stn., 1972 (Aberystwyth). p. 25-45.
21. Bryan, W. B., W. F. Wedin, and R. L. Vetter. 1970. Evaluation of reed canarygrass and tall fescue as spring-summer and fall-saved pastures. Agron. J. 62:75-80.
22. Buckner, R. C. 1960a. Performance of inbred lines, polycross progenies, and synthetics of tall fescue suited for improved palatability. Agron. J. 52:177-180.
23. ———. 1960b. Use of polyethylene tubes for selfing tall fescue. Agron. J. 52:410-411.
24. ———. 1973. The tall fescue breeding program at the University of Kentucky. p. 85-88. *In* South. Pasture and Forage Crop Improvement Conf. Proc. 30th. 29-31 May 1973. Univ. of Kentucky, Lexington.
25. ———, and P. B. Burrus, Jr. 1962. Comparison of techniques for evaluating palatability differences among tall fescue strains. Crop Sci. 2:55-57.
26. ———, and ———. 1965. Kenwell tall fescue. Characteristics and management. Kentucky Agric. Exp. Stn. Circ. 601.
27. ———, ———, and J. R. Todd. 1969. Morphological and chemical characteristics of inbred lines of tall fescue selected for palatability to cattle. Crop Sci. 9:581-583.
28. ———, ———, G. T. Webster, and L. P. Bush. 1975. Breeding pasture, hay, and turf grasses. Kentucky Agric. Exp. Stn. Annu. Rep. 87th. p. 14-15.
29. ———, L. P. Bush, and P. B. Burrus, II. 1973. Variability and heritability of perloline in *Festuca* sp., *Lolium* sp., and *Lolium-Festuca* hybrids. Crop Sci. 13:666-669.
30. ———, and E. N. Fergus. 1960. Improvement of tall fescue for palatability by selection within inbred lines. Agron. J. 52:173-176.
31. ———, Helen D. Hill, and Paul B. Burrus, Jr. 1961. Some characteristics of perennial and annual ryegrass × tall fescue hybrids and of amphidiploid progenies of annual ryegrass × tall fescue. Crop Sci. 1:75-80.
32. ———, ———, Arne W. Hovin, and Paul B. Burrus, Jr. 1963. Cytogenetic and morphological characteristics of progenies of crosses of annual ryegrass × tall fescue hybrids and their derivatives. Crop Sci. 3:453-454.
33. ———, J. R. Todd, P. B. Burrus II, and R. F. Barnes. 1967. Chemical composition, palatability, and digestibility of ryegrass-tall fescue hybrids, 'Kenwell' and 'Kentucky 31' tall fescue varieties. Agron. J. 59:345-349.
34. ———, G. T. Webster, P. B. Burrus II, and L. P. Bush. 1976. Cytological, morphological, and agronomic characteristics of tall × giant fescue hybrids and their amphiploid progenies. Crop Sci. 16:811-816.
35. Burton, G. W., and E. H. DeVane. 1953. Estimating heritability in tall fescue (*Festuca arundinacea*) from replicated clonal material. Agron. J. 45:478-481.
36. Bush, L., and R. C. Buckner. 1973. Tall fescue toxicity. p. 99-112. *In* A. G. Matches (ed.) Anti-quality components of forages. Crop Sci. Soc. Am., Madison, Wis.
37. ———, J. A. Boling, G. Allen, and R. C. Buckner. 1972. Inhibitory effects of perloline to rumen fermentation in vitro. Crop Sci. 12:277-279.
38. ———, R. C. Streeter, and R. C. Buckner. 1970. Perloline inhibition of in vitro ruminal cellulose digestion. Crop Sci. 10:108-109.
39. Carlson, Irving T. 1974. Correlations involving in vitro dry matter digestibility of *Dactylis glomerata* L. and *Phalaris arundinacea* L. p. 732-738. *In* V. G. Iglovikov and A. P. Movsissyants (eds.) Int. Grassl. Congr. Proc., 12th, Moscow, USSR. 11-20 June 1974.
40. Carnahan, H. L., and Helen D. Hill. 1962. Increased meiotic irregularities in inbred programs of a *Festuca arundinacea* clone with apparent inversion heterozygosity. Crop Sci. 2:445-446.
41. Chatterjee, B. N. 1966. Analysis of ecotypic differences in tall fescue (*Festuca arundinacea* Schreb). Ann. Appl. Biol. 49:560-562.
42. Clarke, J., P. Chandrasekharan, and H. Thomas. 1976. Studies in *Festuca* 9. Cytological studies of *Festuca pratensis* var. *apennina* (De Not) Hack. Z. Pflanzenzuchtg. 77:205-214.
43. Cooper, J. P. 1973. Genetic variation in herbage constituents. p. 379-417. *In* G. W.

Butler, and R. W. Bailey (eds.) Chemistry and biochemistry of herbage. Vol. II. Academic Press, New York.
44. Cornelius, P. L., R. C. Buckner, L. P. Bush, P. B. Burrus II, and J. Byars. 1974. Inheritance of perloline content in *Lolium multiflorum* Lam. × *Festuca arundinacea* Schreb. hybrids. Crop Sci. 14:896–898.
45. Cowan, J. R. 1956. Tall fescue. Adv. Agron. 8:283–320.
46. Craigmiles, J. P., L. V. Crowder, and J. P. Newton. 1964. Palatability differences in tall fescue using leaf and plant type. Crop Sci. 4:658–660.
47. CSIRO Division of Plant Industry. 1973. Annu. Rep. for 1972. Canberra, Australia. p. 192.
48. Cunningham, I. J. 1949. A note on the cause of tall fescue lameness in cattle. Aust. Vet. J. 25:27–28.
49. Davies, W. E., B. F. Tyler, M. Borrill, J. P. Cooper, H. Thomas, and E. L. Breese. 1972. Plant introduction at the Welsh Plant Breeding Station. Annu. Rep. Welsh Plant Breeding Stn., 1972 (Aberstwyth). p. 143–162.
50. Djikstra, J. 1974. Seedling growth of allopolyploids from the cross *Lolium multiflorum* × *Festuca arundinacea*. New ways in fodder crops. Fodder Crops Section, Eucarpia Proc. p. 5–6.
51. Donefer, E., E. W. Crampton, and L. E. Arnold. 1966. The prediction of digestible energy intake potential (NVI) of forages using a simple in vitro technique. p. 442–445. *In* G. G. Hill (ed.) Int. Grassl. Congr. Proc. 10th. Helsinki, Finland. 7–16 July 1966. Univ. of Helsinki, Finland.
52. Echeverri, S. 1964. A comparison of progeny testing methods and estimates of combining ability for seed yield and associated variables in tall fescue (*Festuca arundinacea* Schreb.). Ph.D. Thesis. Oregon State Univ. Univ. Microfilms. Ann Arbor, Mich. (Diss. Abstr. 25:1450–1451).
53. Evans, G. M., K. H. Asay, and R. G. Jenkins. 1973. Meiotic irregularities in hybrids between diverse genotypes of tall fescue (*Festuca arundinacea* Schreb.). Crop Sci. 13:376–379.
54. Flores, J. M., R. A. Chapman, and L. Henson. 1969. Susceptibility of detached and attached leaf blades of tall fescue to infection by two species of *Helminthsporium*. Phytopathology 59:1010–1011.
55. ———, and K. I. Matheson. 1973. Progeny testing for forage yield in tall fescue, *Festuca arundinacea*, Schreb. Crop Sci. 13:293–295.
56. Frame, J. 1963. Tall fescue. West of Scotland Agric. Coll. Scottish Agriculture (Autumn, 1963). 2 p.
57. ———. 1972. The agronomic evaluation of Syn I tall fescue in the West of Scotland. J. Br. Grassl. Soc. 27:155–162.
58. ———, R. D. Harkess, I. V. Hunt. 1974. Comparison of productivity of varieties of meadow fescue and tall fescue. West of Scotland Agric. Coll. Exp. Record 40:17–25.
59. ———, I. V. Hunt, and R. D. Harkess. 1970. Potentiality studies of tall fescue. p. 209–214. Int. Grassl. Congr. Proc. 10th, Helsinki, Finland.
60. Gentry, C. E., R. A. Chapman, L. Henson, and R. C. Buckner. 1969. Factors affecting alkaloid content of tall fescue, *Festuca arundinacea* Schreb. Agron. J. 61:313–316.
61. ———, L. Henson, and R. A. Chapman. 1968. The interrelationships of *Rhizoctonia solani* and *Helminthsporium vagans*, nitrogen, and variety on alkaloids in *Festuca arundinacea*. Phytopathology 58:1051.
62. Gillet, M., and J. Jadas-Hecart. 1965. Leaf flexibility, a character for selection of tall fescue for palatability. p. 155–157. *In* Int. Grassl. Congr. Proc. 9th, Sao Paulo, Brazil. 8–19 Jan. 1965.
63. Gross, H. D., L. Goode, W. B. Gilbert, and G. L. Ellis. 1966. Beef grazing systems in Piedmont, North Carolina. Agron. J. 58:307–310.
64. Hanson, A. A. 1972. Grass varieties in the United States. USDA Handb. 170.
65. ———, and H. L. Carnahan. 1965. Breeding perennial forage grasses. USDA Tech. Bull. 1145.
66. Hill, H. D., and R. C. Buckner. 1962. Fertility of *Lolium-Festuca* hybrids as related to chromosome number and meiosis. Crop Sci. 2:484–486.
67. Hill, R. R., and S. B. Guss. 1976. Genetic variability for mineral concentration in plants related to mineral requirements in cattle. Crop Sci. 16:680–685.
68. Hollowell, E. A. 1945. Registration of varieties and strains of grasses. Am. Soc. Agron. J. 37:653–654.

69. Hoveland, C. S. 1967. Goar tall fescue. Auburn Agric. Exp. Stn. Leaflet 75.
70. ———, E. M. Evans, and D. A. Mays. 1974. Cool season perennial grass species for forage in Alabama. Auburn Agric. Exp. Stn. Bull. 397.
71. Hunt, I. V., J. Frame, and R. D. Harkess. 1972. Classification of herbage plant varieties for West of Scotland conditions. West of Scotland Agric. Coll. Publ. 320.
72. Jacobson, D. R., R. H. Singer, J. W. Rust, and D. M. Seath. 1957. A preliminary comparison of the nutritive values of orchardgrass, smooth bromegrass, 'Kentucky 31' fescue, and G 1-43 fescue. J. Dairy Sci. 40:613.
73. Jauhar, P. P. 1975a. Genetic regulation of diploid-like chromosome pairing in the hexaploid species, *Festuca arundinacea* Schreb., and *F. rubra* L. (Gramineae). Chromosoma (Berlin) 52:363-382.
74. ———. 1975b. Genetic control of chromosome pairing in polyploid fescues: its phylogenetic and breeding implications. Rep. Welsh Plant Breeding Stn., 1973 (Aberystwyth). p. 114-127.
75. Jenkin, T. J. 1933. Interspecific and intergeneric hybrids in herbage grasses. XVIII. Various crosses including *Lolium rigidum* sens. ampl. with *L. temulentum* and *L. loliaceum* with *Festuca pratensis* and with *F. arundinacea*. J. Genet. 53:467-486.
76. ———. 1955. Interspecific and intergeneric hybrids in herbage grasses. IX. *Festuca arundinacea* with one other *Festuca* species. J. Genet. 53:81-93.
77. Johnson, I. J. 1952. Evaluating breeding materials for combining ability. p. 327-334. *In* S. S. Atwood (ed.) Int. Grassl. Congr. Proc. 6th. 17-23 Aug. 1952. State College, Pa.
78. Jones, D. I. H. 1974. Some recent developments in techniques for assessing the digestibility and intake characteristics of grasses. Annu. Rep. Welsh Plant Breeding Stn., 1974 (Aberystwyth). p. 128-133.
79. ———, R. J. K. Walters, and E. L. Breese. 1973. The evolution of herbage breeding programmes for improved voluntary intake and other nutritive characteristics. p. 111-120. European Grassl. Fed. 5th General Meeting Proc. Uppsala, Sweden.
80. Joppa, L. R., and C. W. Roath. 1964. A comparison of Kenmont, Alta, and other tall fescue varieties in Montana. Montana Agric. Exp. Stn. Bull. 582.
81. Juska, F. V., A. A. Hanson, and A. W. Hovin. 1969. Evaluation of tall fescue, *Festuca arundinacea* Schreb., for turf in the transition zone of the United States. Agron. J. 61: 625-628.
82. Kawabata, Syutaro, Kanji Gotoh, Yukio Mori, Suguru Saiga, Shigeru Suzuki, Jiro Abe, and Noboru Takase. 1972. New cultivars of tall fescue (*Festuca arundinacea* l.) (in Japanese, English Summary). Hokkaido Natl. Agric. Exp. Stn. Res. Bull. 103.
83. Kendall, W. A., J. R. Todd, and C. W. Templeton, Jr. 1970. Simplification of the dry matter disappearance technique to estimate forage quality. Crop Sci. 10:47-48.
84. Lewis, E. J. 1963. Hybrids between geographic races of *Festuca arundinacea*, Schreb. Annual Rep. Welsh Plant Breeding Stn., 1962 (Aberystwyth). p. 26-27.
85. ———. 1966. The production and manipulation of new breeding material in *Lolium-Festuca*. p. 688-693. Int. Grassl. Congr. Proc. 10th, Helsinki, Finland.
86. Martz, F. A., S. Bell, A. G. Matches, and D. A. Sleper. 1976. Advanced evaluation of forages. Missouri Agric. Exp. Stn. Spec. Rep. 192. p. 35-37.
87. Matheson, K. I. 1965. The breeding behavior of forage yield in two populations of tall fescue (*Festuca arundinacea* Schreb.) Oregon State Univ. Ph.D. Thesis. Univ. Microfilms. Ann Arbor, Mich. (Diss. Abstr. 26:14-15).
88. Miles, D. G., G. Griffith, and R. J. K. Walters. 1964. Variation in the chemical composition of four grasses. Annu. Rep. Welsh Plant Breeding Stn., 1963 (Aberystwyth). p. 110-114.
89. Moll, R. H., W. S. Salhuana, and H. F. Robinson. 1962. Heterosis and genetic diversity in variety crosses of maize. Crop Sci. 2:197-198.
90. Mott, G. O., C. J. Kaiser, R. C. Peterson, Randall Peterson, Jr., and C. L. Rhykerd. 1971. Supplemental feeding of steers on *Festuca arundinacea* Schreb. pastures fertilized at three levels of nitrogen. Agron. J. 63:751-754.
91. Mourtray, J. B., Jr., and R. V. Frakes. 1973. Effects of genetic diversity on heterosis in tall fescue. Crop Sci. 13:1-4.
92. Norris, K. H., and R. F. Barnes. 1976. Infrared reflectance analysis of nutritive value of feedstuffs. p. 237-242. *In* P. V. Fonnesbeck, L. E. Harris, and L. C. Kaarl (eds.) Int. Symp. on Feed Composition. Animal Nutrition Requirements and Computerization of Diets 1st Proc. Utah State Univ. 11-16 July 1976. Utah State Agric. Exp. Stn.
93. ———, ———, J. E. Moore, and J. S. Shenk. 1976. Predicting forage quality by in-

frared reflectance spectroscopy. J. Anim. Sci. 43:889-897.
94. Patil, B. D., and D. I. H. Jones. 1970. The mineral status of some temperate herbage varieties in relation to animal performance. p. 726-730. Int. Grassl. Congr. Proc. 11th, Queensland, Australia.
95. Peterson, R. G., G. O. Mott, M. E. Heath, and W. M. Beeson. 1959. A comparison of tall fescue and orchardgrass when grazed by yearling beef steers and heifers. J. Anim. Sci. 18:1542.
96. ―――, P. H. Weswig, and J. R. Cowan. 1958. Measuring palatability differences in tall fescue by grazing sheep. Agron. J. 50:117-119.
97. Phillips, T. G., J. T. Sullivan, M. E. Loughlin, and V. G. Sprague. 1954. Chemical composition of some forage grasses I. Changes with plant maturity. Agron. J. 46:361-369.
98. Raymond, W. F. 1969. Improving the nutritive value of herbage varieties. Br. Grassl. Soc. Occas. Symp. No. 5:29-36.
99. Robson, M. J., and O. R. Jewiss. 1968. A comparison of British and North African varieties of tall fescue (*Festuca arundinacea*) II. Growth during winter and survival at low temperatures. J. Appl. Ecol. 5:179-190.
100. Ross, J. G., S. S. Bullis, and K. C. Lin. 1970. Inheritance of in vitro digestibility in smooth bromegrass. Crop Sci. 10:672-673.
101. Saiga, Suguru, R. C. Buckner, P. B. Burrus. 1976. The fluorescent seedling character in ryegrass—tall fescue hybrids and its relation to nutritive value index and rust resistance. Jpn. J. Breed. 26:1-5.
102. Shenk, J. S., W. N. Mason, M. L. Risius, K. H. Norris, and R. F. Barnes. 1976. Application of infrared reflectance analysis to feedstuff evaluation. p. 242-248. *In* P. V. Fonnesbeck, L. E. Harris, and L. E. Kearl (eds.) Int. Symp. on Feed Comp., Animal Nutrient Requirements, and Computerization of Diets, 1st Proc. Utah State Univ., 11-16 July 1976.
103. Sleper, D. A., G. B. Garner, K. H. Asay, R. Boland, and E. E. Pickett. 1977. Breeding for Mg, Ca, K, and P content in tall fescue. Crop Sci. 17:433-438.
104. ―――, ―――, M. L. Mitchell, and K. L. Hunt. 1976. Breeding tall fescue. Missouri Agric. Exp. Stn. Special Rep. 192. p. 22-24.
105. ―――, and M. L. Mitchell. 1975. Breeding tall fescue. Missouri Agric. Exp. Stn. Special Rep. 177. p. 18-19.
106. Stone, P. C. 1953. The house cricket as a laboratory insect. Turtox News 31:150-151.
107. ―――. 1963. Preliminary studies on forage diets of crickets. Entomol. Soc. Am. Proc. N. Cent. 18:24-25.
108. ―――, and A. G. Matches. 1966. Cricket growth and survival on forage diets. Entomol. Soc. Am. Proc. N. Cent. 21:133-135.
109. Sulinowski, Stanislaw. 1972. Induced alloploids in grasses of *Festuca* and *Lolium* genera. Part III. *Festuca gigantea* (L.) Vill. (2n=42) *F. arundinacea* Schreb. (2n=42) hybrid derivatives. Genet. Pol. 13:93-105.
110. Sullivan, J. T. 1944. Further comparisons of plants with different chromosome numbers in respect to chemical composition. J. Am. Soc. Agron. 36:537-543.
111. ―――, and W. M. Myers. 1939. Chemical composition of diploid and tetraploid *Lolium perenne*. J. Am. Soc. Agron. 31:869-871.
112. ―――, T. G. Phillips, M. E. Loughlin, and V. G. Sprague. 1956. Chemical composition of some forage grasses. II. Successive cuttings during the growing season. Agron. J. 48:11-14.
113. Thomas, J. R. 1967. Heritabilities and association of seed yield components and seed yield in tall fescue (*Festuca arundinacea* Schreb.). Ph.D. Thesis. Univ. Microfilms. Ann Arbor, Mich. (Diss. Abstr. 27:3759-60).
114. ―――, and R. V. Frakes. 1967. Clonal and progeny evaluations in two populations of tall fescue (*Festuca arundinacea* Schreb.). Crop Sci. 7:55-58.
115. Tilley, J., M. A., and R. A. Terry. 1963. A two-stage technique for the in vitro digestion of forage crops. J. Br. Grassl. Soc. 18:104-111.
116. VanKeuren, R. W., and M. H. Niehaus. 1970. Ohio forage crop evaluation tests for 1969. Ohio Agric. Res. Dev. Ctr. Agron. Dep. series No. 195.
117. Van Soest, P. J. 1963. The use of detergents in the analysis of fibrous feeds. II. A rapid method for the determination of fiber and lignin. J. Assoc. Offic. Agric. Chem. 46:829-835.
118. ―――. 1964. Symposium on nutrition and forage and pastures. New chemical procedures for evaluating forages. J. Anim. Sci. 23:838-845.

119. ———. 1965. Symposium on factors influencing the voluntary intake of herbage by ruminants. Voluntary intake in relation to chemical composition and digestibility. J. Anim. Sci. 24:834–843.
120. ———, R. H. Wine, and L. A. Moore. 1966. Estimation of the true digestibility of forages by the in vitro digestion of cell walls. p. 438–441. *In* G. G. Hill (ed.) Int. Grassl. Congr. Proc. 10th, Helsinki, Finland. 7–16 July 1966. Univ. of Helsinki, Finland.
121. Watson, C. E. 1976. Heritability for perloline, nitrogen, and digestibility characteristics in tall fescue (*Festuca arundinacea* Schreb.) single crosses grown in two locations. Ph.D. Thesis. Oregon State Univ., Corvallis. Univ. Microfilms. Ann Arbor, Mich. (Diss. Abstr. 37:3199B).
122. Webster, G. T., and R. C. Buckner. 1971. Cytology and agronomic performance of *Lolium-Festuca* hybrid derivatives. Crop Sci. 11:109–112.
123. Wedin, W. F., I. T. Carlson, and R. L. Vetter. 1966. Studies on nutritive value of fall-saved forage, using rumen fermentation and chemical analyses. p. 424–428. Int. Grassl. Congr. Proc. 10th, Helsinki, Finland.
124. Williams, M., S. R. Shaffer, G. B. Garner, S. G. Yates, and H. L. Tookey. 1973. Ion-exchange chromatography of a toxic fescue extract assayed by intraperitoneal infusion into cattle. Fescue toxicity Conf. Proc., Univ. of Kentucky, Lexington. 29–31 May 1973.
125. Wilsie, C. P. 1958. Hybrid vigor, yield variance, and growth habit inheritance in divergent crosses. p. 21–33. *In* H. O. Graumann (ed.) Alfalfa Improvement Conf. Proc. 16th, Ithaca, N.Y. 28–30 July 1938.

Chapter 8

Seed Production

HAROLD YOUNGBERG
Agronomy Department
Oregon State University
Corvallis, Oregon

HOWELL N. WHEATON
Department of Agronomy
University of Missouri
Columbus, Missouri

ECONOMIC IMPORTANCE AND VOLUME OF SEED PRODUCTION

The availability of adequate supplies of tall fescue (*Festuca arundinacea* Schreb.) seed is essential to the maintenance of a strong forage system in many regions of the United States. During the years 1950 to 1977, tall fescue seed production in the United States has risen from 8.7 to 45.6 million kg, while the area harvested increased from 35,235 to 182,455 ha.

The major reason for the vast buildup of seed supplies was the development of management techniques that used tall fescue seed fields for both forage and seed. In fact, the bulk of the tall fescue seed produced in this country is produced from fields where the primary product is forage, not seed. The seed is a by-product of the overall forage program. For example, in 1977 Missouri produced 34.0 million kg of tall fescue seed, about 75% of the total United States production, but not a single hectare was planted primarily for seed (Table 8-1).

Tall fescue seed harvest in the United States reached a peak in 1973 at

Table 8-1—Tall fescue seed: area harvested, yield, and production, 1977.†

State	Area harvested	Yield	Production, clean seed
	ha	kg/ha	1,000 kg
Alabama	1,620	168	272
Arkansas	6,685	314	2,097
Georgia	3,645	241	878
Kansas	14,175	157	2,225
Kentucky	6,075	235	1,430
Missouri	134,865	252	34,000
Oregon	4,050	673	2,724
South Carolina	4,050	202	817
Tennessee	6,075	196	1,192
United States total	181,240	252	45,635

† Crop Report, Statistical Reporting Service, USDA. 1977.

Copyright © 1979 ASA-CSSA-SSSA, 677 South Segoe Road, Madison, WI 53711 USA. *Tall Fescue.*

185,000 ha, then fell to 161,000 ha in 1976. In 1973, U.S. seed production was 60.4 million kg, 17% greater than consumption. This large carry-over depressed prices, resulting in reduced production. Thus, tall fescue seed production has expanded with increased forage plantings, until large carry-over supplies reduced prices and reduced the area harvested for seed.

It appears that seed produced from hay and pasture fields in the Midwest will continue to dominate the market for some time into the future. Many growers will produce seed if enough income is generated to exceed harvest costs; other expenses such as fertilizer and labor are charged to the livestock enterprise.

AREAS OF SEED PRODUCTION

Originally, most of the tall fescue seed harvested in this country was produced in the Pacific Northwest and Kentucky. However, recent drastic shifts have occurred in seed production areas. The largest single area for tall fescue seed production is now centered in the Southern Corn Belt and the Northern Cotton Belt.

A. Pacific Northwest

During the 1940's and 1950's, tall fescue was grown for seed in the dryland areas of eastern Oregon, Washington, and Idaho, as well as in the humid western parts of Oregon. This region has a natural advantage for grass seed production with its mild, wet winter and spring months which promote grass growth. The dry, low-humidity summer months favor harvesting clean, high-germination seed.

Tall fescue produces seed well on the fertile, well-drained soils in this region. Most grass seed is grown on farms specializing in production of tall fescue seed or other grass seed as a cash crop. There is little grazing of the tall fescue seed fields.

The use of well-established production techniques and the favorable environment results in high seed yields. Tall fescue seed production in the state of Oregon in 1977 averaged 673 kg/ha, more than 2.5 times the national average. A 3-year average seed production of 1,754 kg/ha has been recorded under experimental conditions (R. E. Fore, unpublished data).

B. Midwest and Other

Outside the Pacific Northwest, tall fescue seed production is principally a by-product of a livestock (primarily beef cow) enterprise. Spring growth in excess of forage needs is set aside and harvested for seed. The summer growth after seed harvest is then either grazed or clipped and baled for hay.

Fall regrowth is used for fall and winter grazing. This is an excellent way to utilize the total tall fescue crop, but it sometimes results in seed yields somewhat lower than if managed primarily for seed. Although there are many instances of seed yields exceeding 1,100 kg/ha, the average is approximately 335 kg/ha when used for seed and pasture (Missouri Farm Facts. 1975. Missouri Crop and Livestock Reporting Service. Columbia, Mo.). Under experimental conditions, seed yields as high as 1,000 kg/ha are obtained in solid stands managed basically for forage production (Kroth, 1973).

CULTURAL PRACTICES AND MANAGEMENT

Tall fescue has a wide climatic adaptation for seed production. Best yields are from fertile, well-drained soils managed for seed. Plants will survive on infertile soils, but little seed is matured. Climatic factors that limit seed production include late spring frosts at the time the inflorescence emerges from the boot and very hot or dry conditions at the time of seed maturation. Supplemental irrigation is required in the West where annual rainfall is less than 45 cm.

A. Planting and Establishment

A clean, weed-free, firm seedbed is desirable for tall fescue establishment because it is not a strong competitor with annual weeds during its first season of growth. A fall-planted crop does not reach maximum seed yield until the second year after seeding. Spring seedings were found to be the best for seed production under conditions at Prosser, Wash. (Van Keuren and Canode, 1963). The most common procedure is to prepare a good seedbed on land that has grown a cultivated crop or a spring-seeded grain crop. Fall plantings are usually made from 15 August to 15 October. Spring seedings are made from March through May.

Row vs. Solid Seeding

Tall fescue may be planted by broadcast or drill equipment. Drilling is preferred because more uniform stands result from the positive control of seeding depth, placing the seed in the most favorable position for germination and emergence. Solid stands are most common and are seeded in 30.5- to 35.5-cm drill rows using 9 to 11 kg of seed/ha. Commercial seed production in cultivated rows is common in dry regions and on less fertile soils. This method has been recommended for production of seed of high quality and for maintaining high seed yields over a longer stand of life. Rows are spaced 76 to 107 cm using seeding rates of 3.4 to 5.6 kg of seed/ha.

Numerous solid stands of tall fescue in Missouri and Oregon still produce 900 to 1,100 kg of seed/ha, although the stands are often 15 years old or older. There are, however, many reports and observations of a decline in seed yields as a stand becomes older (Canode and Van Keuren, 1963). Other workers (Spencer, 1950; Wheeler and Hill, 1957; Rampton, 1949) have reported that tall fescue seed yields from row plantings also tend to decline, but at a much slower rate than those established as solid stands. Spencer (1950) found that tall fescue in 67-cm rows showed a progressive decline in seed from year to year, although of a much smaller magnitude than that shown by broadcast seedings.

Hyer et al. (1950) reported a survey of grower practices in Oregon during 1948. They found that seed yields in solid stands dropped with each year of production, while most fields planted in rows maintained their production level. This study was conducted before post-harvest burning was practiced, with fertilizer applications of 37 kg N/ha on solid stands, while 53 kg N/ha were applied on row stands.

They reported that growing tall fescue in cultivated rows was the standard seed production procedure in the drier climate of eastern Oregon but not in western Oregon at that time. They found that row seedings were more costly to maintain and required five times as much labor as solid seedings during establishment. Even allowing for greater yields, the net cost of seed per kilogram from row seedings was higher than from solid stands.

The use of high fertility levels and post-harvest burning of residue permitted the maintenance of high seed yields in solid stands in Oregon. Rampton (1965) showed that tall fescue seed yields declined over 5 crop-years when planted in 15- and 30-cm drills at low fertility levels. However, using high fertility levels coupled with annual post-harvest burning and fall applications of herbicide for grass seedling control, seed yields were maintained at more than 227 kg/ha over 5 years in the 15- and 30-cm drills and 91-cm rows. These treatments prevented the stands from becoming excessively dense.

The term "sod-bound" has been applied to several forage crops and is associated with the thickening of the stand as a result of the plant's spreading by tillers and rhizomes. Stands planted at lower seeding rates do not become sod-bound as quickly as those seeded at heavier rates. Wheeler and Hill (1957) attribute the decline in seed yields from some older stands of fescue to this sod-bound condition because there is excessive competition for space and few crown shoots are produced from which fertile tillers can develop. Once a solid stand has become sod-bound, heavy fertilization will produce excellent forage growth, but only a limited number of culms.

The use of row plantings, with rows spaced as wide as 61 to 71 cm, has been recommended to avoid sod-bound stands (Spencer, 1950; Wheeler and Hill, 1957; and Rampton, 1949).

Richardson[1] and Furtick[2] studied techniques of renewing the seed yield in sod-bound fields. Richardson[1] concluded that the condition was the re-

---

[1] Richardson, G. L. 1961. Some studies on the causes of sod-binding in Alta fescue (*Festuca elatior* var. *arundinacea* (Schreb.) Wimm.) Ph.D. Thesis. Oregon State Univ., Corvallis.
[2] Futrick, W. R. 1952. Renovating Alta fescue for seed production. M.S. Thesis. Oregon State Univ., Corvallis.

sult of insufficient nutrients combined with failure of the plants to produce enough new rhizomes from which primordia may develop. He increased seed yield of a sod-bound stand from 108 to 304 kg/ha by reducing the stand to rows with a garden tiller and applying 336 kg N/ha in the spring. After 2 or 3 years, the seed yield returns to the level of the sod-bound condition. Furtick[2] tried chemical renovation techniques to increase seed yield without success.

The practice of "skim plowing" has been used on sod-bound fields. The sod of the old tall fescue stand is turned under to a depth of 10 cm in the fall with a sod plow. The ground is rolled to provide a smooth surface for the operation of harvesting equipment the following season. The stand re-establishes itself the year following plowing, and seed is harvested the second year. Seed yields on a plow-renovated stand decline at about the same rate as on a new seeding. These renovation methods have not been widely adopted by seed growers, as other recent techniques have been more successful. The use of 30- or 46-cm row widths, post-harvest burning, heavy fertilizer rates, and annual herbicide applications have become standard production practices in the West to maintain seed yields for 8 to 10 years.

Solid stands of tall fescue grazed during the fall do not show a decline in seed yields with advancing age under Missouri conditions (Peter Booysen, unpublished data). This may be related to greater fall tillering as a result of defoliation which facilitates a greater number of induced tillers.

Weed Control in Stand Establishment

The most important step in weed control in young stands is the selection of a clean field and proper seedbed preparation. Under conditions where grass weeds are a particular problem, a system of a spring seeding on a chemically prepared seedbed has been used in Oregon (Lee, 1965). A seedbed is prepared in the fall, followed by applications of propham (isopropyl carbanilate) or paraquat (1,1'-dimethyl-4,4'-bipyridinium ion) during the winter to control weed seed germination. The tall fescue is planted into this seedbed without additional tillage in March. This is followed by the standard weed control program for established stands.

Lee (1973) developed a technique of establishing weed-free stands from fall plantings. A narrow band of activated charcoal is applied directly over the seeded row during planting; then the entire field is sprayed with diuron [3-(3,4-dichlorophenyl)-1,1-dimethylurea]. The herbicide is inactivated by the charcoal protecting the crop in the seeded row, but it effectively controls weeds between the rows.

Broadleaf weeds can be controlled in new and established seed fields of tall fescue with 2,4-D [(2,4-dichlorophenoxy)acetic acid]; dicamba (3,6-dichloro-*o*-anisic acid) and 2,4-D in combination; bromoxynil (3,5-dibromo-4-hydroxybenzonitrile); or MCPA {[(4-chloro*o*-tolyl)oxy]acetic acid} (Hepworth, 1977). New seedings should not be sprayed with 2,4-D, MCPA, or dicamba-2,4-D combinations before the tall fescue has five leaves (usually 4 to 5 weeks after emergence). Label precautions should be followed in all spray operations.

In the eastern seed-producing areas, solid stands with adequate N fertilization have tended to control most weed problems (Wheaton, 1971).

B. Fertilization of Established Stands

The use of commercial fertilizers is essential for seed production. Nitrogen is the most important nutrient affecting the seed yield of tall fescue. Phosphorus and K applications are made on the basis of soil test results and should be maintained in the medium range. Lime should be applied if the soil pH falls below 5.5 (Gardner et al., 1970; Wheaton, 1971).

Although all major fertilizer elements are necessary for tall fescue seed production, the greatest response is to N. Time of N application is important in seed production. In trials with 'Kentucky 31' tall fescue, Spencer (1950) reported that applications of N in October, January, or February increased seed yield. Applications made in early April did not increase seed yield, though they greatly increased forage production. Nitrogen applied in early spring, shortly before growth starts, causes more lodging and excessive straw production than if the N is applied during the winter.

In Oregon, 56 to 67 kg N/ha are applied to established stands in the fall with 100 to 112 kg/ha in late February or March (Gardner et al., 1970). Wheaton (1971) recommends 78 to 112 kg N/ha applied during December or January for seed production in Missouri. If a late summer N application is made to encourage fall growth for winter grazing, then winter application rate should be reduced.

When tall fescue is used for the dual purpose of forage and seed, determining the proper amount of N for a seed crop is more difficult. Late summer (August) N application is needed to promote fall and winter grazing. It is seldom economical to apply enough N at this time to encourage the full potential of seed production the following summer. Some additional N should be applied in December or January, but the amount is somewhat dependent upon how much was applied in the late summer. The additional N needed is influenced by the amount of fall vegetative growth, grazing intensity, autumn and winter rainfall, and the soil texture and organic matter content.

A rule of thumb is, when no fall N was applied, 90 to 135 kg/ha should be topdressed during the winter; if 67 or 90 kg/ha were used in the fall, then 67 to 90 kg/ha are added during the winter; but if 90 to 112 kg/ha were fall applied, then 34 to 45 kg/ha applied in the winter may be sufficient for seed production (Wheaton, 1971).

Nitrogen applied during the winter dormant period, December through March, appears to produce the greatest response (Kroth, 1973). Certainly N should be available to the plant prior to spring growth. Some increase in seed is indicated with late spring application, but the vegetative growth seems to be stimualted to a greater degree than seed production. Lodging often becomes a problem with N application in late spring.

C. Weed Control in Established Stands

Effective weed control is essential in producing seed that meets market quality standards. Complete removal of crop residues after seed harvest is essential to maximizing the effectiveness of chemicals used in annual weed control. Burning straw on seed fields soon after harvest is a practice that is widely used in the Pacific Northwest. Baling straw and grazing of summer regrowth is common where seed is grown in combination with livestock production. In either case, thorough residue removal is necessary for maximum herbicidal activity.

Annual grasses and broadleaf weeds in established stands of tall fescue can be controlled by fall application of simazine [2-chloro-4,6-bis(ethylamino(-s-triazine], propham, chlorpropham (isopropyl m-chlorocarbanilate), or diuron under Pacific Northwest conditions. Simazine, propham, or chlorpropham is especially effective for control of wild oats (*Avena fatua* L.) and ripgut brome (*Bromus rigidus* Roth). Diuron is more effective in control of annual bluegrass (*Poa annua* L.) (Hepworth, 1977).

Broadleaf weeds in established fields are controlled by spring applications of 2,4-D, 2,4-D plus dicamba, or bromoxynil under Pacific Northwest conditions. These materials must be applied before seed heads appear in the boot. Fall applications may be advisable under certain conditions with weeds that are difficult to control (Hepworth, 1977).

Careful attention must be given to application methods, rates, and timing to avoid crop injury. Chemical effectiveness will vary with the environment and soil conditions. Local recommendations should be followed.

D. Post-Harvest Management

Care of seed fields immediately after harvest is extremely important to seed production in the following year. Tillers for the next season's seed crop are initiated during the early fall. Clipping the regrowth after seed harvest seems to encourage tiller development. Seed production is significantly reduced when tall fescue stubble is not clipped and aftermath removed by some procedure (Rampton, 1965; Kroth, 1970).

Seed head size seems to be related to the time that the tiller is developed in the fall. Early tillers tend to be the most productive. There is some indication that tillers developed later in the winter do not receive sufficient photothermal induction to become reproductive in the spring. Seasonal weather conditions can modify this to some degree; cool spring weather may delay head emergence that could result in larger seed heads. Post-harvest management should provide conditions that will favor maximum early fall tiller development[3] (Templeton et al., 1961).

---

[3] Dunn, S. J. 1958. Seedhead development in tall fescue as influenced by date of application of 3-chloro IPC. Ph.D. Thesis. Oregon State Univ., Corvallis.

Burning of the seed crop residue has many benefits and is widely used in tall fescue seed fields in the Pacific Northwest. When post-harvest field burning was applied to a tall fescue crop in western Oregon in 1949, it controlled blind seed (*Gloetinia temulenta*) and ergot (*Claviceps purpurea*) (Hardison, 1976). The use of fire for residue disposal, disease control, and weed control has continued to the present.

In addition to disease control benefits from post-harvest burning, seed yield increases of 24 and 28% from machine-burned and open-burned tall fescue, as compared to treatments with straw removed, have been reported by Youngberg et al. (1975). Chilcote and Youngberg (1975) also report benefits of early (August), as compared with late (October), burning of perennial grasses. Seed yields in the following year were greatly reduced when burning was delayed until after initiation of fall regrowth.

Spencer (1950) burned tall fescue in Kentucky in early February when accumulated dead leaves permitted a very complete burning of all aboveground vegetation. Seed yields from these burned plots were of the same magnitude as from plots that had remained unclipped.

Tall fescue cultivars vary in tolerance to post-harvest burning. Rampton (1965) found at least one late-maturing cultivar that was severely injured by burning, and caution must be used in applying the practice to new cultivars.

Cattle management becomes an important factor in seed production if tall fescue seed fields are grazed during the fall and winter. Grazing, if any, should not be too heavy during August, September, and October when tiller initiation occurs. After the first of November, grazing pressures may be increased, and all growth should be removed by 15 January. Trampling during wet weather should be avoided at all times.

CERTIFICATION

The maintenance of the genetic identity of a cultivar is an important part of seed increase. Standards published by the Federal Seed Act and the Association of Official Seed Certifying Agencies provide the guidelines for certified seed production in the United States. Specific requirements are established for planting stock, land history, isolation, and tolerances for other cultivars allowed at field inspection. Standards require a field inspection shortly after planting and an annual inspection at heading to assure that the genetic standards are being maintained. State certification requirements should be consulted before a seed field is planted (Table 8-2).

HARVESTING

The seed of tall fescue shatters easily when ripe. It is usually harvested by direct combining of standing grass, or windrowed and then combined. Shattering due to delays in harvest caused by rains, unavailability of har-

SEED PRODUCTION

Table 8-2—Hectares applied for certified tall fescue production, United States, 1977.†

Oregon	2,389
Kentucky	213
Missouri	212
Georgia	53
Total	2,867

† Garrison, Robert H., and Bridget R. Tuttle. 1977. Report of acres applied for certification in 1977 by seed certifying agencies. Clemson, S. C.

vesting machines, or high winds can easily reduce yields 50% or more. Even under favorable conditions, extreme care and skill by the combine operator is necessary to prevent serious losses.

Direct combining of tall fescue is a feasible method of harvesting under the following conditions: 1) the area of fescue seed to be harvested is small (can be completed in 1 or 2 days); 2) a combine is available without delay; 3) seed will not be placed in large bins. Combining should begin when 5 to 15% of the seed are immature. Harvesting with more than 20% immature seed usually results in low yields. Excessive seed moisture associated with early harvest will cause heating in storage, low seedling vigor, and low germination.

When a large area is to be harvested, or when delays in obtaining harvesting equipment are expected, then mowing, windrowing, and curing the seed in the windrow is the best method of harvesting. A swather can complete the cutting and windrowing in one operation. The swathed straw and seed can be easily picked up with a combine when cured (Fig. 8-1). Tall fescue should be mowed at an earlier stage of seedhead maturity when windrowed. Mowing should begin when the straw in the head is yellowing. An occasional seed will shatter from the earliest maturing heads in the field as the stem is tapped below the head.

The mower or windrower should be set to cut high enough to leave 10 to 15 cm of grass stubble standing with the windrow placed on top of the stubble. Air will circulate through the swath and decrease drying time. The tall fescue should be combined as soon as the windrows are thoroughly dry. Field-dry seed will not require special handling and can be stored in bins.

Seed losses can be reduced and yields maximized by proper timing of the harvest operation. Klein and Harmond (1971) studied the maturation characteristics of 'Alta' tall fescue to determine the proper cutting time to maximize seed yields. They found that the optimum cutting time for grass seed crops often preceded the state of maturity that produced the highest germination and frequently occurred after some seed shatter. The optimum cutting time is a balance between the gain in seed weight and germination gain with maturity and losses due to shattering. Klein and Harmond (1971) found that seed moisture related well with maximum yield of pure live seed (Fig. 8-2). The seed moisture content curve during maturity has a characteristic slope and shape and may be used to predict the proper mowing date.

Klein and Harmond (1971) reported that farmers tended to allow tall fescue to become too ripe before cutting for seed. The average seed mois-

Fig. 8-1—Seed harvest from the windrow allows seed to mature and reduces seed loss due to shattering.

ture when farmers cut was 25%. This was approximately 6 days beyond the optimum.

The seed moisture level in the mature standing crop stabilized at approximately 10% under Oregon conditions. Delaying cutting until the seed reached 10% moisture resulted in significant seed losses. They found that cutting the crop when the seed moisture in the standing crop was 43% and allowing the crop to mature in the windrow produced the maximum pure live seed per hectare. Seed losses of 112 kg/ha were observed when cutting was delayed until seed moisture reached 10%.

The combine should be set according to the manufacturer's instructions. Excessive cylinder action is not necessary to remove the seed from the straw. The combine operator should check from time to time to assure that seed is not being carried over in the straw and chaff. The operator should recognize sterile or immature florets that may give the impression that seed is being blown out.

Seed that is directly combined will usually require special handling and care to prevent heating. The direct harvested seed should be cleaned immediately to remove all green material and then air-dried to a moisture level that is safe for bin storage. If only a small amount of seed is to be handled,

SEED PRODUCTION

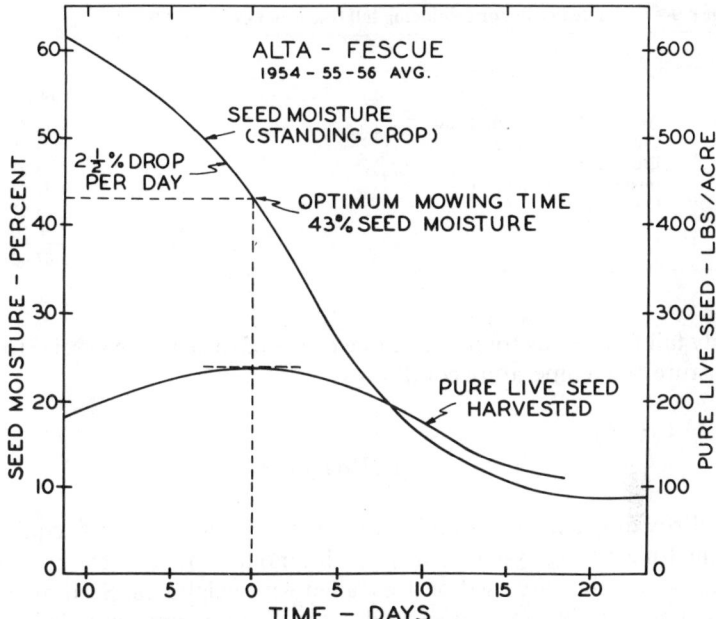

Fig. 8-2—The influence of seed moisture at mowing time on combine harvest yields of Alta tall fescue under Oregon conditions (Klein and Harmond, 1971).

it may be spread out in bins or lofts to dry. If drying bins are used, the circulating air should not exceed 32 C at the flue entrance.

SEED STORAGE

Seed viability in storage is determined by the initial seed quality, seed moisture, and storage temperatures. High moisture and high temperature conditions are most damaging to storage life. In a study conducted under low summer and high winter humidity conditions of western Oregon, Ching et al. (1963) found that tall fescue seed retained its viability very well for 15 months when stored in burlap bags under standard warehouse conditions.

Initial seed quality will also influence seed longevity. One tall fescue lot that germinated 98% dropped to 80% after 5 years of storage, while another lot that had an initial 78% germination lost viability in less than 5 years.

Seed viability decreases very rapidly under high temperature and high moisture storage conditions. Ching et al. (1960) reported that forage crop seeds reached hydroscopic equilibrium with 90% relative humidity at 32 C. Seed viability was lost in 7 weeks' storage under these conditions. At 32 C and 40% relative humidity, the equilibrium moisture was 9.5%. In this

Table 8-3—Estimated costs for producing tall fescue in Oregon, 1975.

Item	cost/ha
Amortized establishment cost	$ 12.55
Cultural operations	43.25
Processing	8.97
Interest on land and capital	31.52
Land taxes	5.01
Overhead	6.10
Total	$107.40

study tall fescue was found to be one of the most sensitive species to adverse moisture and temperature conditions.

MARKETING

Providing adequate supplies of seed to forage producers requires a well defined marketing system. Only small amounts of seed are grown on the farms where they are used or are traded with neighbors. Seed of new cultivars, particularly, are grown under certification programs in specialized seed-growing areas far from the regions of consumption. Some established firms have branch offices in production, as well as consumption, areas. Others use the services of brokers to secure supplies of seeds to meet their needs. Through these lines of communication and commerce, seed supplies are made available to the consumer.

Seed production has been high, with approximately 6% of annual seed production carried into the next marketing year. In response to depressed seed prices, growers have reduced the area harvested for seed and brought supply in line with demand.

ECONOMIC RETURN FROM SEED PRODUCTION

Returns for seed production will vary widely. In those areas where fields are extensively grazed, such as the Midwest, the return from livestock can be very important. The decision whether to harvest seed will depend upon the livestock price and forage availability, as well as the seed price. In the Pacific Northwest where seed fields are not grazed, seed is harvested each year.

Estimated cost for tall fescue seed production in Oregon in 1975 was $107.40/ha (Nelson et al., 1975) or 17 cents/kg, based on average seed yields (Conklin and Fisher, 1973) (Table 8-3).

LITERATURE CITED

1. Canode, C. L., and R. W. Van Keuren. 1963. Seed production characteristics of selected grass species and varieties. Washington Agric. Exp. Stn. Bull. 647.
2. Chilcote, D. O., and H. W. Youngberg. 1975. Techniques and timing of post-harvest grass seed field burning. Oregon State Univ., Dep. Crop Sci., Oregon Ext. Memo ACS 7.
3. Ching, T. M., I. Schoolcraft, P. Rowell, H. Taylor, and B. Davidson. 1963. Change of forage seed quality in commercial warehouses in western Oregon. Agron. J. 55:379-382.
4. ————, H. L. Taylor, and P. T. Rowell. 1960. Change of forage seed quality under different simulated shipping conditions. Agron. J. 52:37-40.
5. Conklin, F. S., and D. E. Fisher. 1973. Economic characteristics of farms producing grass seed in Oregon's Willamette Valley. Oregon Agric. Exp. Stn. Circ. Inf. 643.
6. Gardner, Hugh E., and Rex Warren. 1970. Oregon State University fertilizer guide for tall fescue in western Oregon. Oregon State Univ. Ext. Memo FG 36.
7. Hardison, J. R. 1976. Fire and flame for plant disease control. Annu. Rev. Phytopathol. 14:355-379.
8. Hepworth, Homer M. 1977. Oregon weed control handbook. Oregon State Univ. Ext. Serv., Corvallis. p. 45-52.
9. Hyer, Edgar, A., M. H. Becker, and C. D. Mumford. 1950. The economics of grass seed production in Willamette Valley. Oregon Agric. Exp. Stn. Bull. 484.
10. Klein, Leonard M., and Jesse E. Harmond. 1971. Seed moisture—a harvest timing index for maximum yields. Trans. Am. Soc. Agric. Eng. 14:124-126.
11. Kroth, Earl. 1973. Research report. Southwest Missouri Ctr., Columbia, Mo.
12. Lee, William O. 1965. Herbicides in seedbed preparation for the establishment of grass seed fields. Weeds 13:293-297.
13. Lee, William O. 1973. Clean grass seed crops established with activated carbon bands and herbicides. Weed Sci. 21:537-541.
14. Nelson, G., and M. Becker. 1975. Estimated costs for establishing and producing grass seed crops, Oregon's Willamette Valley. Oregon Agric. Exp. Stn. Dep. Agric. Resour. Econ. Memo.
15. Rampton, H. H. 1949. Alta fescue production in Oregon. Oregon Agric. Exp. Stn. Bull. 427.
16. Rampton, H. H. 1965. Forty years of testing grass and legume varieties for seed yields in Oregon. Oregon Agric. Exp. Stn. Bull. 600.
17. Spencer, J. T. 1950. Seed production of Ky 31 fescue and orchardgrass as influenced by rate of planting, nitrogen fertilization and management. Kentucky Agric. Exp. Stn. Bull. 554.
18. Templeton, W. C., Jr., G. O. Mott, and R. J. Bula. 1961. Some effects of the temperature and light on growth and flowering of tall fescue, *Festuca arundinacea* Schreb. II. Floral development. Crop Sci. 1:283-286.
19. Van Keuren, R. W., and C. L. Canode. 1963. Effects of spring and fall plantings on seed production of several grass species. Crop Sci. 3:122-125.
20. Wheaton, Howell N. 1971. Seed production of tall fescue and other cool season grasses. Missouri Ext. Sci. Tech. Guide 4670.
21. Wheeler, W. A., and D. D. Hill. 1957. Grassland seeds. Van Nostrand, New York. p. 473-480.
22. Youngberg, H. W., D. O. Chilcote, and D. E. Kirk. 1975. Evaluation of a field sanitizer for controlled burning of grass seed fields. Oregon State Univ., Dep. Crop Sci. Memo ACS 10.

Chapter 9 **Stand Establishment and Renovation of Old Sods for Forage**

T. H. TAYLOR
Agronomy Department
University of Kentucky
Lexington, Kentucky

W. F. WEDIN
Agronomy Department
Iowa State University
of Science and Technology
Ames, Iowa

W. C. TEMPLETON, JR.
SEA-USDA U.S. Regional
Pasture Research Laboratory
University Park, Pennsylvania

Compared with ryegrass (*Lolium multiflorum* Lam.) and orchardgrass (*Dactylis glomerata* L.), tall fescue (*Festuca arundinacea* (Schreb.) is relatively slow to establish from seed. Once tall fescue is established in its region of adaptation, stands may last for a very long time. An excellent stand, which served as the initial source for the 'Kentucky 31' cultivar, continues on a hill field in Menifee County, Kentucky. The original material dates back at least to 1887 when Mr. William O. Suiter purchased the farm (Fergus, 1972).

Since the early 1950's, tall fescue seed has been plentiful and moderately priced in the United States. This situation has encouraged its use as a forage grass both alone and in mixtures and as vegetative cover on highway banks, parks, playgrounds, home lawns, and waterways. Because of its wide use over the past 25 years and its ability to set and shatter seed, tall fescue may now be found growing as individual plants, in thin, uneven sods, or as dense sods throughout its region of adaptation in the temperate zone of the United States. Volunteering and spreading are enhanced by seed production of unclipped plants along roadways, power lines, fence rows, undergrazed pastures, and over-mature hayfields. Because of these and other sources of plant dispersal, tall fescue is now quite widely distributed in its region of adaptation.

Copyright © 1979 ASA-CSSA-SSSA, 677 South Segoe Road, Madison, WI 53711 USA. *Tall Fescue.*

STAND ESTABLISHMENT

A. Seed Size and Seeding Rate

The number of tall fescue seeds per kg is approximately 500,000 (Metcalfe, 1973; Association Official Seed Analysts, 1970), and, on the average, a seeding rate of 1 kg/ha results in 50 seeds/m^2. Recommended seeding rates vary from approximately 1 to 28 kg/ha (Rampton, 1945; Spencer, 1950; Metcalfe, 1973). This wide range is partially accounted for by circumstances and uses to be made of the plantings.

B. Germination

The accepted germination period for tall fescue seed is approximately 14 days at 15 to 25 C or 20 to 30 C alternating temperatures. For official tests of germination, it is recommended that seed be held at the first temperature for approximately 16 hours and at the second for 8 hours daily (Association Official Seed Analysts, 1970). The fact that seed analysts recommend only the 15 to 25 C regime for germination of Kentucky bluegrass (*Poa pratensis* L.) and orchardgrass seed suggests that tall fescue is tolerant of higher temperatures during germination.

Chippindale (1949) placed grass seed in soil at different levels of available moisture and observed germination. The data indicate that seed of *F. pratensis* and *F. elatior* L. can germinate in considerably drier soil than can those of orchardgrass and Kentucky bluegrass.

C. Row and Broadcast Plantings

Early stand establishment work on tall fescue in the United States was done with primary consideration for seed production. For this purpose, Rampton (1945) recommended planting 'Alta' tall fescue in Oregon in rows 0.9 m apart, as opposed to broadcast plantings. The suggested seeding rates were 3.4 to 5.6 kg/ha for row plantings and 15.7 to 20.2 kg/ha for solid plantings for forage production; however, when planted in mixture with other plants the recommended rates varied from 2.2 to 17.9 kg/ha. It was indicated that plantings may be made in spring or fall, with the latter being better in most circumstances in Oregon. Spencer (1950), working at the Kentucky station, reported that row seedings of Kentucky 31 fescue were superior in seed prodution to broadcast plantings. A broadcast seeding rate of 3.4 kg/ha produced more seed than a 7.8 kg/ha rate, and the 7.8 kg/ha rate was higher yielding than a rate of 16.8. This was true even in the third year.

Cowan (1956) recommended a well prepared, firm, clean seedbed for sowing tall fescue, indicating that seed may be broadcast or drill-planted. In either case, the seed should not be planted more than 2.5 cm deep. These conditions were deemed desirable for more rapid stand establishment because of the slow development of tall fescue seedlings and their relatively low competitive ability with weeds and other plants for light, space, and moisture (Cowan, 1956).

Tall fescue may be seeded successfully in late summer, early autumn, late winter, or early spring. Late summer seedings are generally preferred over spring sowings. Even though tall fescue may survive on soils varying in pH from 4.5 to 9.5, and on soils low in N, P, and K, it grows much better on soils of high fertility. Adequate levels of N, P, and K should be available to establish new stands (Fergus, 1952; Cowan, 1956).

Hunt et al. (1963) concluded at the end of a 4-year study that better establishment of tall fescue and ladino white clover (*Trifolium repens* L.) resulted in 3 out of 4 years by banding fertilizer below the drilled seed. During one year, under favorable moisture conditions, the broadcast plantings were equal to the band-seeded sowings.

D. Seedling and Stand Development

Using a scale of 1 to 14, Blaser et al. (1952) ranked 14 common cool-season grass species as to rate of germination and development. Italian ryegrass (*Lolium multiflorum* Lam.), the quickest to germinate and develop, received a ranking of 1, tall fescue, which developed more slowly, received a rank of 5, while Kentucky bluegrass ranked 13. Working with four grasses, Hays (1976) suggested that poor field establishment of tall fescue, compared with *L. perenne* L., may be due to poor mobilization of nutrient reserves in the seed with consequential poor seminal root growth.

Seedling and sward development of Kentucky bluegrass and tall fescue were observed on newly cleared hill land for 3 consecutive years at the Kentucky station[1]. Seedlings of tall fescue were considerably larger than those of bluegrass, resulting in quicker ground cover by the tall fescue. Sixty days after planting, tall fescue seedlings growing in association with bluegrass were larger than tall fescue seedlings grown in pure stand, indicating strong competitiveness of tall fescue seedlings with each other. Rye (*Secale cereale* L.), seeded as a companion crop, severely reduced seedling size and survival of the perennial grasses. Tall fescue was able to compete more successfully with rye than was bluegrass. Better grass stands were obtained from September plantings than from March plantings; however, tall fescue was more successfully sown in spring than was Kentucky bluegrass. The suppressive effects of rye on grass stands were still evident more than a year after sowing.

In the lower southeastern U.S., Hoveland and McCormick (1974) concluded that rye could be planted with tall fescue, to increase establishment-year forage yields, without seriously reducing tall fescue stands. Second-

---

[1] Howard, Douglas. 1971. Grass establishment studies on newly cleared brush land in the Hills-of-the-Bluegrass region of Kentucky. M.S. Thesis. Univ. of Kentucky.

year forage yields of tall fescue were reduced 25% when established with rye. Autumn and winter production of the rye-tall fescue mixture was 500% greater than the grass sown alone. Broadcast and row plantings gave similar results.

In South Carolina, Jutras (1968) reported establishment of a 'Kentucky 31' tall fescue-ladino white clover mixture in association with winter oats (*Avena sativa* L.). Seedings were made in late August, and the oat-grass-legume vegetation was cut or grazed during late autumn, winter, and spring. The oat component increased herbage production, reduced soil erosion, and only slightly reduced grass and clover stands.

In an extensive literature review on the topic of agronomic practices aimed at reducing competition between cover crops and undersown pasture plants, Santhirasegaram and Black (1965) concluded that a companion cereal crop should be removed early by cutting or grazing, provided the practice does not damage the pasture species. Insufficient light for growth of pasture seedlings is one of the primary problems resulting when the cereal is harvested for grain.

Vogel and Berg (1968) grew Kentucky 31 tall fescue and weeping lovegrass [*Eragrostis curvula* (Schrad) Nees] in greenhouse experiments on acid, stripmine spoils ranging in pH from 3.5 to 5.4. Lovegrass grew well on spoils of pH 4.1 and below, while tall fescue growth was adequate only at 5.4. Flemming et al. (1974) reported that lovegrass was more tolerant than tall fescue to Al concentrations in the range of 2 to 12 ppm. In this instance, lovegrass was favored over tall fescue for revegetation of acid mine spoils of low pH and high exchangeable Al.

E. New Plantings With Associated Legumes

Large hectarages of tall fescue are sown in association with legumes for hay and pasture. Plantings may be made with or without a companion crop. As indicated earlier, a small grain companion crop may be very competitive to the associated grass, particularly if the grass is sown in spring. When plants are sown in mixtures, competition for growth factors may begin early. Donald (1963) stated that the capacity of a plant to exploit the environment quickly may give success over competitors.

An illustration of this was reported by Charles (1967) who researched establishment of tall fescue and Italian ryegrass. Tall fescue was more sensitive than ryegrass to competition during the establishment phase. The competition could be in the form of a cereal companion crop or an associated legume such as white or red clover (*Trifolium pratense* L.). Frequent grazing or cutting of the companion crop and the grass-clover mixture during establishment improved tall fescue stands and grass yields the following season. Ryegrass-white clover swards produced more than tall fescue-white clover when a companion crop was used. When seeded without a companion crop and grazed frequently, tall fescue-clover outyielded ryegrass-clover. The method of establishment also affected total seasonal yield in the second and third year after sowing, but the effects were lessened.

In a study of adaptation of birdsfoot trefoil (*Lotus corniculatus* L.) cultivars to Virginia conditions, Blaser et al. (1952) sowed trefoil with Kentucky bluegrass, orchardgrass, and tall fescue. In the subsequent year, the relative yield of trefoil growing in association with the grasses was 100, 47, and 73, respectively. These data indicate that tall fescue was more aggressive than bluegrass toward trefoil but less so than orchardgrass.

Park et al. (1961) obtained good stands of white clover and tall fescue from 1 September and 1 November plantings in South Carolina. However, clover stands in the November plantings were reduced by heaving, the legumes in plots with trashy seedbeds being damaged less than in plots with clean seedbeds. Band seeding was recommended over broadcast seeding.

Wagner and Hulburt (1953, 1954) compared band and broadcast seedings 6 weeks after planting and reported taller tall fescue seedlings and more white clover plants with trifoliate leaves on the banded than on the broadcast plots. Eight months after sowing, forage yields from band-seeded plots were 30% higher than from broadcast plots. Hart et al. (1968) compared drilled and broadcast-seeded tall fescue and white clover on three soil types on 29 dates over a period of 2 years. In the spring after seeding, tall fescue yields were generally higher from drilled than from broadcast stands. Available soil moisture immediately after germination had a pronounced effect on both stand establishment and yields of both grass and clover the following year. More clover was present in broadcast than in drill plantings with both species seeded in the same row. This difference was ascribed to less competition from the grasses for light and moisture. Chamblee and Lovvorn (1953) reported that alfalfa (*Medicago sativa* L.) and tall fescue sown in alternate row plantings produced less total forage than broadcast or mixed-in-the-row plantings. Tall fescue suppressed the growth of alfalfa more than orchardgrass did. Their data suggested that with alfalfa tall fescue was more competitive for K than was orchardgrass. In the fourth year alfalfa-orchardgrass produced more than twice as much total forage as did alfalfa-tall fescue. Comstock and Law (1948) also reported lower yields for alfalfa-tall fescue than alfalfa-orchardgrass.

F. Establishing Tall Fescue in Bermudagrass Sods

In the upper South, tall fescue is the primary species used as stockpiled forage for late autumn and winter pasturage (Taylor and Templeton, 1976). Many researchers have attempted growing tall fescue in association with bermudagrass (*Cynodon dactylon* (L.) Pers.), in the hope that such a mixture would be useful in both cool and warm periods of the growing season.

Wilkinson et al. (1968) drilled Kentucky 31 tall fescue into a dormant 'Coastal' bermudagrass sod in October at two locations in Georgia. Excellent stands resulted, and the experiment was observed for three growing seasons. Tall fescue stands were greatly reduced under high N fertilization (980 kg/ha) and a 5.1-cm stubble cut. On the other hand, good stands were maintained on plots fertilized with 420 kg/ha of N and a stubble cut of 10.2 cm.

Taylor et al. (1971) established tall fescue in an old sod of 'Midland' bermudagrass through use of an experimental sod-seeder that planted in 20-cm rows. The seeding was made in late September and resulted in excellent stands. The species association was observed through three growing seasons. Under a system of seven harvests per season and a cutting height of 6.4 cm, tall fescue dominated the grass mixture and produced 67% of the herbage the first year, 72% the second, and 87% the third year. Percentages of tall fescue in sward yields for the third year varied from 73% on plots receiving no N fertilizer to 92% on plots dressed with 300 kg/ha. Seasonal production of the grass mixture was more evenly distributed than was that for either species grown alone. These data illustrate the high degree of adaptation of tall fescue to field conditions in Kentucky compared with that of bermudagrass.

Fribourg and Overton (1973) successfully established tall fescue in Midland bermudagrass sod through use of a commercial sod-seeder. Sowings were made in October at a seeding rate of 17 kg/ha. The experiment was observed over a 3-year period. Tall fescue depressed Midland bermudagrass production by 30% but increased total yearly forage production by approximately 2 metric tons/ha and extended the potential grazing season from 5 to 8 or 9 months.

RENOVATION OF OLD SODS

Grassland renovation technology has developed to the point that adapted legumes may be introduced into old tall fescue sods without destroying the grass by disking or plowing. Advantages of sod renovation without destroying the grass are: (a) higher forage production is realized the establishment year; (b) sufficient viable grass plants remain for a stand, eliminating the need for reseeding the grass; (c) much less soil is lost by erosion, resulting in less land deterioration and pollution of the environment. Moreover, grazing animals perform better on tall fescue-legume pastures than do animals grazing pure tall fescue with or without N fertilizer. A tall fescue-legume mixture may be higher yielding than pure tall fescue receiving N at 225 kg/ha (Taylor et al., 1959). Also, the associated legume(s) improve feed quality and add significant amounts of biological N to the system (Smith et al., 1975; Templeton and Taylor, 1975). These additions not only have great economic significance, but increase human food production with less consumption of fossil fuel.

A. Historical Concepts of Grassland Renovation

As an important forage grass, tall fescue is a relatively new crop. A review of concepts and research findings on introduction of legumes into grass pastures and hay fields before the advent of tall fescue as an important forage plant is, therefore, useful in this discussion. For example, Roberts

(1910) maintained clover in a cool-season grass pasture from 1878 to 1903 by sowing in early spring in alternate years 1 or 2 quarts of mixed clover seed per acre (2.3 or 4.7 liters/ha). At that time of year the ground was open, soft, and moist and covering the seed with a brush drag was possible. The spring grass crop was grazed hard the year of legume seeding. The main objective of maintaining legumes in the pasture was to improve the quantity and quality of the pasturage.

In the 1920's, numerous researchers conducted experiments on the effects of legumes on associated grasses and methods of renovating low producing pastures and hay fields (Montgomery, 1921; Stapledon, 1925; Karraker, 1925; White and Holben, 1925; Wiggans, 1926; Graber, 1927, 1928; Bates, 1929). Those studies showed that reducing the vigor of the sod by tilling and/or burning, liming, and fertilizing to meet the needs of the legumes, sowing adapted legume species, and covering the sown seed resulted in higher producing pastures with a significant increase in the legume component.

Interest in introducing legumes into cool-season grass fields and their role in forage production continued in the 1930's. Fergus (1935) pointed out that legumes improve pastures by (a) directly and indirectly increasing total dry matter production; (b) improving vigor of the sod and decreasing weed growth; (c) increasing protein and mineral concentrations of the herbage. Working in Wales, Thomas (1936) reported that wild white clover was more successful during the first 3 years after sowing on harrowed than on plowed and fitted land. Applying lime and P well in advance of seeding, coupled with hard grazing, enhanced clover establishment. Graber (1936), Fuelleman and Graber (1938), and others continued exploring techniques of grassland renovation with legumes and the effects of such renovation on botanical composition and productivity. Field trials (Kentucky Agricultural Experiment Station, 1939) demonstrated that clover stands could be established, improved, and maintained by "top-seeding", along with suitable grazing and/or clipping, liming, and fertilizing practices.

Further refinements in grassland renovation were made during the decade of the 1940's (Brown, 1944; Ahlgren et al., 1946; Smith et al., 1947). Brown (1944) recommended the use of annual lespedezas (*Lespedeza* sp.) to upgrade permanent pastures in Missouri. Improved cultivars and strains of *L. stipulacea* Maxim. were suggested for the northern half of the state and *L. striata* Hook. & Arn. in the southern part. Grazing schemes to permit natural reseeding of both species were indicated. Ahlgren et al. (1946) renovated permanent bluegrass pastures on sloping land in Wisconsin. Yields of dry matter were significantly higher from renovated pastures than from untreated pastures during each of the 5 years immediately following renovation. Sprague et al. (1947), working on Pennsylvania hill land, concluded that adequate legume seedbeds could be prepared in old Kentucky bluegrass sods by surface tillage. They observed that the sod litter remaining on and near the soil surface reduced soil puddling, runoff, and erosion. Working with hill pastures in West Virginia, Smith et al. (1947) reported that better stands of legumes were obtained on shallow-tilled, disked, harrowed, or cultivated plots than on plowed and fitted land. On the other hand, newly sown grass stands appeared to be superior on plowed and fitted seedbeds.

B. Early Studies on Renovation of Tall Fescue Sods

As we have shown, cool-season, grass-sod renovation technology was fairly well developed by 1950, but there had been little adoption of the practice by farmers. Further refinements in renovation techniques have been developed since 1950, namely, the use of herbicides to kill or suppress the grasses while the legumes were becoming established, and the use of minimum tillage in seedbed preparation.

By about 1955 tall fescue had become a major forage grass in the southeast U.S. and in portions of the Northwest (see Chapter 2). Farmers, extension personnel, and research workers soon realized that older, pure tall fescue swards were not high quality forage, especially as pasturage in summer months. Rampton (1945) and Fergus (1952) strongly recommended that tall fescue be grown with adapted legumes if the forage was to be utilized by livestock. However, these recommendations were not often followed during the seed bonanza in the late 1940's and early 1950's. As the incentive to produce seed subsided, larger acreages of tall fescue were available for use by domestic livestock.

The dry summer of 1953 in Kentucky reduced and, in many cases, eliminated legume stands in tall fescue pastures. The low forage quality of tall fescue became more evident, and in 1954 the Kentucky General Assembly appropriated $100,000 to be used for pasture research and forage extension programs. The special appropriation made additional grassland investigations possible, and renovation research, using legumes, was initiated in 1955 (Taylor et al., 1955) as one way of upgrading tall fescue pastures and hay fields. Already, Jurado-Blanco[2] and Fergus (1952) had introduced legumes into old Kentucky 31 tall fescue sods without destroying the grass. Clipping and disking the sod greatly improved legume establishment over the check or clipped-only plots. Comparatively small differences in legume stands were observed for light and heavy disking of clipped fescue plots. In one experiment, legume seedlings were seriously damaged by crickets (*Gryllus* sp.).

Based on the tall fescue-renovation work of Jurado-Blanco, Fergus, and Taylor-Templeton, the University of Kentucky Agricultural Extension Service published a six-step guide for introducing legumes into grass-dominant sods (Taylor et al., 1958).

Employing on-the-farm tools, Taylor et al. (1959) renovated old tall fescue sods in Christian County, Kentucky. Plantings were made in March for 3 successive years and in late summer for 2 years. March-seeded legumes

---

[2] Juardo-Blanco, Bernardo. 1952. Effect of clipping and tillage on the emergence and establishment of some legume species in Kentucky 31 fescue pastures. M.S. Thesis. Univ. of Kentucky, Lexington.

substantially increased herbage production during the year of seeding. In the year after legume establishment tall fescue-alfalfa swards were higher yielding than was tall fescue receiving 224 kg N/ha. In the second year after establishment, grass-alfalfa swards were highest yielding (11,500 kg/ha), grass-N was intermediate (10,400 kg/ha), and grass-clover was lowest (9,300 kg/ha). Legume establishment and growth were more successful when the seedings were made in late winter or spring than in late summer. Seedbed treatments were no-tillage, disked-moderately, disked-heavily, and plowed. Legumes were planted with a grain drill, equipped with a small seed box. In the spring trials no-tilled plots were as productive as the plowed and fitted plots. Moderate and heavily disked plots were similar in production, but higher than the plowed and no-tillage treatments. In the late summer trials plowed, moderately disked and heavily disked plots were similar in production and higher than the no-tilled plots.

Brown (1961) reported that ladino white clover was successfully established in tall fescue without tillage. Establishment success was coupled with proper liming, fertilizing, and rigid scheme of grazing management.

C. Minimum Tillage and Herbicides for Grass-Sod Renovation

Recent advances in grassland renovation techniques include development of more specialized tillage tools and application of herbicides in specified situations (Dalin, 1960; Robinson and Cross, 1960; Sprague, 1960; Blackmore, 1962, 1965). Dalin (1960) showed in a bluegrass sod renovation experiment that 2-year forage yields were the same on plowed and trenched plots but 60% higher than those on non-disturbed plots. Robinson and Cross (1960) suggested that an ideal machine for introducing legumes into a grassy sward should leave a plant-free track approximately 4 cm wide for the seed. This type of seedbed, in conjunction with close grazing of the grass during early establishment stages of the introduced species, has shown promise in New Zealand.

Sprague (1960) reported that a seedbed prepared in bluegrass sod by treating with a herbicide and moderate disking was superior for legume establishment to either herbicide or disking alone. Using herbicide to kill a portion of the sward, Blackmore (1962, 1965) found that legume seedlings were successful when 25% of the grass, a 5-cm band over the row, was killed. Killing more than 25% of the grass was unnecessary. Taylor et al. (1964) used strip tillage to introduce legumes into an established sod, and suggested that a tilled strip of 0.6-cm width and depth would be satisfactory for small-seeded legumes.

Decker et al. (1964) developed a disk-opener device that was used successfully to establish legumes in bluegrass sod. Sund et al. (1966) observed forage plants over a period of 24 years in Wisconsin and concluded that with proper soil conditions and precision placement, rates of sowing could be reduced markedly.

Using minimum tillage and herbicide, Taylor et al. (1969) established alfalfa and white clover in Kentucky bluegrass sod. Their studies showed that minimum tillage and seed coverage enhanced stand establishment. In some instances, banding a grass herbicide over the seeded row improved legume stands. Seed coverage, 1.2 cm for alfalfa and 0.6 cm for white clover, was found to be the controllable factor which most consistently contributed to successful establishment. Stands were equal or superior on the minimum-tilled plots, i.e., strips tilled 0.6 m wide and 1.9 cm deep, to those of other treatments. Based on success with minimum tillage, a prototype field machine was designed and constructed to till, seed, firm soil over the seed, and apply herbicide in one pass over the land (Smith et al., 1973).

Theron et al. (1972) developed a sodseeding machine that could prepare a seedbed, lime, fertilize, and seed in one pass. The machine has been used to upgrade native range and cultivated pasture.

Using on-the-farm tools, Matches et al. (1973) established legumes in tall fescue sod. Red clover, birdsfoot trefoil, and annual lespedeza provided more summer production than fescue receiving N fertilizer. In general, the tall fescue-legume swards were more productive than grass swards receiving 135 kg N/ha.

Employing minimum tillage and herbicide, workers at the Kentucky Agricultural Experiment Station (T. H. Taylor, W. C. Templeton, Jr., and E. M. Smith, unpublished data) planted red clover into tall fescue and Kentucky bluegrass sods at 3.4, 6.7, and 10.1 kg seed/ha. During the establishment year, yields and clover percentages in the harvested forage were increased by increasing the seeding rates. The use of herbicides to suppress or kill the grass along the seeded row indicated no consistent advantage over no herbicide for the 1974 and 1975 plantings. However, in 1976 the unusually dry April and May which followed the 6 April planting resulted in herbicide having a pronounced effect on seedling size 61 days after planting. The largest red clover plants, given a rating of 100, were those that grew in herbicide-treated tall fescue sod. Clover seedlings grown in untreated tall fescue rated 82, while clover grown in untreated and treated bluegrass sod ranked 37 and 73, respectively. It was suggested that size of the legume seedlings was related, primarily, to the amount of available soil moisture in the root zone of young plants. Following adequate precipitation during the summer months, plant size and yield differences between herbicide and no-herbicide treatments were not significant.

Sward production in the establishment year (2-year average) was higher (8,300 kg/ha with 46% clover) for tall fescue-clover stands seeded with 6.7 kg clover seed/ha than tall fescue receiving 112 kg N/ha (7,000 kg). Tall fescue and tall fescue-legume associations were higher yielding than bluegrass or bluegrass-legume mixtures. Production data (2-year average) for the year after establishment show that the use of herbicide did not improve yields; however, higher seeding rates of clover increased yields somewhat. Sward production of the tall fescue-clover plots was higher than that of grass plots receiving 224 kg N/ha.

Nitrogen in harvested herbages of tall fescue alone without N fertilizer, tall fescue fertilized with 224 kg N/ha, and tall fescue-red clover without herbicide in the clover establishment year amounted to 99, 232, and 240 kg/ha, respectively. In the year following clover seeding, tall fescue-red clover yielded 220 kg N/ha more than did tall fescue without N fertilizer and 88 kg more than grass receiving 224 kg N/ha.

Templeton and Taylor (1975) showed that bigflower vetch (*Vicia grandiflora* var. *kitaibeliana* W. Koch), a winter annual, may be introduced into old tall fescue sods, resulting in increased productivity. When the vetch seed were covered the first spring harvest of tall fescue-vetch contained 48 to 71% legume and was essentially equal or superior in production to tall fescue plots receiving 100 kg N/ha. Two-year average seasonal yields for the grass without N fertilizer, grass plus N at 100 kg/ha, and grass-vetch were 4,500, 6,500, and 7,900 kg/ha, respectively. Herbage-N harvest amounted to 42, 90, and 148 kg/ha for unfertilized tall fescue, tall fescue with 100 kg N/ha, and tall fescue-vetch, respectively.

D. Animal Response to Forage of Renovated Sods

Reliable data on animal response to renovated and non-renovated tall fescue pastures were not available in the 1950's and 1960's. Nonetheless, results from grazing experiments (Blaser et al., 1956; Burns et al., 1973; Heinemann and VanKeuren, 1956) have shown that cattle perform better on tall fescue-clover pastures than on tall fescue-N pastures. These and other observations clearly indicated beneficial effects of the associated legumes.

Smith et al. (1975) reported on a 3-year grazing trial in southern Indiana in which brood cows and calves grazed either tall fescue pastures receiving 168 kg N/ha or clover-renovated tall fescue pastures. The old sod was thoroughly disked, then sown to a mixture of ladino white clover and red clover. Cow grains were 9 g/day on tall fescue-N pastures, compared with 263 g for cows grazing tall fescue-clover pastures. Conception was 72% on tall fescue-N pastures and 92% on tall fescue-clover. Daily liveweight gains of the calves on grass-N pastures and grass-legume pastures were 581 and 826 g, respectively. Calf weaning weights were 159 kg/head for the grass pastures and 198 kg for the grass-clover pastures, a difference of 39 kg in favor of the grass-legume mixture. Average legume composition of the renovated pastures was 30%. These more recent data support the concept that growing clover with tall fescue upgrades the pasture for a brood cow and her calf.

E. Recommendations for Renovating Tall Fescue Sods with Legumes

Farmers, extension personnel, and research workers in the Southeast and in more limited areas of the north-central and northwestern U.S. have had wide and varied experiences with tall fescue during the past 30 years. From experience and research it has been learned that the addition of legumes to a tall fescue pasture or meadow results in quick, marked improvement in herbage quality. Recommendations are now available in several states for renovating tall fescue and other grass sods with legumes (Burns, 1976; Evans et al., 1975; Myers and Triplett, 1974; Smith et al., 1976; Wheaton, 1976).

Success in introducing and maintaining legumes in tall fescue depends upon maintaining a sward environment which is favorable to the seeded legume. Thus, major attention must be directed toward well-being of the legume, rather than the grass. As indicated earlier, tall fescue, in its region of adaptation, is a vigorous, persistent plant. Intensive utilization, through grazing and/or clipping, during the growing season prior to renovation aids materially in reducing costs of renovation and in insuring successful legume establishment.

The following guidelines are offered for renovation and improvement of tall fescue-based grasslands with legumes:

1. The old sod should be weakened by intensive, close grazing and/or frequent clipping during a period of several weeks prior to seeding the legume. Moderate tillage of the sod reduces grass vigor and, unless specialized seeding equipment is used, helps provide an improved environment for germination and early seedling development. Grass growth and competition can also be reduced by appropriate herbicides, but these have the disadvantages of additional costs and, unless the grass has been utilized previously, waste of useful feed.

2. Prior to legume seeding, the field should be limed and fertilized in accordance with needs of the legume. A soil test is recommended to ascertain which plant nutrients should be added for successful legume establishment and growth. If lime is needed it should be broadcast several weeks prior to time of planting the legume seed. Nitrogen should never be applied in renovation seedings as it quickly stimulates grass growth, leading to excessive shading of the legume seedlings and undue competition for water and soil nutrients.

3. Successful renovation seedings may be made by broadcast overseeding, with conventional drills or cultipacker-type seeders, or with more specialized grassland seeders, depending upon previous treatment and condition of the sod and availability of equipment. Coverage of the seed is of vital importance.

4. Legume seed should be planted at the appropriate time and at an adequate rate for the conditions and method of seeding employed. Under many conditions, inoculation of the seed immediately before planting, with appropriate strains of *Rhizobium* sp., is essential for successful legume establishment.

5. After planting, renovated fields should be closely grazed until the legume seedlings develop to the point that the animals are biting off young leaves and shoots. At that time the animals should be removed for a period of 6 to 10 weeks to allow establishment of the legumes. Length of the deferred period of grazing varies with growing conditions and species of legume with shorter periods feasible under favorable growing conditions. Alfalfa normally requires a longer period for establishment before grazing

than do the clovers. After establishment, renovated fields should be managed to favor the particular legume that was planted.

6. Renovated grasslands should be observed often for signs of insect damage and appropriate control measures used as needed. Slugs and certain insects may be especially destructive during the period of germination and early establishment.

SUMMARY AND CONCLUSIONS

Compared with ryegrass or orchardgrass, tall fescue is relatively slow to establish from seed. Recommended seeding rates vary from approximately 1 to 28 kg/ha, depending upon circumstances and uses to be made of the planting. Tall fescue appears to be capable of germination at somewhat higher temperatures and in drier conditions than are bluegrass and orchardgrass. Row seedings are preferable to broadcast plantings for seed production. Plantings may be made in late summer, early autumn, or early spring, with late summer seedings being generally preferred over spring sowing.

Companion crops such as small grains, legumes, and other grasses may suppress tall fescue seedling and stand development. Careful management, i.e., grazing, cutting, or both, of mixed seedings may be required to obtain satisfactory stands of tall fescue. Even after establishment, red clover may be very aggressive toward tall fescue unless management schemes are imposed on the clover to reduce shading of the associated grass. Tall fescue may be successfully established and grown in bermudagrass sods. Again, management schemes appropriate for the grass association are required.

The development of grassland renovation technology using legumes began in the latter part of the last century and has slowly but steadily continued during this century. By the time tall fescue came into general use in the late 1940's and early 1950's, cool-season, grass-sod renovation technology was sufficiently developed to be applied to tall fescue sods on an experimental basis. From the mid-1950's to the present, further renovation techniques have been developed, namely, (a) use of herbicides to kill or suppress the grass while the legumes were becoming established; (b) use of minimum tillage for seedbed preparation.

Compared with sods not receiving high rates of N fertilizer, the addition of adapted forage legumes to tall fescue stands improves productivity. Also, animal performance is enhanced because of the legume. Recommended practices for renovating tall fescue sods are given.

LITERATURE CITED

1. Ahlgren, H. L., M. L. Wall, R. L. Muckenhirn, and J. M. Sund. 1946. Yields of renovated and unimproved permanent pastures on sloping land in southern Wisconsin. J. Am. Soc. Agron. 38:914–922.
2. Association of Official Seed Analysts. 1970. Rules for testing seeds. 60:7, 38–39, 44.

3. Bates, G. H. 1929. The mechanical improvement of grasslands. J. Minist. Agric. (G. B.) 36:321-325.
4. Blackmore, L. W. 1962. Band spraying: A new overdrilling technique. N.Z. J. Agric. 104: 13-19.
5. ―――. 1965. Chemical establishment and renovation of pastures in southern Hawkes Bay and northern Wairarapa in New Zealand. p. 307-312. *In* Zoraide Martins (eds.) Int. Grassl. Congr. Proc. 9th, Sao Paulo, Brazil. 7-20 Jan. 1965. Sao Paulo, Brazil.
6. Blaser, R. E., R. C. Hammes, Jr., H. T. Bryant, C. M. Kincaid, W. H. Skrdla, T. H. Taylor, and W. L. Griffeth. 1956. The value of forage species and mixtures for fattening steers. Agron. J. 48:508-513.
7. ―――, W. H. Skrdla, and T. H. Taylor. 1952. Ecological and physiological factors in compounding forage seed mixtures. Adv. Agron. 4:179-219.
8. Brown, E. M. 1944. Improve permanent pastures with lespedeza, phosphate, lime, and supplementary grazing. Missouri Agric. Exp. Stn. Circ. 285.
9. ―――. 1961. Improving Missouri pastures. Missouri Agric. Exp. Stn. Bull. 768.
10. Burns, J. C., L. Goode, H. D. Gross, and A. C. Linnerud. 1973. Cow and calf gains on ladino clover-tall fescue and tall fescue grazed alone and with Coastal bermudagrass. Agron. J. 65:877-880.
11. Burns, Joe D. 1976. Renovate grass pastures. Tennessee Agric. Ext. Serv. Circ. 714.
12. Chamblee, D. S., and R. L. Lovvorn. 1953. The effect of rate and method of seeding on the yield and botanical composition of alfalfa-orchardgrass and alfalfa-tall fescue. Agron. J. 45:192-196.
13. Charles, A. H. 1967. Effects of method of establishment on tall fescue and Italian ryegrass mixtures in the following year. J. Br. Grassl. Soc. 22:245-251.
14. ―――. 1969. A comparison of the herbage yield of Italian ryegrass and tall fescue mixtures in the second and third year after sowing. J. Br. Grassl. Soc. 24:111-118.
15. Chippindale, H. G. 1949. Environment and germination in grass seed. J. Br. Grassl. Soc. 4:57-61.
16. Comstock, V. E., and A. G. Law. 1948. The effect of clipping on yield, botanical composition, and protein content of alfalfa-grass mixtures. J. Am. Soc. Agron. 40:1074-1083.
17. Cowan, J. Ritchie. 1956. Tall fescue. Adv. Agron. 8:283-319.
18. Dalin, A. D. 1960. New methods of grassland improvement. p. 353-356. *In* C. L. Skidmore, P. J. Boyle, and L. W. Raymond (eds.) Int. Grassl. Congr. Proc. 8th. Reading, Berkshire, England. 11-21 July 1960. Reading, Berkshire, England.
19. Decker, A. M., H. J. Retzer, and F. G. Swain. 1964. Improved soil openers for the establishment of small seeded legumes in sod. Agron. J. 56:211-214.
20. Donald, C. M. 1963. Competition among crop and pasture plants. Adv. Agron. 15:1-118.
21. Evans, J. K., G. Lacefield, T. H. Taylor, W. C. Templeton, Jr., and E. M. Smith. 1975. Renovating grass fields. Kentucky Agric. Ext. Serv. Agron. 26.
22. Fergus, E. N. 1935. The place of legumes in pasture production. J. Am. Soc. Agron. 27: 367-373.
23. ―――. 1952. Kentucky 31 fescue-culture and use. Kentucky Agric. Ext. Serv. Circ. 497.
24. ―――. 1972. A short history of 'Kentucky 31' fescue. p. 136-139. *In* Proc. 29th S. Pasture Forage Crop Imp. Conf., Clemson, S.C. 16-18 May 1972.
25. Flemming, A. L., J. W. Schwartz, and C. D. Foy. 1974. Chemical factors controlling the adaptation of weeping lovegrass and tall fescue to acid mine spoils. Agron. J. 66:715-719.
26. Fribourg, H. A., and J. R. Overton. 1973. Forage production on bermudagrass sods overseeded with tall fescue and winter annual grasses. Agron. J. 65:295-298.
27. Fuelleman, R. F., and L. F. Graber. 1938. Renovation and its effect on the populations of weeds in pastures. J. Am. Soc. Agron. 30:616-623.
28. Graber, L. F. 1927. Improvement of permanent bluegrass pastures with sweet clover. J. Am. Soc. Agron. 19:994-1006.
29. ―――. 1928. Evidence and observations on establishing sweet clover in permanent pastures. J. Am. Soc. Agron. 20:1197-1202.
30. ―――. 1936. Renovating bluegrass pastures. Wisconsin Agric. Exp. Stn. Circ. 277.
31. Hart, R. H., G. E. Carlson, and J. H. Retzer. 1968. Establishment of tall fescue and white clover: Effects of seeding methods. Agron. J. 60:385-388.
32. Hays, P. 1976. Seedling growth of four grasses. J. Br. Grassl. Soc. 31:59-64.

33. Heinemann, W. W., and R. W. VanKeuren. 1956. Fattening steers on irrigated pastures. Washington Exp. Stn. Bull. 578.
34. Hoveland, C. S., and R. F. McCormick, Jr. 1974. Establishment of tall fescue and koleagrass with rye as a companion forage crop. Agron. J. 66:394-396.
35. Hunt, O. J., W. C. Hulburt, and R. E. Wagner. 1963. Development of field research equipment and evaluation of methods of establishing forage crops. USDA Tech. Bull. 1279.
36. Jutras, M. W. 1968. Growing oats in clover-grass seedings. South Carolina Agric. Exp. Stn. Circ. 151.
37. Karraker, P. E. 1925. Note on the increased growth of bluegrass from associated growth of sweet clover. J. Am. Soc. Agron. 17:813-814.
38. Kentucky Agricultural Experiment Station. 1939. Effects of legumes on bluegrass. Kentucky Agric. Exp. Stn. 52nd Annu. Rep. p. 15.
39. Matches, A. G., H. N. Wheaton, and J. B. Travis. 1973. Renovation of tall fescue sods with legumes. Univ. of Missouri College of Agriculture Misc. Publ. 73-5.
40. Metcalfe, D. S. 1973. Forage statistics. p. 69. *In* M. E. Heath, D. S. Metcalfe, and R. E. Barnes (eds.) Forages, 3rd Ed. Iowa State Univ. Press, Ames, Iowa.
41. Montgomery, E. G. 1921. Improving old pastures. Cornell (New York) Agric. Ext. Serv. Bull. 46.
42. Myers, D. K., and G. B. Triplett. 1974. Agronomic tips—No-tillage pasture renovation. Ohio Agric. Ext. Leaflet F-7.
43. Park, J. K., E. H. Stewart, B. K. Webb, and C. W. Gantt. 1961. Establishing stands of fescue and clovers. South Carolina Agric. Exp. Stn. Circ. 129.
44. Rampton, H. H. 1945. Alta fescue production in Oregon. Oregon Agric. Exp. Stn. Bull. 427.
45. Roberts, I. P. 1910. The Robers pasture. Cornell (New York) Agric. Exp. Stn. Bull. 280.
46. Robinson, G. S., and M. W. Cross. 1960. Improvement of some New Zealand grassland by oversowing and overdrilling. p. 402-405. *In* C. L. Skidmore, P. J. Boyle, and L. W. Raymond (eds.) Int. Grassl. Congr. Proc. 8th. Reading, Berkshire, England. 11-21 July 1960. Reading, Berkshire, England.
47. Santhirasegaram, K., and J. N. Black. 1965. Agronomic practices aimed at reducing competition between cover crops and undersown pasture. Herb. Abstr. 35:221-225.
48. Smith, E. M., Robert Fehr, Garry D. Lacefield, and L. Kenneth Evans. 1976. Renovating grass fields with a renovation seeder. Kentucky Agric. Ext. Serv. ID-33.
49. ———, T. H. Taylor, J. H. Casada, and W. C. Templeton, Jr. 1973. Experimental grassland renovator. Agron. J. 65:506-508.
50. Smith, R. M., G. G. Pohlman, F. W. Schaller, and G. R. Brown. 1947. Pastures improved with tillage-treatment-seed. West Virginia Agric. Exp. Stn. Bull. 327.
51. Smith, W. H., V. L. Lechtenberg, D. C. Petritz, and K. G. Hawkins. 1975. Cows grazing orchard, fescue or fescue-legume. J. Anim. Sci. 41:339-340.
52. Spencer, J. T. 1950. Seed production of Ky 31 fescue and orchardgrass as influenced by rate of planting, nitrogen fertilization, and management. Kentucky Agric. Exp. Stn. Bull. 554.
53. Sprague, M. A. 1960. Seedbed preparation and improvement of unplowable pastures using herbicides. p. 264-266. *In* C. L. Skidmore, P. J. Boyle, and L. W. Raymond (eds.) Int. Grassl. Congr. Proc. 8th. Reading, Berkshire, England. 11-21 July 1960. Reading, Berkshire, England.
54. Sprague, V. G., R. R. Robinson, and A. W. Clyde. 1947. Pasture renovation: Seedbed preparation, seedling establishment, and subsequent yields. J. Am. Soc. Agron. 39:12-25.
55. Stapledon, R. G. 1925. The improvement of very poor pasture by ploughing and immediate re-seeding. J. Minist. Agric. (G. B.) 32:13-25.
56. Sund, J. M., G. P. Barrington, and J. M. Scholl. 1966. Methods and depths of sowing forage grasses and legumes. p. 319-323. *In* A. G. G. Hill, V. U. Mustonen, S. Pulli, and M. Latvala (eds.) Int. Grassl. Congr. Proc. 10th. Helsinki, Finland. 7-16 July 1966. Helsinki, Finland.
57. Taylor, T. H., J. M. England, R. E. Powell, J. F. Freeman, C. K. Kline, and W. C. Templeton, Jr. 1964. Establishment of legumes in old *Poa pratensis* L. sod by use of paraquat and strip-tillage for seedbed preparation. p. 792-802. *In* Proc. VII Br. Weed Control Conf. Brighton, England. 24-26 Nov. 1964. Br. Weed Control Counc., 95 Wingore Street, London.

58. ─────, E. M. Smith, and W. C. Templeton, Jr. 1969. Use of minimum tillage and herbicide for establishing legumes in Kentucky bluegrass (*Poa pratensis* L.) swards. Agron. J. 61:761-766.
59. ─────, and W. C. Templeton, Jr. 1976. Stockpiling Kentucky bluegrass and tall fescue for winter pasturage. Agron. J. 68:235-239.
60. ─────, ─────, E. N. Fergus, and W. N. McMakin. 1958. Renovation of pastures. Kentucky Agric. Ext. Serv. Leaflet 210.
61. ─────, ─────, and W. N. McMakin. 1959. Improve grass pastures by growing more legumes. Better Crops Plant Food XLIII (No. 1):32-38.
62. ─────, ─────, ─────, and S. H. West. 1955. Establishment of legumes in an old tall fescue sod. Kentucky Agric. Exp. Stn., 68th Annu. Rep. p. 18-19.
63. ─────, ─────, and E. M. Smith. 1976. Yield and legume content of renovated Kentucky bluegrass and tall fescue sods. Kentucky Agric. Exp. Stn. Misc. Publ. 404.
64. ─────, ─────, and J. W. Wyles. 1971. Quality, yield, and distribution of production of bermudagrass and tall fescue grown alone or in association under different levels of N fertilization. p. 58-59. *In* Proc. 68th Annu. Conv. Assoc. Southern Agric. Workers Conf. 1-3 Feb. 1971. Jacksonville, Fla.
65. Templeton, W. C., Jr., and T. H. Taylor. 1975. Performance of big-flower vetch seeded into bermudagrass and tall fescue swards. Agron. J. 67:709-712.
66. Theron, E. P., J. C. Krog, and J. S. Grove. 1972. The cedara contraseeder. Trop. Grassl. 6:91-95.
67. Thomas, N. I. 1936. The introduction and maintenance of nutritious and palatable species and strains. Welsh Plant Breed. Stn. Bull. Ser. H No. 14:4-57.
68. Vogel, W. G., and W. A. Berg. 1968. Grasses and legumes for cover on acid stripmine spoils. J. Soil Water Conserv. 23:89-90.
69. Wagner, R. E., and W. C. Hulburt. 1953. Better forage stands with less seed. Crops Soils 6(2):8-9.
70. ─────, and ─────. 1954. Better forage stands. Natl. Fert. Rev. 29(1):13-19.
71. Wheaton, H. N. 1976. Renovating grass sods with legumes. Univ. of Missouri Ext. Div., Guide 4651.
72. White, J. W., and F. J. Holben. 1925. Development and value of Kentucky bluegrass pastures. Pennsylvania Agric. Exp. Stn. Bull. 195.
73. Wiggans, R. G. 1926. Pasture studies. Cornell (New York) Agric. Exp. Stn. Memoir 104.
74. Wilkinson, S. R., L. F. Welch, G. A. Hillsman, and W. A. Jackson. 1968. Compatibility of tall fescue and Coastal bermudagrass as affected by nitrogen fertilization and height of clip. Agron. J. 60:359-362.

Chapter 10 Management

A. G. MATCHES
SEA-USDA, and Department of Agronomy
University of Missouri
Columbia, Missouri

INTRODUCTION

Research relating to the management of tall fescue (*Festuca arundinacea* Schreb.) is reviewed in this chapter. Particular attention is given to the influence of time of harvest, frequency of defoliation, height of cutting, N fertilization, and the interaction of these variables on the forage yield and quality of tall fescue when grown alone or in mixtures with legumes or bermudagrass (*Cynodon dactylon* L.). Tall fescue is perhaps the temperate grass most often grown for utilization as winter pasture. One section of this chapter deals specifically with management practices which influence yield and quality of stockpiled and winter-grown fescue. Where appropriate, the responses of tall fescue are compared with those of other temperate grasses. Concluding each section are summary comments highlighting findings reported in the literature, plus my interpretative remarks.

SEASON-LONG MANAGEMENT OF TALL FESCUE

A. Tall Fescue Grown Alone

Different frequencies, timing, and heights of defoliation may have sizable influence upon the seasonal yields of tall fescue. Using one, two, three, five, and six harvests per season, Van Keuren (1972) evaluated tall fescue, orchardgrass (*Dactylis glomerata* L.), and Kentucky bluegrass (*Poa pratensis* L.) over four growing seasons. Average yields of dry matter for tall fescue were 5.6 metric tons/ha when harvested only in November; 10.1 metric tons/ha with a June and November harvest; 9.1 metric tons/ha when cut in June, August, and November; 6.2 metric tons/ha with five or six pasture stage harvests per season. Tall fescue yielded more than orchardgrass in all cutting schedules except the pasture harvests and was higher yielding than Kentucky bluegrass only in the two- and three-cut treatments. For the three grasses combined, crude protein (CP) content of herbage averaged 19.4% with pasture harvests as compared to 9.3% with one annual cut in November.

Copyright © 1979 ASA-CSSA-SSSA, 677 South Segoe Road, Madison, WI 53711 USA. *Tall Fescue.*

In Kentucky research (Templeton et al., 1965), tall fescue yielded 4.9 metric tons/ha and orchardgrass 3.9 tons/ha when harvested at 14-day intervals for a total of 12 harvests per year. With five or six harvests per season, tall fescue and orchardgrass yields were the same. The stage of growth when the initial spring harvest was taken greatly influenced total herbage production. With six harvests and the first harvest at the boot stage, tall fescue and orchardgrass averaged 5.6 metric tons/ha; with an early-bloom initial harvest, however, total yields were 8.0 metric tons/ha. Five cuttings per season, with the initial harvest at the late-bloom stage, also yielded 8.0 metric tons/ha.

Matches (A. G. Matches. 1977. Techniques for evaluating cultivars of temperate grasses. Am. Soc. Agron. Abstr., p. 102) also observed an influence of earliness of spring harvest on total yield of dry matter. He harvested different blocks of tall fescue at weekly intervals. The first block was harvested approximately 1 May, and the last of four blocks had its initial spring harvest 3 weeks later. Aftermath harvests were taken every 6 to 8 weeks. At one location, three tall fescue cultivars averaged total yields of 7.77 metric tons/ha with an early spring harvest as compared to 9.01 metric tons/ha when the initial harvest was delayed 3 weeks. At the second location, total yields for early and late spring harvests were 7.75 and 8.96 metric tons/ha, respectively. Matches also found that cultivars differed in amount of yield increase from delaying the initial spring harvest 3 weeks. The average (two locations) yield increase was 0.80 metric ton/ha for 'Fawn', 1.48 metric tons/ha for 'Kenmont', and 1.40 metric tons/ha for 'Kentucky-31'. Similar yield responses were obtained with different cultivars of reed canarygrass (*Phalaris arundinacea* L.), orchardgrass, and smooth bromegrass (*Bromus inermis* Leyss.) which were also included in this experiment.

Schiller and Lazenby (1975) reported that total yields of dry matter for nine populations of tall fescue declined when cutting frequency was reduced from 16- to 2-week intervals. In contrast, they found that numbers of tillers increased with more frequent cutting. Isley and Chamblee (1971) observed a marked increase in leaf area index with more frequent clipping of tall fescue. Minnesota research (G. C. Marten, personal communication) showed that after 2 years, the percent stands of Kentucky-31 tall fescue were less with two cuttings per year than with three or four cuttings. Percent stands were as follows: two cuttings, 70%; three cuttings, 85%; four cuttings, 95%. Corresponding stands of 'Nordstern' orchardgrass were 61, 81, and 90% with two, three, and four cuttings, respectively. Marten indicated that stand losses of both tall fescue and orchardgrass were not due to winter injury but occurred during the growing season.

In Missouri experiments, three cultivars of tall fescue were subjected to eight systems of defoliation over three growing seasons (A. G. Matches, unpublished data). The eight defoliation systems (Table 10-1) represent different ways that tall fescue has been or is being used in the Southern Corn Belt. Results followed trends similar to the findings of others reported above. Total yields were lowest in plots most frequently defoliated. Delaying the first harvest resulted in higher forage yields, as may be seen in com-

Table 10-1—Average yield (1974-1976) of tall fescue cultivars under eight systems of defoliation at Columbia, Mo. (A. G. Matches, unpublished data).

Cultivar	Yield of dry matter							
	Systems of defoliation†							
	1	2	3	4	5	6	7	8
	kg/ha							
Ky-31	5,977	5,811	6,975	7,126	7,909	7,572	8,875	8,596
Mo-96	5,851	5,410	6,677	7,132	7,505	6,911	8,596	7,847
Kenhy	5,833	5,949	6,995	7,455	7,922	7,432	9,041	7,546
	Harvest dates (1974)‡							
	8 May	8 May	23 May	23 May	23 May	23 May		
	27 June	27 June	10 July	10 July			27 June	
	21 Aug.	21 Aug.	21 Aug.	21 Aug.		21 Aug.		
	7 Oct.		7 Oct.					
	10 Dec.	10 Dec.	10 Dec.	10 Dec.	10 Dec.	10 Dec.	10 Dec.	10 Dec.

† Systems of defoliation: 1. Simulated grazing; 2. Simulated grazing + stockpiling for winter; 3. Hay harvest + stockpiling for summer; 4. Hay harvest + stockpiling for both summer and winter; 5. Hay harvest + stockpiling for winter; 6. Two hay harvests + stockpiling for winter; 7. Seed harvest + stockpiling for winter; 8. Total stockpiling for winter.   ‡ Harvest dates varied some among years due to yearly variation in earliness of spring growth.

parison of treatments 1 and 3 and 5 and 7 in Table 10-1. A significant cultivar-by-defoliation system interaction (P = 0.01) indicated that cultivars responded differently under some of the defoliation systems. Stand ratings taken in March following the last season of harvesting showed marked differences among defoliation systems. Simulated grazing (System 1) had the best stands remaining (91%), whereas the total stockpiling system (System 8) had the poorest stand (37%). Stand estimates for the other systems were: System 2, 87%, System 3, 85%; System 4, 78%; System 5, 76%; System 6, 79%; System 7, 59%. As in the case of forage yield, there appeared to be a cultivar-by-defoliation system interaction.

Dobson et al. (1976) reported that tall fescue grown alone and clipped at monthly intervals yielded 4.5 metric tons/ha clipped at 5 cm as compared to 3.0 metric tons/ha clipped at 10 cm. Hart et al. (1971) in Maryland combined time of initial spring harvest, frequency of cutting, and stubble cutting height into a single experiment with Kentucky-31 tall fescue and 'Potomac' orchardgrass. Initial spring harvests were taken when the first growth was in a non-elongated vegetative stage, when 50% or more of the stems were in boot, and when 50% or more of the heads were at anthesis. Plots were cut weekly to a stubble height of 5 or 10 cm, and cut every 28 days to a 5-cm stubble height. When cut back to 5 cm either weekly or monthly, tall fescue generally produced more forage than did orchardgrass. No differential response to date of first cutting was noted between tall fescue and orchardgrass. Forage yields were highest when the grasses were cut to 5 cm every month instead of every week. With the 10-cm cutting height, more dry matter was left as stubble; consequently, total dry matter production (harvested dry matter + stubble dry matter) was greatest with the 10-cm stubble. Leaving a 10-cm stubble and cutting less frequently (monthly vs. weekly) reduced the number of tillers/dm^2. The greatest reduction in tillers occurred when the first harvest was delayed until flowering. The authors suggested that frequent close grazing beginning before tiller elongation would be the best grazing management regime to use with tall fescue and orchardgrass. With less frequent defoliation, forage yields would be higher, but quality would decline. Even so, less frequent defoliation would seem desirable when tall fescue is grown for hay.

Isley and Chamblee (1971) found that a summer rest treatment of tall fescue from 15 June to 15 August yielded up to 15% more total dry matter than tall fescue that was cut all season. When tall fescue was cut back to a stubble height of 5 cm each time growth reached 15 cm height, yields were 33% greater than yields with a 9-cm stubble height.

Nitrogen fertilization may also influence the response of tall fescue and other grasses to frequency of defoliation. For example, McKee et al. (1967) reported that heavy N fertilization (336 vs. 112 kg N/ha) was associated with reduced yield and thinning of Kentucky-31 tall fescue stands to 10% when spring growth was not clipped before flowering. These decreases seemed to occur when the spring growth became very dense before cutting at the bloom stage. With frequent defoliation, tillering was stimulated, and stands following the spring harvest were not reduced. Interactions between rate of N fertilization and cutting treatment for tiller density were observed. Stands and regrowth yields during the summer were reduced only slightly by the high N rate.

Jung et al. (1974) in West Virginia reported that persistence of tall fescue, Kentucky bluegrass, orchardgrass, and timothy (*Phleum pratense* L.) improved when clipping frequency increased from three to eight harvests per season combined with using a higher rate of N fertilization (336 kg/ha). With high N and only three cuts per season, stands of tall fescue, Kentucky bluegrass, and timothy were nearly eliminated. They attributed poor survival with infrequent defoliation to etiolation and death of basal leaves. At their high rate of N, tall fescue yields were highest with five cuts; however, orchardgrass yields were greatest with three cuts and Kentucky bluegrass yields highest with eight cuts. With the low rate of N (168 kg/ha), all species (including tall fescue) except Kentucky bluegrass and timothy had highest production with the three-cut system.

In other experiments, Jung and Kocher (1974) fertilized 39 perennial grasses in the spring with N levels ranging from 0 to 240 kg N/ha. Grasses were clipped at vegetative stage (four cuts/season) or after head emergence plus two aftermath harvests (three cuts/season). There were large differences in amounts of winter injury among five tall fescue cultivars and three *Lolium-Festuca* synthetics. For example, with no N there was no winter injury with Fawn, Kentucky-31, and 'Alta' tall fescue; 1.0% for 'Kenwell' and 3.2% for Kenmont. But with 120 kg N/ha, there was a range in winter injury from 27.5 to 78.8%; Fawn had the least injury and Kenmont the most. The authors stated that clipping also differentially affected the winter survival of the grasses, and its effect was nearly as great as that due to N fertilization. Tall fescue generally was injured less when the initial spring harvest occurred after head emergence. Winter injury associated with clipping regimes usually increased as rate of N increased. However, their data suggest interactions among cultivars by N rates by clipping regimes. The authors suggested that different reactions among cultivars of tall fescue and other grasses to environmental stresses should receive more attention when testing new cultivars.

B. Tall Fescue-Legume Mixtures

Optimal management of tall fescue-legume mixtures is more complicated than management when tall fescue is grown along. In mixtures, particular attention must be directed to practices which will maintain stands and productivity of both the grass and legume components. Both components may react to environmental conditions and management practices in the same way when grown in mixtures as they do when grown alone. However, practices which are optimal for one component may not be desirable for other forage components of the mixture. Therefore, priorities in managing tall fescue-legume mixtures generally are given to management practices which favor (a) the most desirable species in the mixture or (b) the least dominant but desirable species of the mixture. Whether tall fescue or the legume is the dominant species is highly dependent on the geographic region in which they are grown.

Maintaining stands of tall fescue when grown with alfalfa (*Medicago sativa* L.) sometimes is very difficult to accomplish in the northern states. In Wisconsin trials, nine grasses were grown with alfalfa and harvested two, three, or four times annually at stubble heights of 4 and 10 cm (Smith et al., 1973). Tall fescue was nearly eliminated by cutting twice at the 4-cm stubble height but persisted well in all other treatments. In the two-cutting system, tall fescue was allowed to reach a late stage of maturity (green to ripe seed) before the first harvest was taken. Tall fescue was in the late-stem elongation and pre-anthesis stage in the three- and four-cut treatments, respectively, at time of first harvest. Apparently, tall fescue in the late stages of seed development was not physiologically able to withstand the stress of close defoliation and competition from alfalfa.

In Illinois experiments (Burger et al., 1958), forage yields after the first year for tall fescue grown with alfalfa or ladino clover (*Trifolium repens* L.) were highest when cut three times per season as hay. When cut four times as silage or five times as pasture, yields were nearly equal but lower than yields with three cuttings. Legume percentages were maintained at highest levels in the hay management treatment. Yields and percent legumes increased as stubble height was lowered from 7.5 to 5.0 to 2.5 cm.

Dobson et al. (1976) reported a legume by clipping height interaction for tall fescue-legume plots clipped four to six times annually at either a 5- or 10-cm stubble height. Legumes grown with fescue included cicer milkvetch (*Astragalus cicer* L.), crownvetch (*Coronilla varia* L.), birdsfoot trefoil (*Lotus corniculatus* L.), white clover (*Trifolium repens* L.), and red clover (*T. pratense* L.). When clipped at 5 cm, the tall fescue-white clover

Table 10-2—Forage yield of legume-tall fescue swards compared to tall fescue when clipped at 5 and 10 cm; Blairsville, Ga., 1971 to 1973 (Dobson et al., 1976).

Mixture	Yield of dry matter			
	Cutting height			Average increase for legumes
	5 cm	10 cm	Average	
	Tons/ha			
Tall fescue-milkvetch	4.3	4.4	4.4	0.6
Tall fescue-'Penngift' crownvetch	8.5	7.8	8.2	4.4
Tall fescue-'Chemung' crownvetch	9.3	8.1	8.7	4.9
Tall fescue-'Empire' birdsfoot trefoil	7.7	6.8	7.3	3.5
Tall fescue-'Viking' birdsfoot trefoil	10.7	9.4	10.1	6.3
Tall fescue-'Kenland' red clover	9.1	9.0	9.1	5.2
Tall fescue-'Kenstar' red clover	10.4	10.5	10.5	6.7
Tall fescue-'Regal' white clover	8.3	6.0	7.2	3.4
Tall fescue-'Tillman' white clover	9.5	8.1	8.8	5.0
Tall fescue-'Ladino' white clover	7.0	5.9	6.5	2.7
Tall fescue alone	4.5	3.0	3.8	

mixture averaged 1.6 metric tons/ha more forage than when clipped to 10 cm (Table 10-2). Yield increases of 1.1 metric tons/ha for tall fescue-birdsfoot trefoil and 0.9 metric ton/ha for tall fescue-crownvetch resulted from clipping at 5 cm. Stubble height had little or no influence on yields of tall fescue-red clover and tall fescue-milkvetch.

Tall fescue-crownvetch yields were depressed by a severe cutting schedule in Iowa (Helsel et al., 1974). Tall fescue-crownvetch, with no N fertilizer, yielded 1.6 metric tons/ha when cut monthly from mid-May to mid-October (six cuts) as compared to 2.7 metric tons/ha with an initial cutting in mid-June followed by two monthly pasture cuttings and a final harvest in mid-October.

C. Summary for Season-Long Management of Tall Fescue

Infrequent defoliation (but more than one harvest annually) usually results in highest forage yields for tall fescue. Delaying time of the initial spring harvest to about the early-bloom stage results in greater total forage production than if first harvest is taken at a more immature stage of growth. Harvestable yields tend to be greater when short stubble heights are used. Tillering of tall fescue is stimulated by frequent defoliation. Infrequent defoliation may result in sparse, clumpy stands, whereas stands with frequent defoliation may become dense and turf-like in structure. Nitrogen fertilization may intensify stand losses of tall fescue under some time and frequency of defoliation combinations. Tall fescue cultivars may respond differently to various management practices.

Management of tall fescue-legume mixtures to maintain a proper balance of the two species is more difficult than when tall fescue is grown alone. Interactions between time of cutting, frequency of cutting, stubble height, species, and cultivar of legumes can be expected.

STOCKPILED TALL FESCUE FOR AUTUMN AND WINTER GRAZING

Stockpiling has been defined by Mays and Washko (1960) as, "the practice of allowing forage to accumulate in the field until it is needed for grazing." Throughout the Southern Corn Belt and upper South, many farmers depend on stockpiled tall fescue as their main source of late fall and winter pasture. Cattle management and animal response to stockpiled tall fescue are discussed in Chapters 11 and 12.

A. Period of Accumulating Growth

The period of accumulating growth for stockpiling tall fescue may range from over the entire season to just a few weeks in the fall. During the 1960's, governmental programs to divert land from grain production allowed the seeding of forages on the diverted hectarages. However, these areas could not be grazed or harvested until after the growing season. Consequently, growth of tall fescue was accumulated the entire growing season, but grazing did not begin before early winter.

Matches and Tevis (1973) followed the yield and quality trends of tall fescue growth accumulated on a year-long basis over a 2-year period. In their experiments, maximum accumulation of dry matter occurred during September and averaged over 10 metric tons/ha for the 2 years. From September to the following spring, yields of dry matter declined to as low as 4.5 tons/ha; this decline represented a 55% loss of dry matter from fall to spring. Under central Missouri conditions, they found that it was not unusual to have periods of several days in February and March in which the temperature reached over 10 C, causing a greening-up of the stem bases and lower leaves of tall fescue and at times even a small increase in yield of dry matter. Tall fescue progressed from a vegetative stage in April (20 to 25 cm tall), boot stage to heading in May, flowering to late-dough stage of seed development in June, ripe seed to the shattering of seed in July, and the complete shattering and early decay of seed-bearing stems in August. There was an average of 54% living tillers from June through October, followed by declines to 31% living tillers in November, 16% in December, and 5% thereafter until the end of March. In June, the sward averaged 26% green reproductive stems and heads, but no living reproductive stems were observed in the later sampling dates.

Percent crude protein (CP), crude fiber (CF), and estimated digestible dry matter (by the nylon bag technique) are shown in Fig. 10-1. In general, there was a curvilinear decrease in percent protein from April to mid-July. Crude protein averaged 24 to 29% in April and 8 to 10% in July. After July, levels of CP ranged from 6 to 9% and averaged 7% from November on. Crude fiber trends were the inverse of CP, averaging 22% in April, 30%

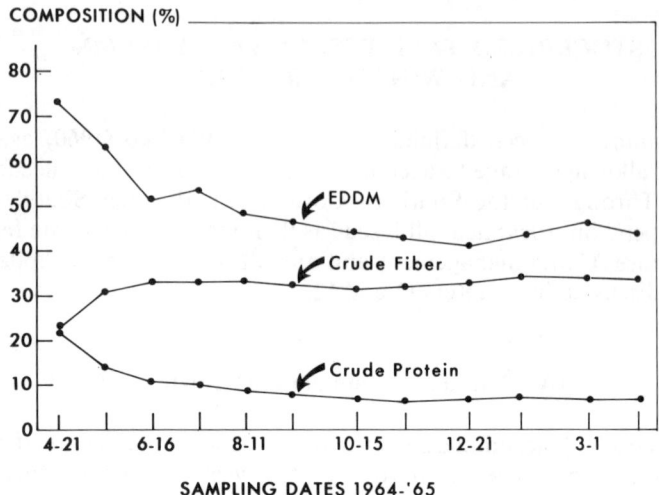

Fig. 10-1—Seasonal trends for estimated digestible dry matter (EDDM), crude fiber, and crude protein for total accumulated growth of tall fescue (Matches et al., 1973).

in May, and 33% from mid-June on. Digestibility decreased sharply from highs of over 75% in April, when the growth was 20 to 25 cm tall, to a range of 51 to 61% in June and then gradually declined to an average of 49% in January. Slight increases in digestibility after January appeared to be related to periods of warm temperatures. These results show that under conditions of total growth accumulation, forage available after seed maturity in July is of low quality. During the fall and winter months, total accumulated winter growth would likely provide adequate nutrients for only dry cows.

Stockpiled tall fescue of improved quality probably could be obtained if the growth of grass was accumulated following the completion of its reproductive cycle, providing the reproductive growth is removed. Removing the reproductive growth should eliminate much of the dilution in quality from decaying seed stems.

Several investigators have examined the influence of different periods of accumulation on the yield and quality of stockpiled tall fescue (Balasko, 1977; Green, 1974; Kroth et al., 1977; Matches and Tevis, 1973; and Van Keuren, 1972). The longer the periods of accumulation, the higher the yields of stockpiled tall fescue, but the inverse results for percent digestibility and CP. In Virginia, Green (1974) accumulated growth of tall fescue from 20 May, 11 June, and 6 August and harvested the stockpiled growth in December. Yields of dry matter averaged 8,780, 8,098, and 5,376 kg/ha when growth was accumulated from May, June, and August, respectively. Matches and Tevis (1973) in Missouri, also accumulated tall fescue growth from 10 May, 21 June, and 2 August. Over the 3 years, stockpiled yields in October averaged 2,616, 1,272, and 775 kg of dry matter/ha, respectively (Table 10-3). In Ohio (Van Keuren, 1972) stockpiled tall fescue yields in

Table 10-3—Yield and quality of accumulated growth of tall fescue and losses between 25 October and 18 January (average for 3 years) at Columbia, Mo. (Matches and Tevis, 1973).

Growth accumulated after:	Harvest date	Yield of accumulated dry matter	IVDMD	Crude protein	Losses between 25 Oct. and 18 Jan.		
					Dry matter	IVDMD	Crude protein
		kg/ha	%			%	
10 May	25 Oct.	2,616	57.9	8.3			
10 May	18 Jan.	2,006	49.6	6.6	23.3	14.3	20.5
21 June	25 Oct.	1,272	66.4	8.5			
21 June	18 Jan.	850	54.4	6.8	33.2	18.1	20.0
2 Aug.	25 Oct.	775	68.4	10.1			
2 Aug.	18 Jan.	510	58.9	7.2	34.2	13.9	28.7

November were 5.6 metric tons/ha with total accumulation, 3.7 metric tons/ha when accumulated from June, and 0.9 metric ton/ha when accumulated from August.

Improved forage quality can be expected with shorter periods of accumulation. In Missouri experiments (Matches and Tevis, 1973), percent in vitro dry matter digestibility (IVDMD) in October averaged 57.9% for tall fescue growth accumulated from May as compared to 68.4% for growth accumulated from August (Table 10-3). Comparative results for CP were 8.3 vs. 10.1% for the long and short accumulation periods. Crude protein percentages of stockpiled tall fescue herbage in November from the Ohio experiments (Van Keuren, 1972) averaged 8.6% with total accumulation, 10.1% for growth accumulated from June, and 12.2% for growth accumulated from August. Balasko (1977) and Green (1974) also observed similar trends in the quality of stockpiled tall fescue for varying periods of accumulating growth. Green found that stockpiling tall fescue after a May hay cut gave 65% dead herbage in December as compared to 57% for stockpiling after 6 August. Total nonstructural carbohydrate (TNC) levels of the herbage were inversely related to percent dead forage. On 4 December, the green forage averaged 17% TNC, whereas the brown herbage averaged 4.5% TNC. Similar results have been reported by Balasko (1977) and Taylor and Templeton (1976). Brown et al. (1963) reported that high TNC levels in the herbage were associated with high digestibility of herbage. Also, they found that tall fescue maintained high digestibility in the fall. They attributed the lack of decline in digestibility with age of fall-grown tall fescue to an increase in soluble carbohydrate content in the absence of an increase in crude fiber and lignin contents.

In Missouri and other midwestern states where tall fescue is grown for seed, the grass may be stockpiled following a seed harvest. Kroth et al. (1977) measured the yield and quality of the residue growth immediately following seed harvest in June, the combined residue plus regrowth in December, and the regrowth in December when the residue growth was harvested and removed in June. Depending on level of N fertilization and location, average yields of residue growth in June ranged from 1.2 to 5.5

Table 10-4—Quality of tall fescue seed crop residue.†

	Harvested		
	Twice		Once
Quality components	June	Dec.	Dec.
		%	
Crude protein (CP)	7.0	10.0	8.6
IVDMD	47	53	45

† Mean of 3 or 4 years from north and south Missouri Centers (Kroth et al., 1977).

metric tons/ha. When the seed crop residue had not been removed, yields of stockpiled growth in December ranged from 3.0 to 6.3 metric tons/ha, whereas December yields following the removal of the seed crop residue in June ranged from 0.9 to 3.9 metric tons/ha. Average CP levels ranged from 7 to 10% and IVDMD from 45 to 53% for the various types of growth (Table 10-4).

B. Effects of Summer Defoliation on Stockpiled Yields and Winter Growth

Few investigations have been conducted to determine the effects of frequency of defoliation during the spring and summer on the yields of stockpiled tall fescue when fall growth is accumulated from a common date in late summer. Green (1974) in Virginia, simulated grazing from March to August by harvesting Kentucky-31 tall fescue every time growth reached a height of 15 to 25 cm. Tall fescue was also harvested as hay with an initial harvest on 20 May or 11 June, plus another harvest on 6 August. The two-management system had no effect on yield of stockpiled grass in December. Stockpiled yields on 4 December for the simulated grazing and two haying treatments averaged slightly over 5 metric tons of dry matter/ha. Ocumpaugh and Matches (1977) observed a similar response for Kentucky-31 tall fescue grown in Missouri. In their investigations, tall fescue was harvested two, three, and five times between April and mid-August; then growth in all cutting treatments was allowed to accumulate until the end of October. During the first few weeks of autumn regrowth, dry matter yields from the frequently defoliated (five cuts) plots were about 200 kg/ha less than plots subjected to two or three harvests during the spring-summer period. However, by the end of October, total yield of accumulated growth was not different among the three preconditioning treatments. Yields of dry matter averaged 1,000 and 1,800 kg dry matter/ha for the first and second years, respectively. These two experiments suggest that for the upper South and Southern Corn Belt frequency of spring and summer defoliation will have little effect on the yield of stockpiled tall fescue when growth is accumulated from August to the end of the growing season.

In the mid-South, tall fescue may grow throughout much of the winter (Chapter 2). Hoveland and Anthony (1971) in Alabama found that during

MANAGEMENT

the cool-season period from September through April, growth of tall fescue generally was lowest during November to March. The distribution of tall fescue growth during this cool-season period was 30% during September and October, 18% from November to March, and 52% during March and April (Hoveland, 1970). Studies on the effect of summer defoliation on the autumn-winter (October–February) production of tall fescue were conducted in Alabama by Berry and Hoveland (1969). Clipping in mid-July reduced autumn-winter growth of Kentucky-31 tall fescue 25% but had no significant effect on 'Goar' tall fescue. However, clipping in early June reduced autumn-winter production nearly 20% for Goar but had little effect on Kentucky-31. They concluded that summer resting of tall fescue was necessary for maximum autumn-winter production of forage in the lower southeastern U.S.

C. Comparison of Tall Fescue with Other Grasses for Stockpiling

Reynolds (1975) in Tennessee harvested Kentucky-31 tall fescue and 'Boone' orchardgrass in May and September and their harvest in September represented stockpiled growth. With 112 kg N/ha, spring yields of both grasses were nearly equal; in September, however, tall fescue yields were usually larger than yields of orchardgrass. Yields of dry matter in September averaged (3 years) 2.60 metric tons/ha for tall fescue and 1.98 metric tons/ha for orchardgrass.

Van Keuren (1972) in Ohio compared tall fescue, orchardgrass, and Kentucky bluegrass stockpiled for the entire season, stockpiled from June to November, or stockpiled from August to November. Over the 4 years, stockpiled yields of tall fescue and Kentucky bluegrass were very similar; they averaged 5.36 and 3.66 metric tons/ha of dry matter with total accumulation and stockpiling from June, respectively. Under the same stockpiling systems, orchardgrass yields were 1.5 and 1.3 kg/ha less. With accumulation of growth from August to November, yields of all species were nearly equal and averaged 0.84 metric ton/ha. In Maryland experiments, tall fescue, because of its superior fall growth, was considered better adapted than orchardgrass for use as stockpiled pasture (Archer and Decker, 1977a).

Taylor and Templeton (1976) compared 'Kenblue' Kentucky bluegrass and Kentucky-31 tall fescue under 0, 50, and 100 kg N/ha applied in August. Growth was accumulated from 15 August to as late as 2 March. On each of five sampling dates between October and March, stockpiled yields of tall fescue exceeded yields of Kentucky bluegrass within each level of N fertilization. Yields on 1 December with 50 kg N/ha averaged 3.09 metric tons/ha for tall fescue and 1.82 metric tons/ha for Kentucky bluegrass.

Five temperate grasses (Table 10-5) were evaluated as stockpiled forage by Wedin et al. (1966, 1967). Growth was accumulated after late July and harvested from mid-September to late-November. With its high rate of N fertilization (67 vs. 269 kg N/ha), tall fescue had the greatest yield of stock-

Table 10-5—Yield of dry matter of five temperate grasses grown at two locations in Iowa during 1963 and 1964.

	Low N†			High N†		
	Total yield	Stockpiled yield		Total yield	Stockpiled yield	
			% of total			% of total
	——— Tons/ha ———			——— Tons/ha ———		
Tall fescue	3.97	0.79	20	8.42	2.80	33
Reed canarygrass	4.75	1.28	27	9.93	2.51	25
Orchardgrass	4.44	0.99	22	8.16	2.02	25
Smooth bromegrass	4.30	0.63	14	7.96	1.39	17
Meadow foxtail	3.56	0.70	19	7.38	1.84	25

† Low N = 67 kg N/ha; High N = 269 kg N/ha (Adapted from Wedin et al., 1966 and 1967).

piled forage and the greatest portion (33%) of total seasonal yield in the fall. The authors suggested that tall fescue appeared to be most promising when used mainly as fall-saved forage. In later studies, Iowa researchers (Bryan et al., 1970) again found tall fescue to be superior to reed canarygrass as a fall-saved forage.

D. Response of Stockpiled Tall Fescue to Nitrogen Fertilization

Fertilization with N in late summer or autumn can result in sizable increases in fall growth. Fifty kilograms per hectare of N applied on 10 September resulted in 50% more dry matter production for Kentucky-31 tall fescue stockpiled from 10 September in Maryland experiments (Archer and Decker, 1977a). Similar results were obtained in West Virginia (Balasko, 1977) where 60 kg/ha of N applied in the spring and each three successive cuttings resulted in summer, winter, and annual yields up to three times greater than plots not fertilized with N. Likewise, Green (1974) in Virginia observed a 60% increase in yield of stockpiled tall fescue with an additional 56 kg N/ha above the base level of 28 kg N/ha. Kentucky data (Taylor and Templeton, 1976) showed a 66% increase in fall production of tall fescue stockpiled from 15 August to 1 November with the first 50-kg/ha increment of N in August, but the second 50-kg/ha increment of N (100 kg/ha total) increased yields only another 27%. Yields of dry matter for the stockpiled tall fescue on 1 November were 1.83, 3.03, and 3.84 metric tons/ha for 0, 50, and 100 kg/ha of N. Similar responses to N fertilization have been reported by other stockpiling studies in Missouri (Kroth et al., 1977); Oklahoma (Fuller et al., 1971); Iowa (Wedin et al., 1966); Tennessee (Hannaway and Reynolds, 1976; and Reynolds, 1975).

Besides stimulating fall growth, N fertilization may also increase CP levels of stockpiled tall fescue (Archer and Decker, 1977a; Balasko, 1977; Green, 1974; Hannaway and Reynolds, 1976; Kroth et al., 1977; Reynolds, 1975; Taylor and Templeton, 1976) and delay senescence (Balasko, 1977; Taylor and Templeton, 1976). Fertilizing with high rates of N, especially in the fall, may sometimes result in reduction of tall fescue stands. High N rates are believed to result in lower nonstructural carbohydrate levels which may serve as energy source for winter survival. Chapters 4 and 5 discuss N fertilization and nonstructural carbohydrate relationships in more detail.

E. Seasonal Variation for Yields and Losses in Stockpiled Tall Fescue

Yields and quality of stockpiled tall fescue differ considerably from year to year. Balasko (1977) contends that besides the length of accumulation period and fertilization rates, lack of late summer and fall precipitation may be a key climatic factor limiting the amount of fall growth. He concluded that excessive accumulated growth may be lower in overall quality than when growth is accumulated in moderate amounts. He attributed lower quality to a higher ratio of senesced to live tissue with heavy accumulation. Brown and Blaser (1965) also reported that inadequate moisture in the autumn limited the potential yield of tall fescue. Brown et al. (1963) suggested that conditions favoring growth may result in soluble carbohydrates being used for synthesis of new growth rather than accumulating soluble carbohydrates in the herbage.

Ocumpaugh and Matches (1977) evaluated the influence of late summer and autumn temperature and precipitation on the fall production of tall fescue. During the 2 years of experimentation, neither average maximum nor average minimum weekly temperatures was correlated to change in autumn yield. However, the amount of accumulated growth was highly correlated ($r = 0.93$ and 0.98 in Years 1 and 2, respectively) with amount of accumulated precipitation from August to early November. The accumulated yields of dry matter/cm of accumulated precipitation was 62 kg/ha in Year 1 and 164 kg/ha in Year 2. Lack of similar rates of accumulation/cm of precipitation was believed due to an uneven distribution of precipitation between years.

Losses in yield and quality of stockpiled tall fescue may occur over the winter (Archer and Decker, 1977a, b; Balasko, 1977; Bryan et al., 1970; Green, 1974; Matches and Tevis, 1973; Ocumpaugh and Matches, 1977; Taylor and Templeton, 1976; Wedin et al., 1967; Van Keuren, 1972). Matches and Tevis (1973) reported from Missouri experiments that losses in dry matter of stockpiled tall fescue between 25 October and 18 January ranged from 23 to 34% (Table 10-3). Corresponding losses in digestibility and CP averaged 15 and 23%, respectively. Winter losses in dry matter of 14 to 82% were observed in later experiments (Ocumpaugh and Matches, 1977). Green (1974) found that delaying utilization of stockpiled tall fescue from 4 December to 4 February resulted in 19 to 29% reduction in yield. Winter yields averaged a decrease of 10% as harvest was delayed from December to January in West Virginia experiments (Balasko, 1977). The IVDMD was 2.8 percentage units lower in January than in December. Van Keuren's (1972) data showed that losses in dry matter between August and November were slightly less for tall fescue than for orchardgrass and Kentucky bluegrass. Tall fescue averaged a yield loss of 0.67 metric ton/ha as compared to 1.05 metric tons/ha for the orchardgrass and bluegrass. The data showed a 2.0-percentage unit decrease in CP for tall fescue (from 14.2 to 12.2%), a 0.9-percentage unit decrease for orchardgrass (from 15.4 to 14.5%), and a 0.3-percentage unit increase for Kentucky bluegrass (from 14.6 to 14.9%) between August and November.

Archer and Decker (1977a, b) attributed dry matter losses to leaf death and decay. During autumn and early winter, proportions of dead leaves in the stockpiled tall fescue increased from approximately 20 to 46%; but the IVDMD of green leaf fraction decreased only slightly (from 89 to 85% IVDMD). The dead leaf component averaged 67% IVDMD 1 year and 60 to 75% IVDMD the second year. In the first year, the IVDMD of orchardgrass and tall fescue herbage combined decreased by 3.38% ($r = -0.85$) for each 10% increase in dead leaves; but in the second year, the decrease was only 1% ($r = -0.32$). Because of this variation between years, the authors suggested caution in predicting the forage quality of stockpiled herbage based only on the percent dead leaves.

Losses in dry matter, IVDMD, and CP of stockpiled tall fescue as influenced by autumn and winter environmental conditions were studied by Ocumpaugh and Matches (1977). Autumn growth stopped after the first week of below-freezing, average minimum temperatures during early November. After freezing temperatures, yields of dry matter began to decline. The average rate of decline in harvestable dry matter was about 25 kg/ha/week for both years. The IVDMD of stockpiled tall fescue remained fairly constant (63 to 68%) from September until the freezes in November. Following freezing temperature, digestibility decreased an average of 1.0 percentage unit IVDMD/week. Decline in CP content was much slower, averaging 0.1 percentage unit/week. The K content of the autumn-stockpiled grass decreased at a linear rate of about 0.1 percentage unit/week throughout most of the autumn-winter period. The minimum requirement of K in the diet of ruminants has been set at between 0.60 and 0.80% (National Research Council, National Academy of Science, 1970). During one winter, K levels of herbage declined to 0.26%, well below the minimum standards for ruminants. Potassium deficiency may result in poor appetite in animals (Devlin et al., 1969), resulting in reduced voluntary intake of herbage by the grazing cattle (Pfander and Rubio, 1972). Therefore, as winter quality of stockpiled tall fescue decreases and if K also becomes deficient and results in lower intake, animal health could be severely jeopardized. Ocumpaugh and Matches (1977) suggested for the Southern Corn Belt that if alternative feed sources are available for late winter feeding, cattle might graze stockpiled tall fescue before the end of December to avoid sizable losses of dry matter, IVDMD, CP, and K.

F. Summary for Stockpiling

Stockpiling is the practice of accumulating growth for fall and winter grazing. Results from most sections of the tall fescue belt indicate that tall fescue is probably the best cool-season grass presently available for stockpiling. It gives higher yields of stockpiled forage of superior quality compared to most other temperate grasses. The longer the periods of accumulating growth, the higher the yields of stockpiled forage, but shorter periods of accumulating growth result in better forage quality. Fall yields of tall fescue can be substantially increased with N fertilization. However, in the northern and southern fringe areas of tall fescue culture, high rates of N have been associated at times with depletion of tall fescue stands. Frequency of summer defoliation generally has little or no effect on amount of fall growth within the areas where tall fescue is best adapted. As in the case of N, frequent summer and fall defoliation may have detrimental effects, resulting in loss of stands in its fringe areas of adaptation.

Declines in dry matter per ha, percent IVDMD, CP, and water-soluble minerals generally occur in stockpiled herbage during winter. These declines seem to be triggered by the first sustained period of below-freezing temperatures in the fall. In some winters, levels of K in stockpiled tall fescue may decline below the minimum levels recommended for ruminants.

NITROGEN-FERTILIZED TALL FESCUE AND TALL FESCUE-LEGUME MIXTURES

Many producers are faced with providing most of their livestock's feed requirements through pasture and preserved forage. Tall fescue alone or tall fescue-legume mixtures probably should not be the only source of feed. For example, tall fescue-legume mixtures are of higher quality feed during summer than tall fescue alone, as documented in Chapters 11 and 12. Yet, when tall fescue is grown alone, stockpiled for winter grazing, and fertilized with N, it may be more desirable than stockpiled tall fescue-legume mixtures. Generally, freezing temperatures cause leaf drop of most legumes; consequently, legumes would likely contribute little to the yield and quality of the stockpiled feed supply.

Variable forage production systems of tall fescue when grown alone, grown with legumes, and when fertilized with N when grown alone or in a mixture with legumes are essential to provide the feed requirements of livestock. Therefore, certain aspects of tall fescue-N and fescue-legume combinations will be further emphasized (the concept was introduced in Chapter 4).

A. Nitrogen-Fertilized Tall Fescue

Nitrogen is perhaps the single fertility component that most often limits the growth of tall fescue. Therefore, fertilizing with N is a management tool available for exerting considerable control over the seasonal and total forage production potential of tall fescue. Numerous research reports show that applications of reasonable amounts of N increases forage yield and CP content of most grasses (Whitehead, 1970). Tall fescue is no exception.

Table 10-6—Response of tall fescue to N fertilization at 12 locations in the United States.

Location and rate of N	Time N applied	Period of production	Yield of dry matter	Kg dry matter/kg N	Remarks	Reference
kg/ha			kg/ha			
Southern Oregon	Split application—winter and following each of six harvests	Total season			Irrigated	Yungen et al., 1977
0			8,160	--		
118			11,190	25.7		
235			13,160	21.3		
353			15,490	20.8		
Colorado	April	Total season			Irrigated	Dotzenko, 1961
0			4,930	--		
90			7,620	29.9		
180			8,300	18.7		
358			9,420	12.5		
717			9,300	6.1		
Nevada	Fall, winter or spring	Total season (average yields for fall, winter, and spring N application periods)			Irrigated	Jensen, 1970
0			7,850	--		
56			9,420	28.0		
112			11,660	34.0		
224			11,430	16.0		
448			13,230	12.0		
896			14,120	7.0		
Oklahoma #1	August	September–February			Non-irrigated	Rommann and McMurphy, 1973
0			521	--		
45			2,886	52.6		
90			3,058	28.2		
180			3,167	14.7		
358			3,330	7.8		
Oklahoma #2	August	March–July			Non-irrigated	Rommann and McMurphy, 1973
0			1,136	--		
45			2,547	31.4		
90			3,925	31.0		
180			5,979	27.1		
358			7,332	17.3		
Southern Missouri	½ March and ½ July	Total season			Non-irrigated	Wheaton et al., 1973
0			948	--		
84			1,574	7.4		
168			2,590	9.8		
336			4,273	9.9		

(continued on next page)

MANAGEMENT

Table 10-6—Continued.

Location and rate of N	Time N applied	Period of production	Yield of dry matter	Kg dry matter/kg N	Remarks	Reference
kg/ha			kg/ha			
Central Missouri		Total season			Non-irrigated	Matches et al., 1973
0	67 kg February		3,087	--		
67	67 kg August		3,750	9.9		
134			4,856	13.2		
Spooner, Wis. #1	Split over season	Total season			Non-irrigated	Rohweder, D. A., personal communication
0			1,230	--		
84			4,840	43.0		
168			6,500	31.4		
336			6,700	16.3		
Spooner, Wis. #2	Split over season	Total season			Irrigated	Rohweder, D. A., personal communication
0			2,380	--		
84			5,200	33.6		
168			7,510	30.5		
336			7,850	16.3		
West Virginia		Total season			Non-irrigated	Colyer et al., 1977
0			4,973	--		
112	Spring		8,690	33.2		
224	Spring		10,075	22.8		
224(2)	½ spring, ½ after first cutting		11,417	28.8		
448	Spring		11,791	15.2		
448(2)	½ spring, ½ after first cutting		11,823	15.3		
448(4)	¼ spring, ¼ after each of three cuttings		12,846	17.6		
Virginia	35% Feb., 20% Apr., 15% May, 15% June, 15% July	Total season			Non-irrigated	Hallock et al., 1965, 1966
112			3,035	†		
224			5,240	--		
448			6,760	22.1		
896			6,680	18.6		
				9.1		
Georgia	Annually, time not given	Total season			Non-irrigated	Dobson and Beaty, 1977
0			360	--		
37			1,240	23.8		
112			2,720	21.1		
336			6,050	16.9		

† Response to N above rate of 112 kg N/ha

Forage yields of tall fescue grown at 12 locations in the United States under varying rates of N fertilization are presented in Table 10-6. This sampling of locations represents a span of tall fescue cultures ranging from states in the far West to the Eastern Seaboard and from areas of humid pasture production to areas where irrigation is required to grow tall fescue successfully. The magnitude of response to different rates of N appears to be highly associated with the geographic region in which tall fescue is grown. However, there do not seem to be distinct response patterns among locations. Yields with no N range from a low of 360 kg/ha in Georgia (non-irrigated) to a high of 8,160 kg/ha under irrigation in southern Oregon. Yields of 15,490 kg/ha were obtained with 353 kg N/ha plus irrigation in Oregon, 14,120 kg/ha with 896 kg N/ha and irrigation in Nevada, and 12,846 kg/ha with 448 kg N/ha and no irrigation in West Virginia. Irrigated tall fescue with 336 kg N/ha yielded 7,850 kg/ha in Wisconsin, and this was only 1,150 kg/ha more than non-irrigated plots at that location. Hoveland and Evans (1970) in Alabama also reported that irrigation did not increase yields of tall fescue fertilized with 179 kg N/ha or a tall fescue-ladino clover mixture which was not fertilized with N.

Conversion ratios of N to dry matter ranged from 6.1 to 52.6 kg of dry matter/kg N. In nearly all cases, growth response per unit of N diminished as rate of N increased. Lower increments of N, especially rates of 112 kg/ha or less, generally yielded the greatest increase in dry matter production per unit of N.

B. Tall Fescue with Legumes and Nitrogen

Total forage yields generally increase when legumes are grown with tall fescue. Dobson et al. (1976) showed that average increases in total yield in Georgia ranged from 0.6 to 6.7 metric tons/ha for 10 legumes grown with tall fescue (Table 10-2). Greatest yield increases were obtained with 'Kenstar' red clover and 'Viking' birdsfoot trefoil. In Arkansas, mixtures of tall fescue-'Victoria' alfalfa, tall fescue-'Orbit' red clover, and tall fescue-'Regal' white clover yielded an average of 30, 14, and 8% more herbage, respectively, than did tall fescue grown without legumes (Offutt and McKee, 1973).

Yields of tall fescue-legume mixtures generally equal or exceed yields of tall fescue along when fertilized with moderate amounts of N (Dobson and Beaty, 1977; Matches et al., 1973, 1975; Templeton and Taylor, 1966a, b; Tesar, 1974; Wagner, 1954a, b). Total yields for tall fescue and tall fescue-legume mixtures fertilized with 0, 67, and 134 kg N/ha in Missouri are shown in Table 10-7 (Matches et al., 1973, 1975). The 67-kg rate was applied in February, and the 134-kg rate was a split application with one-half the N in February and one-half in August. Results from this trial are representative of findings from many other experiments. In all cases, yields of mixtures exceeded those of tall fescue alone. Yields for mixtures of tall fescue with Kenstar red clover and 'Dawn' birdsfoot trefoil not fertilized with N equaled yields of tall fescue fertilized with 134 kg N/ha.

MANAGEMENT

Table 10-7—Average yield of tall fescue when grown alone or when renovated with sod seedlings of legumes in Missouri (Matches et al., 1973, 1975).

Mixtures	Total yield of dry matter			Mixture average
	0	67	134	
		kg/ha		
Tall fescue alone	3,087	3,750	4,856	3,898
Tall fescue & common red clover	4,242	5,059	6,043	5,115
Tall fescue & 'Kenstar' red clover	4,857	5,051	5,668	5,192
Tall fescue & 'Dawn' birdsfoot trefoil	5,138	5,793	5,727	5,553
Tall fescue & 'Summit' lespedeza	4,176	4,664	5,408	4,749
N average	4,300	4,863	5,540	

Table 10-8—Proportion of the total yield obtained after the first harvest for tall fescue grown alone or renovated with sod seedlings of legumes in Missouri (Matches et al., 1973, 1975).

Mixture	Total yield after the first harvest†		
	N (kg/ha)		
	0	67	134
		%	
Tall fescue alone	63‡	38	48
Tall fescue & common red clover	59	40	45
Tall fescue & 'Kenstar' red clover	66	47	49
Tall fescue & birdsfoot trefoil	70	50	50
Tall fescue & 'Summit' lespedeza	65	44	49
N average	65	44	48

† Two to four regrowth harvests/year depending on the amount of rainfall.   ‡ Plots were invaded by volunteer legumes (trefoil, red clover, and lespedeza) during the last year of experimentation.

Seasonal distribution of forage availability may be of greater importance to the livestock producer than is total yield. In the Southern Corn Belt, availability of forage for grazing during mid-summer when growth rates of many cool-season grasses are normally very low may be of more value to the producer than higher yields during the spring and fall. Growing legumes with tall fescue is one means of providing a more even distribution of forage availability throughout the growing season. For example, under irrigation in Texas, tall fescue grown alone made little growth during May, June, and July; in a tall fescue-alfalfa mixture, however, alfalfa yields held up well during these months (Brooks and Holt, 1954). The combined yield during May–July for Kentucky-31 tall fescue grown alone was 420 kg/ha as compared to 1,861 kg/ha when grown with alfalfa. In South Dakota experiments (Johnson and Nichols, 1969a), tall fescue fertilized with 112 kg N/ha and cut twice during the growing season produced 72 and 28% of its total yield in the first and second harvests, respectively. A tall fescue-alfalfa mixture with no N had a 60 and 40% distribution of production. Total yields of tall fescue and tall fescue-alfalfa were nearly the same. In the previously reported Missouri test (Table 10-8) tall fescue mixtures with Kenstar red clover and birdsfoot trefoil that were not fertilized with N had more than 66% of their season-long production after the first harvest. In comparison, tall fescue fertilized with 67 kg N/ha in the spring had only 38% of its total yield after the first harvest. An additional 67 kg N in August resulted in 48% of the production after the initial spring harvest. Templeton and Taylor (1966b) found that spring-applied N at rates up to 270 kg/ha did not increase annual yields of tall fescue-white clover but resulted in larger spring and early summer yields with proportionally less growth in later summer and autumn. Nitrogen rates of 67 and 135 kg/ha were detrimental to clover when applied in March but less so for May and August applications (Templeton and Taylor, 1966a). Kroth and Mattas (1974) reported that for one 5-year experiment tall fescue-lespedeza (*Lespedeza stipulacea* Maxim.) yields were not different with 0, 34, and 112 kg N/ha, but 48% of the dry matter production occurred after the first harvest with no N, as compared to 24% with 112 kg N. In another 5-year experiment, total yields were higher with than without N fertilization, but the influence of N on seasonal distribution was the same as for the first experiment. Nitrogen rates above 34 kg/ha resulted in a depletion of lespedeza stands.

C. Summary Comments

Advantages exist for both growing tall fescue alone and fertilizing with N or growing tall fescue-legume mixtures. The choice depends on many factors; some of the more pertinent ones are: 1) cost of N and legume seed; 2) availability of improved or adapted legume cultivars; 3) expected longevity of legume stands; 4) period when forage availability is most needed; 5) how forage will be utilized—pastured, hayed, stockpiled, etc.; 6) level of forage quality required for the type of livestock being fed; 7) climate, soil fertility, and other farm operation limitations which might prevent the successful growing of either tall fescue or legumes; 8) animal health hazards such as bloat or fescue toxicosis; 9) the managerial capabilities of the producer. In advising producers, the above factors should be thoroughly considered before specific recommendations are formulated. Seemingly, for many livestock enterprises, some balance in hectares of N-fertilized tall fescue and tall fescue-legume mixtures would appear more appropriate than only one or the other.

INTERSEEDING TALL FESCUE INTO BERMUDAGRASS

Several researchers have proposed seeding tall fescue into bermudagrass as a means of extending grazing into the winter, but very little research has been reported on this topic. An Oklahoma publication (Rommann et al., 1973) describes how to establish tall fescue into bermudagrass. Kentucky-31 and Alta tall fescue were listed as two cultivars that had been successfully established into bermudagrass sods, and October was preferred for seeding tall fescue. By drilling tall fescue into bermudagrass sod treated with paraquat, they reported fescue stands of 84% as compared to 51% stands with drilling and not using a herbicide. Discing and broadcasting tall fescue seed resulted in a 77% stand.

Tall fescue was seeded into dormant 'Coastal' bermudagrass, fertilized at five levels of N, and clipped at 5.1 and 10.2 cm over a 3-year period in Georgia tests (Wilkinson et al., 1968). Tall fescue persisted and contributed to yield at N levels up to 420 kg N/ha/year at the 5.1-cm stubble height, and at N levels up to 560 kg N/ha/year at the 10.2-cm stubble height. Overall, N fertilization decreased tall fescue content, whereas the higher clipping increased tall fescue content of the harvested forage. Percent stands without irrigation for tall fescue in the second year were as follows:

280 kg N/ha: 5.1 cm stubble = 39%, 10.2 cm stubble = 58%;
420 kg N/ha: 5.1 cm stubble = 17%, 10.2 cm stubble = 46%;
980 kg N/ha: 5.1 cm stubble = 3%, 10.2 cm stubble = 6%.

Tall fescue content was high in the spring and fall but low in summer. Seeding tall fescue into bermudagrass failed to greatly increase total dry matter production over the Coastal bermudagrass grown alone; there was a more even distribution of forage production during the year with tall fescue seeded into bermudagrass.

Decker et al. (1974) evaluated tall fescue, orchardgrass, crownvetch, and several annuals sod-seeded into 'Midland' bermudagrass. Tall fescue was successfully established in bermudagrass; however, the annuals, wheat (*Triticum aestivum* L.), rye (*Secale cereale* L.), and hairy vetch (*Vicia villosa* Roth) were superior yielding. Their results further substantiated that tall fescue could be grown with bermudagrass; however, they encountered difficulty in maintaining a uniform mixture of tall fescue and bermudagrass.

Overton and Fribourg (1975) in Tennessee evaluated the persistence of Kentucky-31 tall fescue overseeded in 25.4- and 50.8-cm rows in a Midland bermudagrass sod. Tall fescue stands were less persistent in the wide rows. Tall fescue stands remained excellent (above 70%) with 0 or 202 kg N/ha but declined to a low of 32% with 605 kg N/ha. When clipped back to a stubble height of 5 cm whenever growth reached 10 cm, tall fescue maintained stand above 85% as compared to stands around 50 to 60% with cutting back to a stubble height of 2.5 cm when growth reached 5 cm. Cutting back to 2.5 cm when growth reached 30 cm reduced tall fescue stand to below 50%.

All the research cited above shows that although tall fescue can be successfully established into sods of bermudagrass, maintaining a specific balance in stands of tall fescue and bermudagrass may be quite difficult.

TALL FESCUE CULTIVARS AND GRASS SPECIES— A COMPARISON OF YIELD AND QUALITY

Influence of various management practices (defoliation, N fertilization, legumes, period of accumulating growth, etc.) on the yield and quality of tall fescue and other species have been discussed in other sections of this chapter. In this section are given examples of forage yield and laboratory determined quality values for different cultivars of tall fescue. Also, comparative data for other grasses which might be grown are discussed.

A. Tall Fescue Cultivars—Yield and Quality

Forage yields of cultivars of tall fescue grown at 11 locations in the United States are shown in Table 10-9. Within a location, cultivars were fertilized and managed uniformly; among locations, however, fertilization and defoliation practices were different.

Kentucky-31 was grown in all tests. As indicated in earlier chapters, it is the cultivar most widely grown in the United States. The maximum yield difference between Kentucky-31 and the highest yielding cultivar within locations was 1,050 kg/ha (+10%) in Iowa and 1,132 kg/ha (+15%) in the Alabama TVA test. Generally, higher yielding cultivars exceeded yields of Kentucky-31 by less than 10%. Overall, most cultivars of tall fescue have very similar yields. If particular cultivars have been bred for cold or drought tolerance or resistance to specific diseases their superior genetic potential will be reflected in higher yields and better maintenance of stands in seasons in which conditions exist for which they were bred.

Information about the comparative quality of tall fescue cultivars is rather limited. Bates (undated) reported that 'Kenhy', Kenwell, and Kentucky-31 averaged 17.3% CP; varieties were not different. In Iowa experiments (Carlson, 1974), Kentucky-31, Kenhy, and Fawn averaged 61, 63, and 58% IVDMD, respectively, in mid-August. In mid-September, they averaged 67% IVDMD; cultivars differed by less than 2.0 percentage units.

The percent digestibility of Kentucky-31 and Kenwell was not different in Virginia feeding trials (Staff, Virginia Forage Research Station, 1969), but dry matter intake, as percent of body weight of cows, was 2.73% for Kentucky-31 vs. 2.24% for Kenwell. For the cultivars shown in the Montana tests (Table 10-9), Joppa and Roath (1964) did not detect differences in the percent forage consumed by sheep. Also, for the average of two locations, cultivars did not differ in percent CP. Digestible dry matter was determined with the in vitro nylon-bag technique in Alabama for Kentucky-31 and Goar (Hoveland et al., 1970). Digestibilities of Kentucky 31 and Goar, respectively, were 73 and 74% on 23 February, 74 and 71% on 20 March, and 58 and 49% on 25 April.

Chemical and IVDMD data for five tall fescue cultivars sampled several times in the spring, summer, and fall are given in Table 10-10 (F. A. Martz, A. G. Matches, D. A. Sleper, and R. L. Belyea, unpublished data). For most components, cultivar differences were small. Noteworthy are the seasonal trends for tall fescue in general. During the spring, acid detergent fiber (ADF), and neutral detergent fiber (NDF) were less than summer values but greater than concentrations observed during the fall. The inverse was found for IVDMD. The greatest digestibility occurred in the fall (66.7%) and spring (63.3%) and the lowest during the summer (59.3%). Crude protein was nearly 14% in the spring and fall and 11% during the summer. In another experiment, three cultivars of tall fescue averaged 64% IVDMD when harvested in the first week of May (A. G. Matches, unpublished data, 1971-74). During July, IVDMD ranged from 45 to 52%

MANAGEMENT

Table 10-9—Forage yields of tall fescue cultivars at 11 locations in the United States.

						Yield of forage					
						Location of test					
Cultivar	Montana†	Wyoming†	Oklahoma	Texas	Iowa	Missouri	Illinois	Alabama TVA	Alabama Tuskegee	Alabama Monroeville	Florida
						kg dry matter/ha					
Alta	5,605	7,040		4,327			8,048	7,937	3,811	2,982	
Electra		6,659									
Festal (N4-5a)		6,748									
Fawn		6,950	3,566		11,143	8,225	7,982	8,643	3,261	2,993	4,840
Goar	5,000	6,412		5,022							4,420
Kenhy		6,232	3,148		11,995	8,302					
Kenmont	5,896	6,860									
KL 39		6,076									4,960
Kentucky 31	5,336	6,950	3,241	4,742	10,941	8,269	8,206	7,511	3,890	2,791	4,360
Kenwell		6,950	3,181					6,580	3,296	2,242	850
Marade		5,426									
Oregon 4-36	5,089										
S-170		6,547									
Reference	Joppa and Roath, 1964	Moyer and Seamands, 1975	Bates, undated	Brooks and Ho t, 1954	Carlson, 1974	A.G. Matches, unpublished data	Graffis et al., 1974	Hoveland et al., 1970			Dunavin and Bertrand, 1973

† 12% moisture

Table 10-10—Herbage composition of tall fescue cultivars from spring, summer, and fall samplings in southwest Missouri (F. A. Martz, A. G. Matches, D. A. Sleper, and R. L. Belyea, unpublished data).

Season	Tall fescue cultivar	Dry weight						
		ADF	N	Ca	P	NDF	IVDMD	CP
					%			
Spring	Kenhy	30.8	2.17	0.31	0.29	55.1	64.6	13.56
	Kenmont	30.7	2.22	0.35	0.30	54.9	64.6	13.88
	Mo-96	30.9	2.17	0.31	0.29	55.4	63.2	13.56
	Fawn	32.9	2.17	0.29	0.30	57.9	61.2	13.56
	Kentucky-31	32.1	2.21	0.35	0.32	55.9	62.9	13.81
	Average	31.5	2.19	0.32	0.30	55.8	63.3	13.67
Summer	Kenhy	33.1	1.79	0.36	0.30	58.3	60.6	11.19
	Kenmont	32.7	1.84	0.39	0.32	58.3	60.4	11.50
	Mo-96	30.1	1.84	0.36	0.32	58.0	58.6	11.50
	Fawn	33.7	1.78	0.37	0.29	58.1	57.2	11.13
	Kentucky-31	33.4	1.77	0.38	0.33	58.0	59.7	11.06
	Average	32.6	1.80	0.37	0.31	58.1	59.3	11.28
Fall	Kenhy	31.0	2.20	0.28	0.34	53.4	68.1	13.75
	Kenmont	29.9	2.13	0.32	0.34	54.0	66.6	13.31
	Mo-96	29.8	2.20	0.30	0.32	51.5	67.6	13.75
	Fawn	30.1	2.40	0.30	0.35	55.5	65.0	15.00
	Kentucky-31	33.1	2.21	0.32	0.35	54.3	66.2	13.81
	Average	30.8	2.23	0.30	0.34	53.7	66.7	13.92

and in October and November increased to a range of from 54 to 62%. Vartha et al. (1977) reported for Kentucky-31 tall fescue a concavely curvilinear decline in percent IVDMD from a high of 64% on 20 May to a low of 46% in mid-July. Average digestibility from mid-June to August was below 50% IVDMD. In September, digestibility reached a high of 62% IVDMD.

Animal performance differences have also been observed among different cultivars of tall fescue. In a Missouri experiment (Martz et al., 1976), average daily gains of yearling Hereford steers and heifers over 3 years of experimentation were: Kenhy, 0.58 kg; Kenmont, 0.45 kg; 'Mo-96', 0.57 kg; Fawn, 0.44 kg; Kentucky-31, 0.40 kg. The nearly 40% yield advantage for Mo-96 and Kenhy over Kentucky-31 occurred during the spring and fall grazing periods, but gains on cultivars were not different during the summer.

B. Quality Trends

In another section it was shown that with frequent defoliation, yields of tall fescue were less; but digestibility and CP content of the forage are greater. Comparative values for IVDMD and CP are available from Ocumpaugh and Matches (1977). The IVDMD and CP contents, respectively, of spring- and summer-cut tall fescue forage averaged 48 and 9.8% with two cuttings; 54 and 12% with three cuttings; 67 and 16% with five cuttings. Average yields of dry matter were 7,305, 6,396, and 4,419 kg/ha with two, three, and five cuttings, respectively.

Several workers have reported the rate of decline in digestibility for the spring growth of grasses as they mature. Pritchard et al. (1963) stated that tall fescue digestibility decreased at approximately 0.5 percentage units/day starting at a level of 70% in late May and decreasing below 50% IVDMD before July. Porter et al. (1975) found that the digestibility of tall fescue and orchardgrass hay declined as much as 0.5 percentage units each day that harvest was delayed after mid-May. Although N fertilizer increased CP from 13 to 23% in hay harvested on 10 May and from 7 to 12.9% in hay harvested 16 June, N had no effect on its digestibility. Rates of decline for IVDMD were slightly different for tall fescue, orchardgrass, reed canarygrass, and smooth bromegrass in research conducted over 3 years and at two locations in Missouri (A. G. Matches, unpublished data, 1971-74). From early April to early June, the daily rate of decline for IVDMD in percentage units was as follows: tall fecue—0.63; orchardgrass—0.84; reed canarygrass—0.56; and smooth bromegrass—0.79.

C. How Tall Fescue Compares with Other Grasses

Tall fescue compares favorably with other grasses in quality. Eleven grasses were compared for their CP content at anthesis as influenced by 0 or 112 kg N/ha or when grown with alfalfa (Johnson and Nichols, 1969b). Tall fescue, orchardgrass, smooth bromegrass, and reed canarygrass were not significantly different in CP content when fertilized with N or when grown with alfalfa, but reed canarygrass without N was slightly higher in CP than was tall fescue (11.0 vs. 8.2% CP). With N fertilization, fescue averaged 11% protein; with alfalfa, 10.7% protein.

The growth pattern of tall fescue is quite different than those of many warm-season grasses. Berg (1971) found that tall fescue and orchardgrass began growth much earlier in the spring and continued to grow later in the fall than three cultivars of switchgrass (*Panicum virgatum* L.). The tall fescue and orchardgrass produced 3,200 to 5,000 kg/ha more dry forage over the entire season than switchgrass. But during midsummer, the switchgrass was most productive and would provide summer grazing. Rountree et al. (1974) reported that forage yields of several warm-season grasses [Midland bermudagrass, switchgrass, indiangrass (*Sorghastrum nutans* (L.) Nash)] were similar to total yields of Alta tall fescue and 'Ioreed' reed canarygrass. Total yields of caucasian bluestem (*Andropogon caucasicus* Trin.) were higher than those of either tall fescue or reed canarygrass. More important than yield was the distribution of forage production. Tall fescue had only 42% of its season's production between 1 June and 31 August, whereas the warm-season grasses had 90% or more of their forage production between June and September. Tall fescue averaged 14.5% CP and 63% IVDMD. All warm-season grasses were lower than tall fescue, ranging from 8 to 11.5% in CP and from 46 to 58% in IVDMD. These data show that although the quality of the warm-season grasses may be lower than that of tall fescue, their high summer distribution of forage production makes them ideal for fitting into pasture systems. Tall fescue can provide good spring and fall grazing with the warm-season grasses bridging the summer gap left by tall fescue (Matches et al., 1975).

D. Summary for Species and Cultivars

Differences in herbage yield exist among cultivars of tall fescue; generally, however, yield differences are not great. Commercially available tall fescue cultivars usually show only small differences in herbage composition (CP, ADF, NDF, IVDMD, and mineral composition) when harvested at comparable stages of maturity. Even when cut at a vegetative stage of growth, lowest digestibility and highest fiber content usually occur during the hot summer months.

Differences in rate of gain made by cattle have been observed among cultivars of tall fescue. The rate of decline in IVDMD for tall fescue in the spring is about 0.5 to 0.6 percentage units/day.

Composition of tall fescue herbage is very similar to the composition of other cool-season grasses. However, tall fescue herbage is higher in digestibility and CP than many warm-season grasses when both are cut at similar stages of maturity. Tall fescue generally makes most of its growth during the spring and fall. Certain warm-season grasses, because of their superior summer production, offer potential for developing pasture systems to bridge the summer slump left by tall fescue.

LITERATURE CITED

1. Archer, K. A., and A. M. Decker. 1977a. Autumn-accumulated tall fescue and orchardgrass. I. Growth and quality as influenced by nitrogen and soil temperature. Agron. J. 69:601–605.
2. ————, and ————. 1977b. Autumn-accumulated tall fescue and orchardgrass. II. Effects of leaf death on fiber components and quality parameters. Agron. J. 69:605–609.
3. Balasko, J. A. 1977. Effect of N, P, and K fertilization on yield and quality of tall fescue forage in winter. Agron. J. 69:425–428.
4. Bates, Richard P. Undated. Three-year summary 1974–1977 forage yields and crude protein from perennial cool-season grasses. The Samuel Roberts Noble Foundation, Inc., Route One, Ardmore, Okla. R-162.
5. Berg, C. C. 1971. Forage yield of switchgrass (*Panicum virgatum*) in Pennsylvania. Agron. J. 63:785–786.
6. Berry, R. F., and C. S. Hoveland. 1969. Summer defoliation and autumn-winter production of *Phalaris* species and tall fescue varieties. Agron. J. 61:493–494.
7. Brooks, Lester, E., and Ethan C. Holt. 1954. Cool-season grasses in the Wichita Valley. Texas Agric. Exp. Stn. Prog. Rep. 1716.
8. Brown, R. H., and R. E. Blaser. 1965. Relationship between reserve carbohydrate accumulation and growth rate in orchardgrass and tall fescue. Crop Sci. 5:577–582.
9. ————, ————, and J. P. Fontenot. 1963. Digestibility of fall grown Kentucky 31 fescue. Agron. J. 55:321–324.
10. Bryan, W. B., W. F. Wedin, and R. L. Vetter. 1970. Evaluation of reed canarygrass and tall fescue as spring-summer and fall-saved pasture. Agron. J. 62:75–80.

11. Burger, A. W., J. A. Jacobs, and C. N. Hittle. 1958. The effect of height and frequency of cutting on the yield and botanical composition of tall fescue and smooth bromegrass mixtures. Agron. J. 50:629-632.
12. Carlson, Irving. 1974. Strain test of cool-season perennial forage grasses, 1973-74. Iowa State Univ. Shelby-Grundy Exp. Farm. Annu. Prog. Rep.-1974. p. 5-7.
13. Colyer, Dale, F. L. Alt, J. A. Balasko, P. R. Henderlong, G. A. Jung, and Vinh Thang. 1977. Economic optima and price sensitivity of N fertilization for six perennial grasses. Agron. J. 69:514-517.
14. Decker, A. M., H. J. Retzer, and R. F. Dudley. 1974. Cool-season perennials vs. cool-season annuals sod-seeded into Midland sward. Agron. J. 66:381-383.
15. Devlin, T. J., W. K. Roberts, and V. V. E. St. Omer. 1969. Effects of dietary potassium upon growth, serum electrolytes and intrarumen environment of finishing beef steers. J. Anim. Sci. 28:557-562.
16. Dobson, James W., and E. R. Beaty. 1977. Forage yield of five perennial grasses with and without clover at four nitrogen rates. J. Range Manage. 30:461-465.
17. ─────, C. D. Fisher, and E. R. Beaty. 1976. Yield and persistence of several legumes growing in tall fescue. Agron. J. 68:123-125.
18. Dotzenko, A. D. 1961. Effect of different nitrogen levels on the yield, total nitrogen content, and nitrogen recovery of six grasses grown under irrigation. Agron. J. 53:131-133.
19. Dunavin, L. S., and J. E. Bertrand. 1973. Fescue in review—Varieties, grazing and climatic aspects. Soil Crop Sci. Soc. Fla. Proc. 33:9-12.
20. Fuller, William W., William C. Elder, Billy B. Tucker, and Wilfred E. McMurphy. 1971. Tall fescue in Oklahoma. Oklahoma State Univ. Agric. Exp. Stn. Prog. Rep. P-650.
21. Graffis, D. W., D. A. Miller, O. W. Pile, M. R. Bell, R. Bell, and D. E. Millis. 1974. Forage crops variety trials in Illinois—1974. Illinois Agric. Exp. Stn. Bull. AG-2009.
22. Green, James Terrill, Jr. 1974. Accumulating canopies of tall fescue (*Festuca arundinacea* Schreb.) as influenced by nitrogen and cutting management. Ph.D. Thesis. Virginia Polytechnic Institute and State Univ. (Libr. Congr. Card No. Mic. 75-15874).
23. Hallock, D. L., R. H. Brown, and R. E. Blaser. 1965. Relative yield and composition of Kentucky 31 fescue and Coastal bermudagrass at four nitrogen levels. Agron. J. 57:539-542.
24. ─────, ─────, and ─────. 1966. Response of 'Coastal' and 'Midland' bermudagrass and 'Kentucky 31' fescue to nitrogen in southeastern Virginia. Virginia Agric. Exp. Stn. Res. Rep. 112.
25. Hannaway, David B., and John H. Reynolds. 1976. Seasonal changes in minerals, protein and yield of tall fescue forage as influenced by nitrogen and potassium fertilization. Tennessee Agric. Exp. Stn. Prog. Rep. 98. p. 14-17.
26. Hart, Richard, G. E. Carlson, and D. E. McCloud. 1971. Cumulative effects of cutting management on forage yields and tiller densities of tall fescue and orchardgrass. Crop Sci. 63:895-898.
27. Helsel, Zane R., W. F. Wedin, and Wayne Fruehling. 1974. Crownvetch-grass management study. Iowa State Univ. Annu. Prog. Rep-1974. OEF 74-10. p. 18-19.
28. Hoveland, C. S. 1970. Dormancy and seasonal growth of *Phalaris* species in Alabama. p. 608-611. *In* N. J. T. Normand (ed.) XI Int. Grassl. Congr. Proc. Surfers' Paradise, Queensland, Australia, 13-23 Apr. 1970. Univ. of Queensland Press, St. Lucia.
29. ─────, and W. B. Anthony. 1971. Winter forage production and in vitro digestibility of some *Phalaris aquatica* genotypes introduction. Crop Sci. 11:461-462.
30. ─────, and E. M. Evans. 1970. Cool-season perennial grass and grass-clover management. Auburn Univ. Agric. Exp. Stn. Circ. 175.
31. ─────, ─────, and D. A. Mays. 1970. Cool-season perennial grass species for forage in Alabama. Auburn Univ. Agric. Exp. Stn. Bull. 397.
32. Isley, Charles H., and Douglas S. Chamblee. 1971. Influence of defoliation systems on the morphological development and growth of tall fescue. p. 61. *In* Assoc. South. Agric. Workers, Inc., Proc. 68th Ann. Convention.
33. Jensen, E. H. 1970. Nitrogen fertilization of tall fescue. Nevada Agric. Exp. Stn. Rep. 76.
34. Johnson, James R., and James T. Nichols. 1969a. Crude protein content of eleven grasses as affected by yearly variation, legume association and fertilization. Agron. J. 61:65-68.
35. ─────, and ─────. 1969b. Production, crude protein and use of eleven irrigated grasses and alfalfa-grass combinations on clay soils in western South Dakota. South Dakota Agric. Exp. Stn. Bull. 555.
36. Joppa, L. R., and C. W. Roath. 1964. A comparison of Kenmont, Alta, and other tall fescue varieties in Montana. Montana Agric. Exp. Stn. Bull. 582.

37. Jung, G. A., and R. E. Kocher. 1974. Influence of applied nitrogen and clipping treatment on winter survival of perennial cool-season grasses. Agron. J. 66:62–65.
38. ———, J. A. Balasko, F. L. Alt, and L. P. Stevens. 1974. Persistence and yield of ten grasses in response to clipping frequency and applied nitrogen in the Allegheny Highlands. Agron. J. 66:517–521.
39. Kroth, Earl M., and Richard Mattas. 1974. Yields and soil test values resulting from topdressing forage crops. Missouri Agric. Exp. Stn. Res. Bull. 1005.
40. ———, ———, Louis Meinke, and Arthur Matches. 1977. Maximizing production potential of tall fescue. Agron. J. 69:319–322.
41. Martz, F. A., S. Bell, M. Mitchell, A. G. Matches, and D. A. Sleper. 1976. Advanced evaluation of forages. Univ. of Missouri Spec. Rep. 192. Southwest Ctr. Res. Rep. 1976.
42. Matches, Arthur G., and Jesse B. Tevis. 1973. Yield and quality of tall fescue stockpiled for winter grazing. p. 54–57. *In* Research in agronomy-1973. Univ. of Missouri Misc. Publ. 73-5.
43. ———, G. B. Thompson, and F. A. Martz. 1975. Post-establishment harvesting and management systems of forages. p. 111–135. *In* No-Tillage Forage Symp. Proc. Columbus, Ohio. 14–16 Oct. 1975. Ohio State Univ. and Ohio State Agric. Res. and Dev. Ctr.
44. ———, H. N. Wheaton, and J. B. Tevis. 1973. Renovation of tall fescue sods with legumes. p. 58–59. *In* Research in agronomy-1973. Univ. of Missouri Misc. Publ. 73-5.
45. Mays, D. A., and J. B. Washko. 1960. The feasibility of stockpiling legume-grass pasturage. Agron. J. 52:190–192.
46. McKee, W. H., Jr., R. H. Brown, and R. E. Blaser. 1967. Effect of clipping and nitrogen fertilization on yield and stands of tall fescue. Crop Sci. 7:567–570.
47. Moyer, J. L., and W. J. Seamands. 1975. Tall fescue. Wyoming Agric. Ext. Serv. B626.
48. National Research Council-National Academy of Sciences. 1970. Nutrient requirements of beef cattle. No. 4. 4th Revised Ed. NAS, Washington, D.C.
49. Ocumpaugh, William R., and A. G. Matches. 1977. Autumn-winter yield and quality of tall fescue. Agron. J. 69:639–643.
50. Offutt, M. S., and C. E. McKee. 1973. Performance of three legumes grown in mixture with tall fescue. Arkansas Farm Res. 22:3.
51. Overton, Joseph R., and Henry A. Fribourg. 1975. Persistence of Kentucky 31 tall fescue in Midland bermudagrass sod. Tennessee Agric. Exp. Stn. Prog. Rep. 96. p. 5–7.
52. Pfander, W. H., and E. Rubio. 1972. Composition and supplements for fall-grown fescue. J. Anim. Sci. 35:233.
53. Porter, T. K., V. L. Lechtenberg, and C. L. Rhykerd. 1975. Effect of harvest date and nitrogen fertilization on digestibility and composition of grass hay. p. 61–65. *In* 1975 Indiana Beef-Forage Res. Day, Purdue Univ.
54. Pritchard, G. I., L. P. Folkins, and W. J. Pigden. 1963. The in vitro digestibility of whole grasses and their parts at progressive stages of maturity. Can. J. Plant Sci. 43:79–87.
55. Reynolds, John H. 1975. Yield and chemical composition of stockpiled tall fescue and orchardgrass. Tenn. Farm Home Sci. 94:27–29.
56. Rommann, L. M., and W. E. McMurphy. 1973. Tall fescue establishment and management. Oklahoma State Univ. Ext. Facts No. 2559.
57. ———, ———, and B. D. Boyer. 1973. Tall fescue in bermudagrass. Oklahoma State Univ. Ext. Facts No. 2564.
58. Rountree, Billie H., Arthur G. Matches, and F. A. Martz. 1974. Season too long for your grass pasture? Then use several grasses in several pastures. Crops Soils 26:7–10.
59. Schiller, F. M. A., and Alec Lazenby. 1975. Yield performance of tall fescue (*Festuca arundinacea*) populations on the northern tablelands of New South Wales. Aust. J. Exp. Agric. Anim. Husb. 15:391–399.
60. Smith, Dale, A. V. A. Jacques, and J. A. Balasko. 1973. Persistence of several temperate grasses grown with alfalfa and harvested two, three, or four times annually at two stubble heights. Crop Sci. 13:553–556.
61. Staff, Virginia Forage Research Station. 1969. Managing forages for animal production. Virginia Polytechnic Inst. Res. Div. Bull. 45. p. 35, 77.
62. Taylor, T. H., and W. C. Templeton, Jr. 1976. Stockpiling Kentucky bluegrass and tall fescue forage for winter pasturage. Agron. J. 68:235–239.
63. Templeton, W. C., Jr., and T. H. Taylor. 1966a. Some effects of nitrogen, phosphorus, and potassium fertilization on botanical composition of a tall fescue-white clover sward. Agron. J. 58:569–572.

64. ————, and ————. 1966b. Yield response of a tall fescue-white clover sward to fertilization with nitrogen, phosphorus and potassium. Agron. J. 58:319-322.
65. ————, ————, and J. R. Todd. 1965. Comparative ecological and agronomic behavior of orchardgrass and tall fescue. Kentucky Agric. Exp. Stn. Bull. 699.
66. Tesar, M. B. 1974. Nitrogen on grasses compared to alfalfa-grass mixtures in northern Michigan. Michigan State Univ. Agric. Exp. Stn. Res. Rep. 256. (Progress Rep. CS-LC-7301). p. 101-104.
67. Van Keuren, R. W. 1972. All-season forage systems for beef cow herds. p. 39-44. *In* Twenty-seventh Annu. Meeting Soil Cons. Soc. Am. Portland, Oreg. 6-9 Aug. 1972. Soil Cons. Soc. Am., Ankeny, Iowa.
68. Vartha, E. W., A. G. Matches, and G. B. Thompson. 1977. Yield and quality trends of tall fescue grazed with different subdivisions of pasture. Agron. J. 69:1027-1029.
69. Wagner, R. E. 1954a. Influence of legume and fertilizer nitrogen on forage production and botanical composition. Agron. J. 46:167-171.
70. ————. 1954b. Legume nitrogen versus fertilizer nitrogen in protein production of forages. Agron. J. 46:233-237.
71. Wedin, W. F., I. T. Carlson, and R. L. Vetter. 1966. Studies on nutritive value of fall-saved forage, using rumentation and chemical analysis. p. 424-428. *In* N. J. T. Normand (ed.) X Int. Grassl. Congr. Proc., Univ. of Helsinki, Finland. 7-16 July 1966. Valtioneuvoston Kirjepaino, Helsinki.
73. Wheaton, H. N., Douglas Lowe, Dan Devine, and Ron Weaver. 1973. Tall fescue yield response to nitrogen, phosphorus and potassium. p. 66. *In* Research in Agronomy-1973. Univ. of Missouri Misc. Publ. 73-5.
74. Whitehead, D. C. 1970. The role of nitrogen in grassland productivity. Commonw. Agric. Bur. Pasture and Field Crops. Bull. 48. Hurley, Berkshire, England.
75. Wilkinson, S. R., L. F. Welch, G. A. Hillsman, and W. A. Jackson. 1968. Compatibility of tall fescue and Coastal bermudagrass as affected by nitrogen fertilization and height of clip. Agron. J. 60:359-362.
76. Yungen, J. A., T. L. Jackson, and W. S. McGuire. 1977. Seasonal response of perennial forage grasses to nitrogen application. Oregon Agric. Exp. Stn. Bull. 626.

Chapter 11 Tall Fescue in Forage-Animal Production Systems for Breeding and Lactating Animals

R. W. VAN KEUREN
Agronomy Department
Ohio Agricultural Research and Development
Wooster, Ohio

JOHN A. STUEDEMANN
Animal Science Department
Southern Piedmont Research Center
Watkinsville, Georgia

INTRODUCTION

Tall fescue (*Festuca arundinacea* L.) has received increased attention as a forage for breeding and lactating ruminants, primarily because of its wide adaption to soil conditions, productivity, persistence, response to fertilization, and ease of management. Despite its generally lower palatability and intake as compared with other cool-season grasses and some problems with toxicity, tall fescue is gaining greater usage among livestock producers. Both its advantages and limitations must be considered in using tall fescue in a forage program for breeding and lactating animals, whether they be beef cows, sheep, or dairy cattle.

TALL FESCUE FOR BEEF COWS

For breeding and lactating animals, tall fescue has gained greater acceptance for beef cows than for dairy cattle or sheep. This is primarily because beef cows have lower critical nutrient requirements than dairy cattle or sheep and are increasing in areas where tall fescue is well adapted. Tall fescue is being used for beef cows as summer and winter pasture, combination grazing and hay programs, and in year-round programs with other forages for total feed systems.

Copyright © 1979 ASA-CSSA-SSSA, 677 South Segoe Road, Madison, WI 53711 USA. *Tall Fescue.*

A. Tall Fescue as Summer Pasture for Beef Cows and Calves

Limited data are available on the utilization of tall fescue as summer pasture for beef cows and calves. Figure 1 shows beef cows and calves on tall fescue summer pasture in northern Georgia. In a 2-year North Carolina study, Burns et al. (1973) compared tall fescue, tall fescue-ladino clover (*Trifolium repens* L.), tall fescue rotated with 'Coastal' bermudagrass (*Cynodon dactylon* (L.), Pers.) in midsummer, and tall fescue-ladino clover rotated with midsummer bermudagrass. Spring-born calves, averaging 75 kg (165 lb) going on pasture, made satisfactory gains the first 28 days grazing on tall fescue. This was only slightly less than calves on tall fescue-ladino clover pasture. However, the tall fescue pastured calves subsequently made comparatively poor average daily gains, about 522 g (1.15 lb)/day, and weaned at an average of 28.5 kg (63 lb) less than those pastured on tall fescue-ladino clover. Calves grazing the tall fescue and tall fescue-ladino clover had average weaning weights of 169 kg (373 lb) and 197 kg (435 lb), respectively, with the tall fescue grazed calves consistently grading lower than those that grazed the tall fescue-ladino clover. Calf scores were 9.3 (low good) and 11.0 (high good), respectively. Calves that grazed tall fescue or tall fescue-ladino clover early in the season then grazed bermudagrass had somewhat better gains than did calves on tall fescue all season. The tall

Fig. 11-1—Beef cows and spring-born calves on tall fescue summer pasture in northern Georgia.

fescue-ladino clover gave the best results followed closely by the tall fescue-ladino clover plus bermudagrass. All cows accompanying the calves gained weight, with the tall fescue pastured cows gaining less than those grazing on the other three pasture systems. All cows reached a suitable condition for breeding females.

Tall fescue-ladino clover was compared with orchardgrass (*Dactylis glomerata* L.)- ladino clover for beef cows and calves in a Tennessee study (Anderson and Safley, 1967) with creep feeding as another variable. The pasture period was from early or mid-April to mid-October (Table 11-1). The calves went on pasture ranging in weight from 53 to 81 kg (117 to 179 lb). Calves on newly established tall fescue and orchardgrass pastures with a third of the composition as ladino clover had very satisfactory adjusted average daily gains (Adj. ADG) without creep, about 960 g (2.1 lb)/day. Calves on either grass with a lower proportion of legume gained about 760 g (1.7 lb) or less per day. Excellent weaning weights and grades were obtained from the pasture with good legume stands without creep, ranging from an average of 202 to 254 kg (445 to 560 lb) and grading from 11.6 (high good) to 12.8 (choice). Creep feeding (cracked shelled corn and crimped oats) only partially compensated for the difference in legume content in the pastures. The best daily gains, 1,071 g (2.36 lb) and 1,081 g (2.38 lb), were from calves creep-fed on tall fescue and orchardgrass pastures with good legume stands, respectively. The cows had good weight gains on either grass with good legume stands and on orchardgrass alone. In the first year on tall fescue, cows had small weight gains, and the second year they lost weight. Creep-feeding calves had no effect on weight gain of cows. The ladino clover stand decreased sharply in both grasses the second grazing year.

In an earlier Tennessee study (Anderson and Safley, 1966), pasture renovation to increase the legume stand in tall fescue resulted in better calf daily gains and cow weight gain than before renovation. The legume stand increased from 6 to 12% before renovation to 30 to 52% after renovation. Before renovation, the calves gained an average of 708 g (1.56 lb) adjusted weight daily, graded 12.2 (low choice), and weighed 189 kg (416 lb) at weaning, as compared with 844 g (1.86 lb)/day gain, an average grade of 11.6 (high good), and an average weaning weight of 217 kg (478 lb) for the same pastures after renovation. The cow gains on the pastures (April to October) before renovation were highly variable. They averaged a 50-g (0.11 lb) loss/day in two lots, and a 236-g (0.52 lb)/day average gain in two lots. No reasons were suggested for the variability. Following renovation, cows gained weight on all lots, averaging 259 g (0.57 lb)/day (May to October).

In a Missouri study, preliminary results indicate that spring-born calves gained slightly better on tall fescue summer pasture with no N fertilization than did calves on summer pastures receiving annual N applications of 112 kg/ha (100 lb/acre) or 224 kg/ha (200 lb/acre). ADG were 790 g (1.74 lb), 699 g (1.54 lb), and 745 g (1.64 lb), respectively (Thompson, 1974). Average weaning weights were 194 kg (428 lb), 182 kg (402 lb), and 192 kg (424 lb), respectively, for the three N fertilization rates. Ladino clover was broadcast

Table 11-1—Representative beef cow and calf production on summer, winter, and year-around pastures.

A. Summer pasture

1. Cows and calves†

Performance, 20 Apr.–6 Oct. (3-year average)

Species	Calf performance		Cow performance	
	ADG, g	Weaning wt., kg	ADG, g	Conception, %
Tall fescue	581	159	9	72
Orchardgrass	799	195	263	90
Tall fescue-ladino clover-red clover	826	193	263	92

2. Cows and calves‡

15 Apr.–15 Nov. (two locations, 4-year average) Carrying capacity, cows units

Species	Grazing days/ha	Cow units/ha
Tall fescue	692	3.2
Orchardgrass	519	2.4
Kentucky bluegrass	395	1.8

3. Creep feeding§

Apr.–Oct. Calf performance

Species	Ladino clover		Adj. ADG, g		Weaning wt., kg		Grade		Condition	
	Year	Creep	No creep	Creep	No creep	Creep	No creep	Creep	No creep	Creep	No creep
Tall fescue	1	37	36	1,071	976	255	254	11.8	12.8	9.6	8.8
Orchardgrass	1	29	30	1,081	953	247	215	13.4	11.6	10.0	7.4
Tall fescue	2	9	12	1,044	763	242	202	14.2	11.8	10.6	8.2
Orchardgrass	2	8	9	1,012	754	261	203	14.0	11.8	10.2	8.2

(continued on next page)

Table 11-1—Continued.

Species	Performance	

B. Winter pasture

	15 Nov.–15 Apr. (7-year average)	
	Carrying capacity	
	Grazing days/ha	Cows/ha
Pregnant cows¶		
Tall fescue (small round bales and regrowth)	509	3.3

C. Year-around Pasture

Cow and calf performance (4-year average)

Cows and calves#	Stocking rate cows/ha	Average wt. pregnant cows/kg		Calf crop, %		Calf performance			
		Protein supplement		Protein supplement		Wt. produced kg/ha	Adj. ADG g	Weaning wt., kg	Weaning grade
		High	Low	High	Low				
Bermudagrass-dallisgrass-white clover alternated with tall fescue-white clover	0.82	420	414	98	90	128	704	185	11.0
	1.23	417	414	88	95	197	708	188	10.8
	1.96	396	361	88	85	239	617	155	10.1

† Lechtenberg et al., 1975; Indiana.   ‡ Van Keuren, 1970b; Ohio. 1949 kg N/ha/year, two applications, tall fescue and orchardgrass; 76 kg N/ha/year, one application, Kentucky bluegrass.   § Anderson and Safley, 1967; Tennessee. Creep ration; corn 2/3, oats 1/3, 1st year represents new stand; Calf grades: low good—9, good—10, high good—11, low choice—12, and choice—13.   ¶ Van Keuren and Parker, 1974b; Ohio. No supplement, 74 kg N/ha, Mar. and Aug.   # Ray et al., 1969; Arkansas. No fertilization.

with no seedbed preparation on the tall fescue pastures twice during the prior 4-year period. Seedings were made in winter. This resulted in legume percentages of 12 to 20% of the forage, dry weight basis, for the no-N pastures, and lower percentages in the N-fertilized pastures. The presence of the legume probably accounted for the slightly better average daily gains of calves on the no-N pastures as compared with the two receiving N fertilizer. The cows on the high-N fertilized pastures gained less weight than cows on the two other treatments. These preliminary results also showed a higher conception rate from the no-N pastures than from the fertilized pastures (Redmon, 1974). In this study, the "put and take" procedure of Mott and Lucas (1952) was followed.

In a later report, Stricker et al. (1976) showed a 4-year adjusted ADG of 804 g (1.77 lb), 745 g (1.64), and 740 g (1.63 lb) for the three N fertilizer rates averaged for creep and no-creep feeding. Creep-fed calves gained 826 g (1.82 lb)/day, and no-creep fed calves 699 g (1.54 lb)/day, with adjusted 205-day weights of 210 kg (463 lb) and 186 kg (409 lb), respectively.

Tall fescue summer pasture fertilized annually with N at 74 kg/ha (66 lb/acre), 224 kg/ha (200 lb/acre), or 11.5 metric tons/ha (5 tons/acre) of broiler litter had similar 3-year average calf weaning weights. This was about 167 kg (368 lb) 205-day adjusted weight (Stuedemann et al., 1975). The weights were similar within and between years for the higher rate of N and for broiler litter treatments, but decreased markedly each successive year for the low N treatment. The successive decrease in calf weaning weight for the low N treatment paralleled an increase in nematode eggs recovered from the calves, as well as an increase in actual worm counts. The level of parasitism was inversely related to the quantity of available forage which, in turn, was directly related to the level of fertilization. No parasite problems were observed in cows. The pastures were stocked similarly at 0.40 ha/cow (1 acre/cow) on a year-round basis. Cow weights reflected the N treatments with the lowest gains on the low-N treatment and the highest gains on the broiler litter treatment (highest rate of N). Cow conception rate was considered acceptable for all treatments.

An Indiana study (Lechtenberg et al., 1975) showed spring-born calves gained 218 g (0.48 lb) more per day on orchardgrass than on tall fescue (Table 11-1). With ladino and red clover (*Trifolium pratense* L.) interseeded into tall fescue, the average daily gains were better than on either grass. The 3-year average daily gains were 799 g (1.76 lb), 581 g (1.28 lb), and 826 g (1.82 lb) on orchardgrass, tall fescue, and tall fescue-legume pasture, respectively. Average weaning weights were 195 kg (429 lb), 159 kg (351 lb), and 193 kg (426 lb). The cows had similar gains for the average of the 3 years on orchardgrass and tall fescue-legume, 263 g (0.58 lb)/day, as compared with 9 g (0.02 lb)/day on tall fescue. Conception rates were 90 and 92% for orchardgrass and tall fescue-legume, respectively, as compared with 72% on tall fescue.

Tall fescue has a higher carrying capacity than does orchardgrass or Kentucky bluegrass (*Poa pratensis* L.) in Ohio (Van Keuren, 1970 b) with 692 grazing days/ha (280 days/acre) for tall fescue, 519 (210) for orchardgrass, and 395 (160) for bluegrass with beef cows and calves (Table 11-1). The tall fescue and orchardgrass received 149 kg/ha (133 lb/acre) of N annually in two applications, and the bluegrass received 74 kg/ha/year (66 lb/acre) in one application.

Studies in southern Illinois (Hinds et al., 1974) showed higher calf weaning weights and grades with tall fescue-legume pasture than with tall fescue, as well as better cow conception rates. The legumes were alfalfa (*Medicago sativa* L.), red clover, and Korean lespedeza (*Lespedeza stipulacea* Maxim.). Continuous grazing of tall fescue from mid-April to late October resulted in slightly higher calf weaning weights and better cow conception rates than from a split management system in which a crop of hay was harvested from half of the pasture in early June. The hay was removed and stockpiled for feeding in the latter part of the grazing season.

"Fescue toxicity" was a problem in Oklahoma grazing studies with beef cows on pure stands of tall fescue (Fuller et al., 1971). Symptoms of toxicity began to appear in early to mid-May when daily temperatures remained above 24 C. Symptoms included rapid breathing and excessive salivation, with lameness, loss of tails, and emaciation in advanced stages during the hotter summer months. Calves would also become unthrifty and show a dull haircoat. Adding legume to the tall fescue pasture, however, not only increased forage yields but also alleviated toxicity problems, except for the rough, dull haircoat. Fescue toxicity is discussed in Chapter 13.

In summary, tall fescue used for summer grazing of cow-calf herds should be seeded with a legume for higher calf daily gains, weaning weights, grades, and better cow-conception rates. Creep feeding partially compensated for the lack of legume in pure tall fescue pastures. Creep feeding does not appear to be necessary if good quality grass-legume pastures are provided.

B. Tall Fescue for Wintering Beef Cows

The conserving of grass for grazing in winter after growth ceases due to low temperatures is an old agricultural practice, according to Hughes (1955 a). In early British studies, autumn-conserved herbage was used for wintering cattle with hay fed in increasing amounts toward the end of winter to supplement the grazing (Hughes, 1955 b). Hughes reported that more effective utilization of the forage was obtained where the grazing was restricted by using an electric fence. He noted gradual deterioration during the winter of the conserved standing herbage, with a loss of dry matter and crude protein. Alder and Redford (1958) used a mixture of tall fescue, meadow fescue (*Festuca elatior* L.), and white clover (*Trifolium repens* L.) for winter grazing and reported high yields and good acceptability of tall fescue by both cattle and sheep. Hughes (1955 b) noted that because of tall fescue's fibrous nature and erect growth when it was conserved in situ for winter utilization it retained its acceptability to animals grazed in winter. Baker et al. (1965) reported that although tall fescue had the reputation as a coarse grass not generally acceptable to livestock, in their studies livestock readily ate it in early spring, autumn, or winter, when tall fescue would actually be preferred to orchardgrass.

Spring-Calving

Subsequent studies in the United States confirmed the value of tall fescue as winter pasture for cattle. Indiana studies (Peterson et al., 1964; Kaiser et al., 1966) showed that mature pregnant Hereford cows could be wintered satisfactorily on tall fescue conserved as field-stored small round bales and fall regrowth. Under winter climatic conditions encountered in southern Indiana, some woods or an open shed was adequate protection against the weather. In some trials, a tall fescue seed crop was harvested and the second growth of tall fescue was baled (Peterson and Garrigus, 1964; Kaiser et al., 1966). Peterson and Garrigus (1964) did not improve wintering performance of pregnant cows by providing a small amount of energy [0.45 kg (1 lb)/day/cow ground ear corn (*Zea mays* L.)] and protein supplement as compared with providing protein supplement only. Feeding 1.35 kg (3 lb) of a 20% protein range cube per head daily to brood cows during January through April significantly ($P < 0.01$) improved calf ADG as compared with a check group (bales and regrowth only), 860 g (1.90 lb) and 800 g (1.76), respectively (Kaiser et al., 1966). However, weaning weights at 205 days were similar.

Restricting the area available to the wintering herd (Fig. 2) markedly increased the carrying capacity of the forage, with no effect on calf birth weight and a small effect on cow condition score and weight change, as compared with cows on unrestricted winter feed (Wilson et al., 1965). Energy was not a serious limitation on the tall fescue winter program for spring-calving brood cows (Wilson et al., 1967), but for fall-calving cows supplemental energy was needed during the second to fifth month of lactation (late winter and early spring) (Drewry et al., 1968). In the latter study, feeding 1.36 kg (3 lb/day/cow) ground shelled corn resulted in less weight loss, 11.4 kg (25 lb)/cow as compared with the check group for the 84-day trial period. Conception rates were similar; 82 and 79%, respectively.

Ohio studies under climatic conditions generally similar to southern Indiana confirmed the Indiana research (Van Keuren, 1970 a, 1972; Van Keuren and Parker, 1974 b). 'Kentucky 31' tall fescue fields were established for winter utilization at several locations in southern Ohio. Initially seeded with tall fescue and ladino clover, the clover contributed about 10% of the forage the first year and then largely disappeared. After the second year, the fields received annual applications of 74 kg/ha N (200 lb/acre ammonium nitrate) in early spring (March to April) and again in August. The first growth was harvested as fully headed hay in early to mid-June and left in the fields as small round bales. Subsequent regrowth was accumulated until the beginning of wintering in mid-November. The total dry matter accumulated each year for 7 years at one location averaged 8.5 metric tons/ha

Fig. 11-2.—Pregnant beef cows on Kentucky 31 tall fescue winter pasture of first crop field-stored small round bales and regrowth in southeastern Ohio. Utilization is controlled on this pasture by dividing winter pasture into six lots. (Ohio Agric. Res. and Dev. Center photo).

(3.8 tons/acre); 4.5 metric tons from baled forage, and 4.0 metric tons from regrowth. Winter pasture provided feed for an average of about 3.3 mature pregnant beef cows per ha (0.74 acre/cow), 509 cow-days per ha (207/acre) from mid-November to mid-April (151 days), Table 11-1). No supplement was fed; only salt, mineral, and water were supplied. The average percent crude protein for the hay was 9.7% (range 7.3 to 13.2) and for the regrowth, 11.0% (7.9 to 13.4). The pastures were subdivided into smaller lots with electric fence to control consumption. The winter-pasture herds were compared with barn-fed herds. The winter-pasture groups had available an average of 1.4 to 1.6 kg of forage dry matter/day/45 kg of body weight (3 to 3.5 lb/100 lb body weight) as compared with 0.9 kg/day/45 kg body weight (2 lb/100 lb body weight daily) fed daily to the barn-fed group receiving grass-legume hay. The greater amount of forage for the winter-pasture cows represented a combination of over-consumption and forage loss, although it was observed that both the bales and regrowth were generally cleaned up satisfactorily. Both groups gained weight during the winter, and then lost weight after calving, with the barn groups generally showing a slightly higher condition than the winter-pasture groups at the end of the wintering period. However, calving dates, calf birth weights, and calving percentages were similar. The winter pasturing program markedly saved labor and equipment costs, requiring only 25% as much labor as the barn-wintered herds. The tall fescue utilized in the winter grazing programs had excellent stands at the end of the six winters of grazing (Van Keuren, 1972), and subsequently after nine winters (R. W. Van Keuren, unpublished data), despite occasional wet periods when the sod was severely damaged by the animals.

The in vitro dry matter digestibility (IVDMD) and percent soluble carbohydrates of the regrowth decreased markedly by late winter (Van Keuren and Parker, 1974 b). The baled hay left in the fields did not change in composition other than the outer deteriorated layer. The Ohio winter pastures began to show a severe winter-burn by February and March. The data suggest that for best forage utilization and animal nutrition, the regrowth should be grazed in late fall and early winter, probably best utilized by January.

Subsequent studies in Ohio (R. W. Van Keuren and C. F. Parker, unpublished data) indicated that with first- and second-calf heifers, and particularly with dairy-beef crosses, the quality of hay is very important. Even with good quality hay in the bales, it may occasionally be necessary to provide supplemental energy and protein in late winter and early spring when the young cows are lactating, making some growth yet, and coming into the period where the condition for breeding is important.

Wheaton (1974) in Missouri reported that the beef cow carrying capacity of tall fescue-ladino clover winter pasture with no N was 385 cow-days/ha (156 lb/acre); with 112 kg of N/ha (100 lb/acre), 492 cow days/ha (199); and with 224 of N kg/ha (200 lb/acre), 531 cow days/acre (215). The data suggested more of an advantage for N fertilization of winter pastures than of summer pastures as an alternative for using clover as a source of N.

In an Oklahoma study (Fuller et al., 1971), pregnant beef cows were wintered on tall fescue pasture supplemented with 1.6 kg/head/day (3.5 lb) of bermudagrass hay and wheat straw (*Triticum aestivum* L.). Weight gain was about 50 kg (109 lb)/cow from 1 November until calving in mid-February through late March. The calves averaged 34 kg (75 lb) in weight at birth, 2.3 kg (5 lb) above the average for cows on dry range grass. The researchers suggest using tall fescue winter pasture in a rotation system to provide protein for cows grazing dry range grass.

Hereford and Angus heifers on several wintering programs in Kansas with grain and soybean meal had the following ADG: tall fescue winter pasture, 499 g (1.10 lb); native hay, 381 g (0.84 lb); bermudagrass hay, 477 g (1.05 lb); a hybrid sudangrass (*Sorghum sudanense* (Piper) Stapf.) hay, 536 g (1.18 lb); hay of two sorghum × sudangrass crosses (*Sorghum bicolor* (L.) Moench × *S. sudanense*), 636 g (1.40 lb) and 676 g (1.49 lb), respectively (Chyba and Boren, 1974b). Grazing tall fescue was a good alternative to drylot roughage feeding in terms of labor requirements and animal performance.

Fall-Calving

As pointed out by Parker and Van Keuren (1975), fall-calving has some advantages over spring-calving and may fit into many beef cow-calf operations. However, fall-calving cows require a higher level of nutrition during the winter than do spring-calving cows. In the Ohio studies, legumes (red clover, alfalfa, and ladino clover) were incorporated into the winter pasture programs, either by no-tillage drilling or over-seeding of legumes into established tall fescue, or by using tall fescue-legume mixtures in new seedings. A newly established stand of Kentucky 31 tall fescue-alfalfa was used in a wintering program for a fall-calving herd. The cows calved on summer pasture at an average calving date of 10 September, with an average calf birth weight of 36 kg (80 lb). The herd was placed on the winter pasture 1 November. The hay had been harvested twice and baled as large round bales, averaging 527 kg (1,161 lb), and left in the pasture. The herd wintered on the hay plus fall-accumulated regrowth with no supplemental feed. The hay averaged 13.3% crude protein, the regrowth 10.9%. The calves were not creep-fed. The winter pasture was divided into smaller lots with an electric fence. Cows remained in good condition throughout the winter. The herd went on orchardgrass-alfalfa summer pasture 18 April and calves were weaned on 23 June, averaging 271 kg (596 lb). The average daily gain from birth until weaning (286 days), was 822 g (1.81 lb), with a 205-day adjusted weight of 201 kg (442 lb). This preliminary report indicated that fall-calving can be combined with a winter pasture program in this northern region.

In a 4-year Kentucky study, fall-calving Angus cows were wintered at a stocking rate of 0.81 ha/cow (2 acres/cow) on tall fescue-ladino clover pasture supplemented with corn silage, limited hay, and limited ground-shelled corn, and with and without creep-feeding [G. M. Hill, N. W. Bradley, J. M. Boling, D. R. Lovell, W. C. Templeton, Jr., and D. D. Kratzer. 1975. Forage and creep effects on fall-born calf performance. J. Anim. Sci. 40:193 (Abstr.)]. The average adjusted calf weaning weights for calves weaned at about 290 days were 229 kg (504 lb) for calves on tall fescue-ladino clover with no-creep and 271 kg (598 lb) on tall fescue-ladino clover with creep for 238 days. The calves on creep consumed, on the average, 524 kg (1,155 lb) of feed (cracked yellow corn).

Feeding Large Bales and Stacks

The earlier studies in the United States with field-stored tall fescue hay for wintering beef cows involved using small, round, tied bales made with an Allis Chalmers 'Rotobaler' (Peterson et al., 1964; Van Keuren, 1970a). Recently, a major development has occurred in hay handling involving equipment that makes round bales, generally in the range of 500 to 700 kg (1,000 to 1,500 lb), and stacks of various sizes, from 1 to 10 metric tons. These have been used for handling and storing hay for wintering both cattle (Van Keuren et al., 1973; Smith et al., 1974) and sheep (Van Keuren and Parker, 1973; Parker and Van Keuren, 1973). Tall fescue hay has been fed in several ways, as bales left in the field, stored adjacent to the pasture and fed back, or fed in drylot. Feeding losses of small, round bales left in the field have been small (Van Keuren et al., 1973), but the larger bales or stacks should generally be fed in limited quantities, in racks, or with accessibility to the hay controlled by an electric wire for best utilization (Lechtenberg et al., 1974). Waste with hay stacks fed without a rack can be high (Renoll et al., 1971). These recent developments in hay handling and storage will have a great impact on winter feeding of beef cow herds.

In summary, tall fescue is outstanding in a wintering program for beef cows in northern and central regions such as Kentucky, northern Missouri, southern Illinois, southern Indiana, and southern Ohio. It is productive, very acceptable to the livestock during this period, and withstands treading damage during winter feeding. In northern areas, such as Ohio and Indiana, setting aside a field for wintering is probably the best practice, harvesting several hay crops and field-storing for feeding to supplement the fall and winter regrowth. With good quality hay, and particularly if a legume is included, grain and protein supplement probably will not be needed, except for young cows.

C. Tall Fescue in Year-Round Programs for Beef Cows

Tall fescue is widely used in year-round programs for beef cows in the southeastern U.S. from eastern Kansas and Oklahoma eastward and from southern Iowa across to Virginia southward. In this region, it may be used for both summer and winter programs as in Missouri and Kentucky, or as a winter grazing grass, alternated with warm-season grasses for summer grazing in the more southern states. It is not widely used in the Gulf Coastal region because summer temperatures limit production and persistence. In more northern areas, such as Kentucky, northern Missouri, southern Ohio, southern Indiana, and West Virginia, winter feed depends on summer and fall growth. In more southerly regions, such as Arkansas, northern Alabama, Georgia, and South Carolina, growth of tall fescue occurs during the winter to provide winter grazing for beef cow herds. In the Pacific Northwest and in the intermountain irrigated valleys, tall fescue is used primarily as summer pasture, or grown with subclover for year-long grazing in western Oregon and Washington (Bedell, 1970).

In year-round programs, tall fescue can be alternated with bermudagrass during the year, as in Arkansas (Ray et al., 1969, 1972), Mississippi (Morrison and Gill, undated mimeo), Alabama (Hoveland et al., 1969), South Carolina (Jutras et al., 1975), and other states. It can be interseeded into bermudagrass to extend the grazing season, as in Oklahoma (Rommann et al., 1973), or used as year-round pastures, as in Missouri (Redmon, 1974), southeastern Kansas (Hyde and Kilgore, 1973; Chyba and Boren, 1973), and in the southeastern Piedmont and mountain area (Stuedemann et al., 1975). Tall fescue can also be used for winter pasture with orchardgrass or Kentucky bluegrass used for summer pasture as in Ohio (Van Keuren, 1972) and Virginia (Hammes, 1975). Tall fescue may be grown in pure stands, or in a mixture with a legume(s) such as white clover (Ray et al., 1969, Arkansas; Hammes, 1975, Virginia), ladino clover (King et al., 1953, South Carolina; Burns et al., 1973, North Carolina; Anderson and Safley, 1966, 1967, Tennessee; Thompson, 1974, Missouri), alfalfa (Parker and Van Keuren, 1975, Ohio), red clover (Hammes, 1975, Virginia), and subclover (Bedell, 1970, Oregon). The value of including a legume with grasses has long been recognized by both agronomists and animal nutritionists, and the interseeding of legumes into established tall fescue stands is currently receiving widespread attention. The value of legumes in winter programs needs further evaluation as a replacement for N fertilizer.

An Arkansas study was initiated to determine the forage management needed to provide optimum year-round nutrition for beef cow and calf herds (Ray et al., 1969). Upland pastures of warm-season grasses, common bermudagrass, dallisgrass (*Paspalum dilatatum* Poir), white clover, and common lespedeza (*Lespedeza striata,* (Thunb.) H. and A.) were alternated with bottomland pastures of primarily tall fescue and white clover, with some dallisgrass, using three stocking rates: light (1.22 ha/cow, 3.0 acres), moderate (0.81 ha/cow, 2.0 acres), and heavy 0.51 ha/cow, 1.25 acres), (Table 11-1). Within each stocking rate, high and low levels of protein supplement were fed, each with and without trace minerals. The Hereford cows were bred to calve in late fall and early winter. The heavy stocking rate tended to weaken the tall fescue stand, but with a 50% ground cover of white clover, this was not regarded as serious. In both the upland and bottomland pastures, the stand of the permanent grasses and of white clover improved each year, despite no fertilizer applications. Heifers raised on these pastures after weaning but before first calving benefited from summer protein supplementation, especially heifers on the heavily stocked pastures. From the standpoint of land utilization, the heavy stocking rate was most desirable. Calf birth weights were not affected by grazing pressure, protein supplement level, or trace mineral supplementation. For young cows suckling their first calves and grazing heavily stocked pastures, a protein supplement (about 0.45 kg (1 lb) cottonseed meal per head daily) from 1 October through 31 March was of benefit. Cows on heavily stocked pastures calved an average of 10 days later than did cows on pastures stocked at moderate or light grazing pressures and their calves averaged about 32 kg (70 lb) lighter and a grade lower at weaning than calves in the other two herds.

In a subsequent study (Ray et al., 1972) on similar pastures the value of a forage creep for the calves, which included wheat during January to April followed by hybrid sudangrass during the summer until weaning in August, was compared with fertilizing the perennial pastures with a complete fertilizer (N, P, and K) annually. This study also evaluated the practice of creep-feeding grain to calves and of grain supplementation to the cows during the winter months, January to April. All cows were fed 0.9 kg (2 lb) of cottonseed meal per head daily from 1 October through 10 April each year. Some hay was fed in winter to supplement the pasture. A heavy stocking rate of 0.51 ha/cow-unit (1.25 acres) was used, with a moderate stocking rate of 0.81 ha/cow-unit (2.0 acres) as a control. Cows on moderately grazed pasture weaned an average of 204 kg of calf/ha (182 lb/acre) as compared with 327 kg/ha (292 lb/acre) produced by cows on heavily grazed pastures. Forage creep did not significantly improve weaning weight or grade of the calves nor did it reduce intake of creep-fed grain. Much of the creep forage was not utilized by the calves. Total costs of grain creep were more than the additional return realized from creep feeding. Feeding grain to the cows seemed to be of limited value, since it did not improve calf birth weights, calf gains, or percent calf crop, and the cows tended to accumulate unneeded condition. Stocking rates and pasture treatments did not significantly affect calf weaning weights. Conception rates, ranging from 83 to 98%, were generally similar for all treatments, as was percent calf crop weaned which averaged 84% for all herds for the 4 years.

Another Arkansas study (Ray et al., 1968) compared fall-calving (10 September to 1 December) vs. winter-calving (10 December to 1 March); "high" quality forage (a sequence including oat pasture, orchardgrass, johnsongrass (*Sorghum halepense* (L.) Pers.)-lespedeza, and sudangrass, with supplemental silage in winter) vs. "low" quality forage (bermudagrass in summer, tall fescue in fall and winter, with supplemental grass hay in winter); mixed-breed cows vs. purebred Herefords; sex of calves. Calves produced by cows on the lower quality forage program gained less, graded lower, and had a lower market value than did calves from the cows on the high quality forage, but the higher value of the latter calves did not offset the extra costs involved. The cows on the high quality forage were usually in better condition and were heavier than cows on lower quality forage. Calves from mixed-breed cows had a higher weight per day of age and market value but lower grade than did calves from purebred cows. Season of birth significantly affected weight per day of age but not grade and market value. However, the weight gains were higher for winter-born than for fall-born calves for the first 3 years of study but was the reverse for the last 2 years. Creep-fed calves gained faster, graded higher, and had a higher market value than those not on creep, but creep feeding costs generally were greater than the increased market value of creep-fed calves.

An economic study using linear programming was made of Arkansas beef cow/forage data to determine the least-cost method of producing and utilizing feed for beef cows on thin soils of the Arkansas River Valley upland area (Halbrock et al., 1971). The authors concluded that (1) the establishment of tall fescue was the first step in developing a low-cost program for beef cows; (2) deferring tall fescue production and using it for winter grazing was one of the best ways to decrease feed costs, with deferred winter grazing of tall fescue accounting for much of its importance in forage programs; (3) fertilizing the tall fescue used for deferred grazing was the most profitable use of fertilizer. Tall fescue used for deferred grazing should be fertilized at high levels, 90–67–34 kg/ha (80–60–30 lb/acre) or more of N, P_2O_5, and K_2O, before any fertilizer is used on other forage crops, according to this study.

Drilling winter oats (*Avena sativa* L.) into newly established tall fescue-white clover in South Carolina increased the carrying capacity of beef cows and provided soil erosion control (Jutras, 1968). A study underway in South Carolina on year-round programs (Jutras et al., 1975) showed no significant increase in calf weaning weights from feeding cows a corn and cottonseed meal winter supplement on a combination of tall fescue-ladino clover and bermudagrass pastures with stockpiled tall fescue in winter. In a Mississippi study (Morrison, E. G., and W. J. Gill. Undated. Brood herd management in a fall calving program, 1966–1970. Mississippi Brown Loam Branch Exp. Stn. Mimeo report) several year-round forage combinations using warm-season grasses, with tall fescue-'Regal' white clover or urea-silage in winter for spring- and fall-calving herds were compared. For spring-calving cows, the tall fescue winter forage required supplemental feeding with a urea-silage. With fall-calving cows, tall fescue-Regal white clover in winter and bermudagrass-dallisgrass in summer generally gave calf adjusted weaning weights and daily gains comparable with three other systems compared.

An Alabama study (Hoveland et al., 1969) showed that an economical beef cow-calf forage system with fall-calving in the Piedmont could be based on 'Serala' lespedeza (*Lespedeza cuneata* (Dumont) G. Don.) for summer grazing and 'Goar' tall fescue for fall, winter, and spring grazing, supplemented with Coastal bermudagrass hay in mid-winter.

In another Alabama study (King et al., 1971) year-round systems with fall-calving were compared at the Black Belt Substation in Central Alabama with several combinations of bermudagrass, dallisgrass, white clover, tall fescue, and caley peas (*Lathyrus hirsutus* L.). A combination of 0.6 ha (1.5 acre)/cow-unit of dallisgrass-white clover-caley peas for summer pasture and 0.2 ha (0.5 acre) of tall fescue accumulated for winter grazing or 0.4 ha (1 acre) of each were both satisfactory systems, with the former yielding a slightly higher dollar return for operator's labor and management per cow than the latter. A third Alabama study (Cope et al., 1972) compared warm-season grasses with legume or with N fertilizer, followed by several winter programs using Coastal bermudagrass hay, and one with tall fescue for winter grazing (summer growth cut as hay and fed when tall fescue grazing was inadequate). The latter system included winter-grazed rye (*Secale cereale* L.) as creep-grazing for calves. The tall fescue-rye creep treatment gave the highest calf winter average daily gain, adjusted weaning weight, and grade of the four systems. The calves were born from November to March. The use of legumes in the summer pastures was more profitable than using N fertilizer.

The value of creep-feeding of fall-born calves from Hereford cows on year-round tall fescue was evaluated in a Kansas study (Chyba and Boren, 1974a). The calves consumed an average of 460 kg (1,013 lb) of creep feed, 3.9 kg (9.5 lb)/day, and were weaned at an average of 232 kg (511 lb) with a 767-g (1.69-lb)/day gain as compared with 179 kg (393 lb) weaning weight and 604 g (1.33 lb) ADG for no creep. However, because of the unfavorable price relation between grain costs and beef prices, creep feeding was not profitable. In another Kansas study (Ibbetson et al., 1972), a year-round tall fescue program with Angus cows and fall-born calves, proved to be satisfactory with good-to-excellent calf weaning weights, 198 kg (436 lb) and 251 kg (553 lb) for 2 years. Trace mineral salt was fed and some hay was necessary during periods when snow prevented grazing the tall fescue pasture.

In a year-round program with tall fescue for beef cows in Missouri, Wheaton (1974) reported that the carrying capacity of tall fescue with no N fertilizer was 0.97 kg (2.4 acre)/cow, 373 cow-days/ha (151/acre); with 112 kg N/ha (100 lb/acre), 0.79 ha (1.96 acre)/cow, 460 cow-days/ha (186/acre); and with 224 kg N/ha (200 lb/acre), 0.73 ha (1.8 acre)/cow, 502 cow-days/ha (203/acre).

In summary, tall fescue fits well in year-round forage programs for beef cows, and probably best when alternated with other grass species. Tall fescue fits well with a program of warm-season grasses in the summer in the southern U.S. or with cool-season grass pastures, such as orchardgrass, in summer in more northern regions with tall fescue hay used for wintering, or as a combination of winter grazing and hay. Legumes should be used with it where possible in the summer pastures. In the southern states (Georgia, Alabama, Mississippi, and South Carolina) the value and place of legumes are inconclusive. In this region legumes tend to go dormant in late summer and in winter, insect and disease problems are more prevalent, and problems of grass competition and establishment are greater than in the mountains or more northern regions. Legumes should also probably be used in winter forage programs utilizing tall fescue, but more data are needed.

TALL FESCUE FOR SHEEP PRODUCTION

Tall fescue is not used frequently for sheep as summer pasture, but has more acceptance for winter grazing or winter feed programs. Research, however, indicates that with proper management and when used in combination with legumes, tall fescue is a better grass for sheep than is generally recognized. Tall fescue has received limited research attention as a forage for sheep in the United States, Canada, Great Britain, Australia, New Zealand, and South Africa. In the United States, sheep production is primarily in the upper Midwest and the West.

A. Summer Pasture for Sheep

Tall fescue may have better acceptability by sheep than is generally assumed, particularly if the forage is kept at a short succulent stage and if grown with palatable legumes. In an early Kentucky study, yearling wethers consumed similar amounts of orchardgrass and Kentucky 31 tall fescue, but wether lambs consumed significantly more orchardgrass than tall fescue (Forbes and Garrigus, 1950). The differential consumption of lambs and yearlings was attributed to the relatively tender mouths of the lambs as compared with yearlings and to the fact that midsummer tall fescue is much coarser, stiffer, and harsher than orchardgrass. A mixture of grasses and legumes, including tall fescue, was found satisfactory as irrigated pasture for sheep in South Africa (Mostert, 1956). However, when used in mixtures tall fescue may become dominant as shown in a Canadian study (Wilson and Clark, 1961). In comparison with several other pasture mixtures, tall fescue was used as a component of an orchardgrass-tall fescue-reed canarygrass (*Phalaris arundinacea* L.)-alfalfa irrigated pasture mixture in Alberta, Canada, for ewes. Tall fescue assumed dominance of the sward during the 6 years of the study, with the orchardgrass component decreased because of selective grazing and greater winter-killing as compared with the tall fescue. Reed canarygrass also became a minor component because of the competitiveness of the tall fescue. These results suggest that combining tall fescue with other cool-season grasses is not desirable and that such mixtures are difficult to manage. This mixture was very productive for 3 years, but as the amount of alfalfa decreased, daily gains of the sheep and gains per hectare also decreased, although carrying capacity increased (Clark and Wilson, 1966).

In a grazing trial with young wethers in New South Wales, orchardgrass, perennial ryegrass (*Lolium perenne* L.), and 'Demeter' tall fescue were compared individually and with white clover (Gallagher et al., 1966). Grass and clover were significantly ($P < 0.01$) superior to grass alone in liveweight gains and in wool production. Gains were higher on orchardgrass and ryegrass than on tall fescue, but gains on the three grasses with clover were similar. The best gains were about 136 g/day (0.30 lb). In a subsequent study, the grass-clover mixtures were again superior to the three grasses alone in liveweight gains from weaned crossbred wether lambs (5 to 9 months old), with lambs on the grass-clover averaging 154 g/day (0.34 lb) and on the grasses 145 g/day (0.32 lb), which were significantly different ($P < 0.05$) (Grimes et al., 1967) (Table 11-2). On the three grasses, the ADG were 136 g (0.30 lb)/, 150 g (0.33 lb)/, and 159 g (0.35 lb)/day, respectively, on orchardgrass, tall fescue, and ryegrass; each significantly different (P

Table 11-2—Sheep performance on summer and winter pasture.

Species	Performance				
A. Fattening lambs†	Liveweight gain/day, Oct.–Dec.				
	g				
Orchardgrass	136				
Tall fescue	150				
Perennial ryegrass	159				
White clover	177				
Above three grasses					
grasses alone	145				
grasses + clover	154				
B. Winter pasture‡	Pre-lambing 19 Nov.–15 Feb. Pregnant ewes			Lambs, average wt.	
	Initial wt.	Final wt.	Gain	Birth wt.	30-day wt.
	kg				
1. Barn group: grass-legume hay plus corn-soybean oilmeal during late pregnancy and after lambing	70	79	9.5	4.49	11.3
2. Winter pasture: tall fescue second crop bales and regrowth	70	79	9.5	4.54	10.5

† Grimes et al., 1967; New South Wales, Australia. Dorset Horn × Merino wethers, 5 months old at beginning of trial. Urea fertilizer on grasses. ‡ Parker et al., 1971; Ohio. Four weeks before lambing, pasture ewes put in barn on first crop tall fescue silage and the same concentrate as the barn group.

<0.05). ADG on clover alone was 177 g/day (0.39 lb). In contrast with the previous study, the lambs on orchardgrass grew considerably slower than those on ryegrass and tall fescue throughout the experiment. In another comparison using young weaned crossbred lambs 2 to 3 months old with an average weight of 16 to 18 kg (35 to 40 lb) to graze Demeter tall fescue receiving high and low rates of N fertilizer, no differences in estimated digestible organic matter intake, body composition, or in liveweight gains were found between N treatments (Grimes, 1967). Liveweight gains of about 160 g/day were obtained (0.35 and 0.36 lb/day, respectively, for the high and low N rates). Irrigation was used to keep the pastures growing and regular mowing was followed to keep the herbage in a short vegetative stage. A set stocking rate was followed with the same number of animals on each pasture.

In western Oregon and Washington, the annual legume subterranean clover grows well with perennial ryegrass and 'Alta' tall fescue to provide nutritious forage for grazing cattle and sheep (Bedell, 1968, 1970). The clover-ryegrass is more often used as sheep forage and clover-tall fescue for beef cattle, but both kinds of stock often graze both mixtures. An Oregon study showed that the clover-tall fescue mixture was best utilized as pasture for sheep and cattle or cattle alone than for sheep alone (Bedell, 1971). Cattle grazing subclover-tall fescue pasture preferred the grass to the clover, the converse was true of sheep (Bedell, 1973). A cattle-sheep ration of 2:1 was a better combination for companion grazing than was a 1:1 ratio because at the latter ratio the sheep grazing pressure was too great to satisfactorily maintain the subclover which must set seed annually to persist.

In a study with Merino lambs grazing four temperate grasses in New South Wales, lambs grazing orchardgrass had lower liveweight gains than did lambs grazing hardinggrass (*Phalaris tuberosa* (Hack.) Hitchc.), 'Oregon' tall fescue, or perennial ryegrass (Hamilton et al., 1970). Lambs grazing tall fescue had satisfactory growth rates during the entire trial; lambs grazing on the other grasses had periods of poor or intermediate growth. Demeter tall fescue and alfalfa have been used successfully as a summer pasture for fattening lambs in southeastern South Australia (Lawton, 1974).

In general, it appears that tall fescue used as summer pasture for sheep should be grown as the only grass and with a legume. It should be grazed closely and frequently to keep it vegetative and more palatable, but a 5- to 10-cm (2- to 4-inch) stubble should be left for rapid recovery and constant overgrazing should be avoided. In all of the trials, only fair-to-good average daily gains of lambs were obtained, 0.14 to 0.18 kg (0.30 to 0.40 lb). Tall fescue may serve best in summer as a maintenance pasture for dry ewes.

B. Tall Fescue for Wintering Sheep

As with cattle, tall fescue is more palatable to sheep during fall, winter, and spring than during the summer. This fact, plus the value of extending the grazing season and providing less labor intensive winter feed, has resulted in tall fescue being studied for sheep production in the same areas where tall fescue wintering programs are being used for beef cow herds. In general, tall fescue has proved to be a satisfactory forage either for winter grazing, as a combination of grazing and field-stored bales, or as silage for wintering pregnant ewes.

Studies in Missouri compared Alta tall fescue as fall-deferred winter pasture for pregnant ewes (Thompson et al., 1955). During the last 60 days of gestation, half of the flock received a concentrate feed plus the pasture. The concentrate-fed ewes gained more during pregnancy, bore heavier lambs, and produced more milk and wool, but the difference over the group not receiving concentrates was not sufficient to justify the additional feed. The subsequent performance of the lambs during fattening was not affected by ewe wintering treatment. The tall fescue winter pasture was considered adequate for wintering pregnant ewes. These results were verified in another similar Missouri trial in which the ewes grazed Kentucky bluegrass until 21 November, followed by tall fescue-ladino clover winter-grazing (Ross et al., 1967).

Winter-grazing of tall fescue-meadow fescue-white clover pastures by pregnant ewes satisfactorily replaced hay and silage in British studies (Alder and Redford, 1958). The pasture was stockpiled from late July to early August and grazed between November and February. Nitrogen was applied

Fig. 11-3—Pregnant ewes on Kentucky 31 tall fescue winter pasture of second crop field-stored small round bales and fall regrowth in northern Ohio. Note electric fence used to divide pasture for controlling utilization. (Ohio Agric. Res. and Dev. Center photo).

in August. Sheep readily accepted the forage. Subsequent close-grazing of the pasture from April through July was followed to keep the forage palatable and to counteract the adverse effect of the long autumn rest period on the white clover. Under this management, sward composition remained satisfactory.

In Ohio studies, tall fescue was found to be satisfactory for in-field forage conservation for wintering pregnant ewes as an alternative to the common practice of wintering in the barn on a hay-grain ration (Parker and Van Keuren, 1970; Van Keuren and Parker, 1971; Parker et al., 1971; Van Keuren and Parker, 1974a). In northern Ohio, where the studies were conducted, mid-winter temperatures generally average about 0 C (32 F), occasionally decreasing to −10 to −20 (4 to 14 F), with periods of snow cover. The forage growing season is from April to October. Late summer and fall-accumulated tall fescue used for late fall and winter grazing or as a combination of field-stored round bales and fall regrowth was satisfactory without supplementation for wintering mixed aged pregnant ewes (Fig. 3) from November until 2 to 4 weeks before lambing, generally in February. The fields were divided so that each pasture was utilized in 2 to 3 weeks. The ewes gained weight and produced satisfactory lambs, comparing favorably in results to barn-fed ewes on several diets, including the standard hay and concentrates ration commonly used by Ohio sheep producers (Table 11-2). To provide hay and regrowth of adequate quality and protein level for pregnant ewes, the first crop each season was harvested in early June as hay or silage. The subsequent regrowth was harvested as round tied bales of about 18 to 28 kg (40 to 60 lb) in August and left in the field for the wintering period. The bales and subsequent fall regrowth had crude protein values from 10 to 15% and dry matter yields of 3.4 metric tons/ha (1.5 tons/acre) for each harvest from vigorous stands of tall fescue receiving two applications of N totaling 148 kg/ha (133 lb/acre) annually. The tall fescue silage was satisfactory with concentrate supplementation for the ewes for feeding during late pregnancy and early lactation before going on summer pasture. Studies of the regrowth during the winter growing period showed a decrease in nutritive value of the grass (Van Keuren and Parker, 1974a). Utilization by midwinter (1 January) of the accumulated fall regrowth would be advisable from the standpoint of ewe nutrition and for best utilization of the forage. Excellent acceptability of the tall fescue was noted under the Ohio program. During periods of snow cover, the sheep consumed the bales with no difficulty and also generally continued to graze under the snow. Snow depths up to 46 cm (18 inches) and sub-freezing temperatures were experienced with no problems, despite the fact that no housing or protection was provided. A major savings in labor and equipment for hay handling and storage as well as manure disposal resulted from wintering in the field.

The superior keeping qualities of tall fescue and orchardgrass for late autumn and early winter for sheep was observed in New Zealand studies (Vartha and Clifford, 1971). Levy (1970) considered tall fescue an undesirable pasture species in New Zealand. It was found to be productive and persistent as a winter forage in fat lamb production in New South Wales (Burch, 1973).

Feeding tall fescue hay in bunks in drylot at night with access to tall fescue winter pasture during the day decreased hay feeding waste as compared with leaving the bales in the winter pasture in Missouri studies (McClure and Ross, 1974). A ground corn-biuret supplement was fed to both groups of late-lambing (May to June) ewes during the study (3 February to 5 April). Ewes fed hay in drylot ate less hay and more grass than did those with access to bales in the pasture. The group fed hay in drylot in racks consumed an average of 0.60 kg (1.3 lb) of grass and 0.31 kg (0.7 lb) of hay daily; the pasture group consumed about 0.45 kg (1.0 lb) of both grass and hay. Both groups consumed about 0.3 kg (0.7 lb) of supplement daily per ewe. All ewes subsequently produced strong lambs and had good milk production, with no significant difference in birth and weaning weight of lambs or ewe fleece weight. The tall fescue hay was first growth, baled in July, and the subsequent regrowth was accumulated until grazing in mid-winter. Both forages would be more mature and lower in quality than that used in the Ohio studies where no supplement was needed. In another Missouri study similar to the above, salt was added to the corn-biuret to limit daily consumption (Morrison et al., 1974). Generally, the results were similar to the earlier study. Feeding hay in drylot resulted in less wastage, but required more labor.

In summary, tall fescue fits sheep production for winter feeding much better than for summer pasture. Research shows that used for winter grazing or in combination with hay and winter grazing, tall fescue is satisfactory for pregnant ewes before lambing. If high quality feed is obtained by early harvests for hay and by not delaying utilization of regrowth too late into the winter, supplemental feed is probably not necessary until late pregnancy.

TALL FESCUE FOR DAIRY CATTLE

Despite its many desirable agronomic characteristics and the millions of acres seeded, primarily in the southeastern U.S., dairymen need to evaluate the use of tall fescue in their forage programs rather carefully because of problems reported with intake and palatability, as well as variability in milk production. Both favorable and unfavorable results have been obtained with tall fescue for dairy cattle, reflecting such factors as differences in management and utilization, amount of legume present, weather conditions, time of year grazed, supplements fed, and other factors. Cowan (1956) reviewed the earlier literature and concluded that tall fescue was acceptable and palatable to livestock, including dairy cattle, when properly managed and grown on soil with good fertility.

A. Summer Pasture for Dairy Cattle

For lactating cows, milk production is the most critical measurement of forage value. The response of dairy cattle to tall fescue as compared with other forages has been variable. Several grass-legume mixtures were compared as summer pasture for lactating dairy cows in Virginia (Thompson and Holdaway, 1954 (Table 11-3). In the first year when ladino clover was predominant in the mixture, tall fescue-ladino clover pasture produced significantly ($P < 0.05$) higher average daily 4% fat corrected milk (FCM) per cow, 15.9 kg (35.0 lb), than did orchardgrass-alfalfa, 14.8 kg (32.5 lb) or orchardgrass-ladino clover, 14.9 kg (32.9 lb). Cows on all three pastures decreased in average daily milk production the third year, with those on the tall fescue-ladino clover having the lowest average daily FCM, 10.8 kg (23.7 lb), significantly lower ($P < 0.05$) than the other two mixtures. The decrease in milk production on the tall fescue-ladino clover pastures reflected a decrease in the amount of ladino clover and the increased growth of the tall fescue. The milk production level obtained from all the pastures are only average as compared with general milk production in the United States.

Pure tall fescue fertilized with 74 kg/ha (66 lb/acre) of N annually (half in April and half in July or August) was inferior to Kentucky bluegrass-ladino clover, tall fescue-ladino clover, and orchardgrass-ladino clover as

Table 11-3—Dairy cattle performance on summer pasture.

Species	Performance			
	Avg. daily 4% FCM, kg*			
A. Grass-legume mixtures#	1st year	2nd year	3rd year	Average
Orchardgrass-alfalfa	14.8 a	16.4 b	12.3 a	14.5
Orchardgrass-ladino clover	14.9 a	18.2 a	12.4 a	15.2
Tall fescue-ladino clover	15.9 b†	15.7 b‡	10.8 b‡	14.1
B. Tall fescue-ladino clover††	1st year	2nd year	3rd year	Average
Average daily 4% FCM, kg	17.3	15.1	14.9	15.8
Total standard cow-day/ha	309	262	213	262
Average daily carrying capacity/ha	1.48	1.19	1.19	1.28

C. Grass vs. grass-legume‡‡	3-year average	
	Persistence of milk production average for 4th week	Estimated TDN kg/ha
Orchardgrass-ladino-clover	98.3	1,935
Tall fescue-ladino clover	89.2	2,005
Bluegrass-ladino clover	87.2	1,980
Tall fescue	81.4	2,609

D. Grass species§§	3-year average				2-year average
	DM %	DMD %	Crude protein digestibility%§	Crude protein %	Daily intake¶ kg
Orchardgrass	27.7	67.1	74.4	22.6	11.0
Smooth bromegrass	27.0	66.0	77.1	25.6	10.2
Bluegrass	36.1	65.8	71.7	19.2	10.1
Tall fescue	27.1	65.7	73.9	21.3	9.1

* Means within years followed by the same letters are not significantly different, $p < 0.05$.
† Heavy ladino clover stand.   ‡ Reduced ladino clover stand.   § Dry matter and crude protein digestibilities determined by fecal chromogen technique.   ¶ Forage dry matter intake by fecal chromogen and chromic oxide techniques, in kg/450 kg body weight.   # Thompson and Holdaway, 1954; Virginia.   †† King et al., 1953; South Carolina, Ky-31 tall fescue, 129 kg N/ha annually.   ‡‡ Seath et al., 1954; Kentucky. 74 kg N/ha on tall fescue.   §§ Lassiter et al., 1956; Kentucky. 112 kg N/year in three applications.

pasture for lactating Jerseys and Holsteins in a Kentucky study (Seath et al., 1954). During a comparative test period in which the cows were on pasture plus a 16% protein supplement, cows on orchardgrass-ladino clover averaged 14.3 kg (31.6 lb) 4% FCM daily, on Kentucky bluegrass-ladino clover 13.9 kg (30.7 lb), on tall fescue-ladino clover 13.3 kg (29.3 lb), and on tall fescue 11.2 kg (24.6 lb). A major factor influencing milk production was the amount of clover in the pastures. Volunteer clover appeared in the tall fescue pasture the second and third years of the study with a resulting increase in average daily milk production over the first year. Competitive growth of the tall fescue the second and third year in tall fescue-ladino clover pastures decreased the ladino clover to one-third or less of the original stand, resulting in a corresponding decrease in milk production. Except for tall fescue-ladino clover, the highest average daily FCM production was obtained the third year, about 13.6 kg (30 lb) on tall fescue plus volunteer white clover, about 16 kg (35 lb) on Kentucky bluegrass-ladino clover, and almost 17 kg (37 lb) on orchardgrass-ladino clover. The tall fescue-ladino clover yielded virtually the same as the tall fescue plus volunteer white clover the third year, a decrease from the first year high.

A Virginia study with Holstein dairy cattle (Blaser et al., 1969) compared Kentucky 31 and 'Kenwell' tall fescue as pasture with grain fed at three grain:milk ratios (0, 1:9, 1:3). Nitrogen fertilizer at the rate of 224 kg/ha (200 lb/acre) was applied annually in split applications. The average daily 4% FCM production for 3 years (1963-65) for 0, 1:9, and 1:3 grain to milk ratios for Kentucky 31 tall fescue was 13.4 kg (29.5 lb), 13.7 kg (30.1 lb), and 17.5 kg (38.5 lb). The milk production from the Kenwell tall fescue was lower, probably resulting from "fescue-foot," which was noted for several cows on the Kenwell tall fescue pastures. Grain concentrate feeding did not affect the incidence of the "fescue-foot."

B. Winter Pasture for Dairy Cattle

Several studies have evaluated tall fescue as fall, winter, and early spring grazing for lactating dairy cows. An Ohio study was conducted to evaluate this grass for late fall grazing for lactating Jersey cattle (Pratt and Haynes, 1950). Bluegrass-ladino clover was superior to Kentucky 31 tall fescue-ladino clover in one trial; meadow fescue-birdsfoot trefoil (*Lotus corniculatus* L.)-ladino clover was superior to tall fescue-ladino clover in a second trial; barn-feeding of hay, silage, and grain was superior to tall fescue-ladino clover pasture plus supplemental hay, silage, and grain in a third study. Lower palatability and intake of the Kentucky 31 tall fescue was suggested as the reason for lower milk production from tall fescue compared with Kentucky bluegrass in a reversal trial. An examination of the weather records for this period (Ohio Agric. Res. Dev. Ctr. Weather Reports, September-November 1949) indicated a warm, dry fall with below average precipitation for September, October, and November. October was 2.4 C (4.4 F) above the 62-year mean. These conditions may account for lower acceptability of the tall fescue.

In a South Carolina study on the value of tall fescue and ladino clover as a permanent winter pasture for lactating cows, a mixed dairy herd of Guernseys, Holsteins, and Brown Swiss grazed Kentucky 31 tall fescue-ladino clover pasture (about 25% clover) and received a grain-protein-mineral supplement (King et al., 1953). The above daily 4% FCM per cow for 3 years was 15.8 kg (34.8 lb) (Table 11-3). The cattle were on pasture year-round, but grazed during the day only during colder winter periods. Holsteins and Guernseys in a similar concurrent trial on Alta tall fescue-ladino clover pasture (average 63% clover) averaged 16.7 kg (36.7 lb) 4% FCM daily for 2 years, as compared with 14.8 kg (32.6 lb) for a barn-fed control group. The researchers concluded that the percentage of ladino clover in the stands determined to a large extent the value of tall fescue-ladino clover mixtures for milking cows. They suggested that tall fescue best fits into a dairy program under South Carolina conditions by supplying grazing during the fall, with additional grazing as it becomes available throughout the winter and spring. Excess forage in May and June may be stored as grass silage.

Tall fescue as a late fall and early winter pasture (19 November to 17 December) resulted in decreased milk production as compared with a temporary winter pasture of ryegrass-oats-crimson clover (*Trifolium incarnatum* L.) in Georgia (McCullough et al., 1952, 1953). No difference was found in spring grazing (21 March to 8 May) in Alabama in average daily FCM of cows grazing (1) tall fescue, (2) oats, (3) tall fescue plus alfalfa hay, and (4) oats plus alfalfa hay, with a mean daily FCM production of 15.8 kg (34.7 lb), 101.1% of that of the standardization period (Rollins et al., 1960). In the fall (30 October to 13 November), the mean daily FCM for tall fescue, 20.9 kg (46.0 lb), was not significantly different from cows on alfalfa fed in drylot. Grain was fed in all studies at a grain to FCM ratio of 1:4. The alfalfa hay with a mean of 23% crude protein was chopped. In a Kansas study on tall fescue and smooth bromegrass daily average FCM production of Holsteins in late lactation in October was 10.9 kg (24.1 lb) and 11.6 kg (25.5 lb), respectively (Boren et al., 1968). The cows received a 16% crude protein supplement at a ratio of 1:3 grain to FCM, an average of 4 kg (9 lb)/cow daily. The cows in both groups gained 24.5 kg (54 lb) in weight during the 21-day trial. They consumed an average of 14.5 kg (32 lb) of tall fescue dry matter daily, as compared with 18.6 kg (41 lb) of smooth bromegrass.

C. Persistency of Milk Production, Intake, and Digestibility

Persistence of milk production is an indication of the effect of the forage on maintaining or stimulating milk flow. Several researchers have indicated that tall fescue decreased milk persistency and that this may be the result of decreased intake or lower digestibility of tall fescue as compared with other forages. The high acceptability and intake of white and ladino clover probably explains why the presence of legumes with tall fescue improves animal performance as compared with tall fescue alone.

Dairy cattle were more persistent in milk production while grazing a temporary winter pasture of ryegrass-oats-crimson clover than on a Kentucky 31 tall fescue winter pasture in Georgia (McCullough et al., 1952, 1953). A markedly greater dry matter intake of the temporary pasture was found as compared with the tall fescue pasture, although a greater amount of tall fescue forage was available per cow per acre. King et al. (1953) noted that lactating dairy cows readily grazed pure stands of ladino clover and mixtures of tall fescue and white clover, but were reluctant to consume tall fescue alone. In this study, tall fescue failed to increase or maintain milk production in many of the grazing periods, suggesting to the researchers that this was due to its lower palatability and possibly lower digestibility as compared with white clover or mixtures with ladino clover. Thompson and Holdaway (1954) attributed the superior stimulus of tall fescue-ladino clover in the first year of their study, as compared with orchardgrass-legumes, to the preponderance of ladino clover in the former mixture.

Tall fescue had the lowest persistence of milk production rating of four kinds of pasture, while orchardgrass-ladino clover had the highest in a Kentucky study (Seath, 1952; Seath et al., 1954). The milk persistency rating of tall fescue pasture was sharply improved with the encroachment of volunteer white clover. The persistency rating of a Kentucky bluegrass-ladino clover pasture decreased, apparently from a drastic decrease in the white clover stand from winter-killing. In these studies, the orchardgrass-ladino clover caused the cows to be the most persistent in milk production, apparently because they consumed orchardgrass better than tall fescue and because of the higher ladino clover content than in either the tall fescue-ladino clover or Kentucky bluegrass-ladino clover pastures. Fertilizer-N stimulated rapid tall fescue growth but did not make the forage as palatable as that in the three grass-ladino clover mixtures (Table 11-3). Feeding alfalfa silage or extra grain supplement did not significantly change level of milk production on the several pasture combinations of orchardgrass, Kentucky bluegrass, and ladino clover, but milk production improved somewhat when cows were fed alfalfa silage while grazing tall fescue-ladino clover pasture or when cows were fed grain while grazing tall fescue (Seath et al., 1954). Kentucky 31 tall fescue was inferior to the pure stands of orchardgrass, smooth bromegrass, and Kentucky bluegrass pastures in persistency of milk production and in body weight gains (Seath et al., 1956). The 3-year average of persistence of milk production was 91.6% for Kentucky bluegrass, 90.8% for smooth bromegrass, 87.6% for orchardgrass, and 72.2% for Kentucky 31 tall fescue. The cows gained some weight on Kentucky bluegrass, maintained weight on smooth bromegrass and orchardgrass, and showed significant losses in body weight on tall fescue. The Kentucky researchers found high apparent dry matter and crude protein digestibilities. which, although differing between species and years, were all satisfactory nutritionally (Lassiter et al., 1956) Table 11-3. The daily dry matter intake was lower for tall fescue as compared with the other three grasses. These researchers suggested that the lower dry matter intake may be the major reason for lower persistency of milk production of cows on tall fescue and its most limiting factor regarding its milk producing properties, although apparently not the only factor.

In an Ohio study, cows decreased rapidly in milk production and body weight while grazing tall fescue which was attributed to its lack of palatability (Pratt and Davis, 1954). Apparent daily forage dry matter intake per cow was lower on tall fescue fall pasture, 14.5 kg (32 lb), than on smooth bromegrass, 18.6 kg (41 lb) (Boren et al., 1968). Kentucky 31 tall fescue had a higher relative persistence of milk production, 78% (3-year average over three grain levels) as compared with Kenwell tall fescue, 61% (Blaser et al., 1969).

D. TDN Yield

Kentucky 31 tall fescue pasture fertilized with 74 kg/ha (66 lb/acre) of N annually produced the most calculated total digestible nutrient (TDN) yield in the Kentucky study as compared with three grass-ladino clover mixtures (Seath et al., 1954) Table 11-3. The N stimulated the tall fescue yield, but apparently did not result in forage as palatable as the three grass-legume mixtures. In another Kentucky study with four grasses grown in pure stands, with three applications of 37 kg/ha (33 lb/acre) of N annually, tall fescue produced the least calculated TDN per acre (Seath et al., 1956). The 3-year average yield of TDN per year was 2,560 kg/ha (2,284 lb/acre) for smooth bromegrass, 2,430 kg (2,168 lb) for Kentucky bluegrass, 2,185 kg (1,949 lb) for orchardgrass, and 1,943 kg (1,733 lb) for tall fescue. The tall fescue generally carried more cows per acre, but with a rapid decrease in milk production and large decreases in body weight.

Tall fescue-ladino clover silage was comparable to corn silage for milk production in a South Carolina study (King and LaMaster, 1950). The tall fescue-ladino clover (30% clover) was ensiled in May with 29 kg/metric tons (58 lb/ton) of molasses. This silage was compared with corn silage in rations with 'Kobe' lespedeza hay and a concentrate mixture. Daily silage consumption per cow averaged 20.7 kg (45.6 lb) of corn silage and 21.7 kg (47.8 lb) of tall fescue-ladino clover silage, which were similar in apparent palatability. Average daily 4% FCM per cow was similar, 15.8 kg (34.9 lb) for the corn silage group, 15.7 kg (34.6 lb) for the tall fescue-ladino clover silage group. Liveweight gains for the experimental period were 22 kg (48 lb) and 14 kg (30 lb), respectively, for the corn silage and grass silage groups.

Allowing lactating cows 4, 8, or 20 hours access daily to tall fescue-ladino clover pasture did not influence milk production (Morrison et al., 1952).

E. Tall Fescue as Pasture for Dry Cows and Heifers

Several researchers have studied the use of tall fescue as pasture for dry cows and heifers (Pratt and Haynes, 1950; King et al., 1953) or have reported it as being used by dairymen as a permanent pasture for dry cows and replacement heifers (Spooner and Jeffery, 1964). In an Ohio study (Pratt and Haynes, 1950) Jersey heifers gained from 91 to 182 g (0.2 to 0.4 lb) daily on tall fescue-ladino clover pasture, as compared with 772 g (1.7 lb) when grazing Kentucky bluegrass-ladino pasture and meadow crop mixtures (species not indicated). King et al. (1953) in South Carolina noted better utilization of tall fescue-ladino clover by dry cows and dairy heifers than by lactating cows during summer and suggested using heifers as follow-up animals to the milking herd. Jacobson et al. (1970) obtained widely varying seasonal average daily gains of yearling dairy heifers between years, between tall fescue cultivars, and between cool-season grass species. Average daily gains of heifers on Kentucky 31 tall fescue ranged from 250 to 881 g (0.55 to 1.94 lb), averaging 558 g (1.23 lb) for 12 years; average daily gains on orchardgrass ranged from 540 to 763 g (1.19 to 1.68 lb), with a 12-year average of 608 g (1.34 lb). Several tall fescue cultivars and strains generally gave lower average daily gains than did Kentucky 31. Results were occasionally poor on tall fescue pastures, which were almost always associated with at least some symptoms of fescue toxicity. Of the tall fescue strains, Kentucky 31 had the least incidence of toxicity in this Kentucky study.

In summary, dairymen considering using tall fescue for lactating cows should consider the limitations of tall fescue as a dairy forage, as well as its desirable characteristics. It's major limitation seems to be intake, with a resulting reduced milk production and persistency of milk production. By seeding with a legume and maintaining a vigorous stand of legume in the pasture (at least 25 to 30% or more of the total dry matter), this limitation can be generally overcome for average-producing dairy cows, 14 to 16 kg/day (30 to 35 lb). No research data are available on high-producing cows, 36 to 45 kg daily (80 to 100 lb), grazing tall fescue-legume mixtures, but it is doubtful that tall fescue fits into a forage program for high-producing dairy cows because of intake limitations. Pasture management is critically important to maintain the forage in a nutritional and palatable stage for lactating cows and maintain a legume in the stand.

As with beef cattle and sheep, tall fescue is generally more acceptable for fall, winter, and spring grazing than for summer pasture. Research has shown that it may have a place in areas where winter-grazing is practiced. However, performance is generally less than on winter annual pastures. Tall fescue pastures may be used for dry cows and replacement heifers, but daily gains of heifers may be less than those from other more palatable grasses.

SUMMARY

Tall fescue is being used extensively for beef cow and calf production in many areas of the United States as summer or winter pasture or both, or alternating with other grasses. Its best use for beef cows seems to be in wintering programs for winter grazing with field-stored tall fescue hay or tall fescue hay fed to supplement the grazing. Legumes should be used with tall fescue summer pasture to improve calf gains, as well as improve cow condition and reproduction. For sheep, tall fescue has a place in wintering of pregnant ewes, either for grazing or grazing and hay. For dairy cattle, tall fescue-legume pastures can provide adequate forage for average-producing dairy cows. Good management must be followed, however, to keep the tall fescue palatable and maintain a legume in the pasture.

LITERATURE CITED

1. Alder, R. F., and R. A. Redford. 1958. Further observations on grassland management for meat production. J. Br. Grassl. Soc. 13:239–246.
2. Anderson, J. M., and L. M. Safley. 1966. Effects of two methods of renovating fescue pastures on pasture composition and beef cow-calf production. Tennessee Agric. Exp. Stn. Bull. 407.
3. ————, and ————. 1967. Orchardgrass-ladino clover and fescue-ladino clover pasture for beef cows and calves with and without creep feeding. Tennessee Agric. Exp. Stn. Bull. 422.

FORAGE-ANIMAL PRODUCTION SYSTEMS

4. Baker, H. K., J. R. A. Chard, and W. E. Hughes. 1965. A comparison of cocksfoot and tall fescue dominant swards for out-of-season production. J. Br. Grassl. Soc. 20:84-94.
5. Bedell, T. E. 1968. Seasonal forage preferences of grazing cattle and sheep in western Oregon. J. Range Manage. 21:291-297.
6. ―――. 1970. Nutritive characteristics of sheep and cattle diets on subclover-grass pastures grazed yearlong. Oregon Agric. Exp. Stn. Spec. Rep. 307.
7. ―――. 1971. Nutritive value of forage and diets of sheep and cattle from Oregon subclover-grass mixtures. J. Range Manage. 24:125-133.
8. ―――. 1973. Botanical composition of subclover-grass pastures as affected by single and dual grazing by cattle and sheep. Agron. J. 65:502-504.
9. Blaser, R. E., H. T. Bryant, R. C. Hammes, Jr., R. L. Boman, J. P. Fontenot, and C. E. Polan. 1969. Managing forages for animal production. Virginia Polytech. Inst. Res. Div. Bull. 45.
10. Boren, F. W., D. A. Stiles, G. L. Kilgore, and A. D. Dayton. 1968. Smooth bromegrass and tall fescue compared as dairy pasture. Kansas Agric. Exp. Stn. Rep. Prog. 141:13-14.
11. Burch, G. J. 1973. The role of winter forage in fat lamb production on the New England tablelands. J. Aust. Inst. Agric. Sci. 39:60-61.
12. Burns, J. C., L. Goode, H. D. Gross, and A. C. Linnerud. 1973. Cow and calf gains on ladino clover-tall fescue and tall fescue, grazed alone and with Coastal bermudagrass. Agron. J. 65:877-880.
13. Chyba, L. J., and F. W. Boren. 1973. Beef production potential of fescue pasture with a cow-calf program. Kansas Agric. Exp. Stn. Rep. Prog. 204:5-6.
14. ―――, and ―――. 1974a. Creep feeding fall calves on fescue pasture. Kansas Agric. Exp. Stn. Keeping Up with Research 10.
15. ―――, and ―――. 1974b. Roughage sources evaluated for growing heifers. Kansas Agric. Exp. Stn. Keeping Up with Research 11.
16. Clark, R. D., and D. B. Wilson. 1966. Sheep production on irrigated pasture as influenced by forage mixture and fertilization. Can. J. Anim. Sci. 46:97-106.
17. Cope, J. T., Jr., C. C. King, T. B. Patterson, and S. C. Bell. 1972. Forage and feed systems for beef brood cow herds. Alabama Agric. Exp. Stn. Bull. 435.
18. Cowan, J. R. 1956. Tall fescue. Adv. Agron. 8:283-320.
19. Drewry, K. J., C. J. Kaiser, L. A. Nelson, and R. Peterson, Jr. 1968. Grain supplementation of fall-calving beef cows grazing tall fescue winter pasture and round bales. Indiana Agric. Exp. Stn. Res. Prog. Rep. 343.
20. Forbes, R. M., and W. P. Garrigus. 1950. Some relationships between chemical composition, nutritive value and intake of forages by steers and wethers. J. Anim. Sci. 9:354.
21. Fuller, W. W., W. C. Elder, B. B. Tucker, and W. E. McMurphy. 1971. Tall fescue in Oklahoma: A review. Oklahoma Agric. Exp. Stn. Prog. Rep. P-650.
22. Gallagher, J. R., B. R. Watkin, and R. C. Grimes. 1966. An evaluation of pasture quality with young grazing sheep. I. Live-weight growth and clean wool production. J. Agric. Sci. 66:107-111.
23. Grimes, R. C. 1967. The growth of lambs grazing tall fescue receiving high and low levels of nitrogen fertilizer. J. Agric. Sci. 69:33-41.
24. ―――, B. R. Watkin, and J. R. Gallagher. 1967. The growth of lambs grazing on perennial ryegrass, tall fescue and cocksfoot, with and without white clover, as related to the botanical and chemical composition of the pasture and pattern of fermentation in the rumen. J. Agric. Sci. 68:11-21.
25. Halbrook, W. A., D. C. Denton, A. E. Spooner, and M. L. Ray. 1971. Farm plans for beef cattle producers on shallow upland soils of the Arkansas River Valley. Arkansas Agric. Exp. Stn. Bull. 760.
26. Hamilton, B. A., K. J. Hutchison, and F. G. Swain. 1970. The growth of Merino lambs grazing four temperate grasses. Aust. Soc. Anim. Prod. Proc. 8:455-459.
27. Hammes, R. C., Jr. 1975. Fall stockpiled fescue in forage systems for spring and fall calving cow herds. Virginia Polytech. Inst. Northern Virginia Forage Conference Mimeo (Middleburg). 13 Mar. 1975.
28. Hinds, F. C., G. F. Cmarik, and G. E. McKibben. 1974. Fescue for the cow herd in Southern Illinois. Ill. Res. 16(1):6-7.
29. Hoveland, C. S., W. B. Anthony, R. R. Harris, E. L. Mayton, and H. E. Burgess. 1969. Serala sericea, Coastal bermuda, Goar tall fescue grazing for beef cows and calves in Alabama's Piedmont. Alabama Agric. Exp. Stn. Bull. 388.

30. Hughes, G. P. 1955a. The production and utilization of winter grass. J. Agric. Sci. 45: 179–201.
31. ––––. 1955b. Tall fescue and lucerne drills for winter and summer grass. J. Br. Grassl. Soc. 10:135–138.
32. Hyde, R. M., and G. L. Kilgore. 1973. Tall fescue production. Kansas Agric. Ext. Publ. C-470.
33. Ibbetson, R. W., L. G. Helmer, F. W. Boren, L. J. Chyba, and H. B. Perry. 1972. Dairy and beef cattle investigations, 1971–72. Kansas Agric. Exp. Stn. Rep. Prog. 195.
34. Jacobson, D. R., S. B. Carr, R. H. Hatton, R. C. Buckner, A. P. Graden, D. R. Bowden, and A. W. Miller. 1970. Growth, physiological responses and evidence of toxicity in yearling dairy cattle grazing different grasses. J. Dairy Sci. 53:575–587.
35. Jutras, M. W. 1968. Growing oats in clover-grass seedings. South Carolina Agric. Exp. Stn. Circ. 151.
36. ––––, B. C. Morton, R. L. Edwards, R. M. Rauton, J. W. Hubbard, and W. E. Johnston. 1975. Grazing systems research. South Carolina Agric. Exp. Stn. Prog. Rep.
37. Kaiser, C. J., L. L. Wilson, R. C. Peterson, R. R. Garrigus, and M. E. Heath. 1966. Supplementing beef cows winter grazing tall fescue round bales and regrowth. J. Anim. Sci. 25:881.
38. King, C. C., Jr., W. B. Anthony, S. C. Bell, L. A. Smith, and H. Grimes. 1971. Beef cow grazing systems compared on Eutaw clay. Alabama Agric. Exp. Stn. Bull. 424.
39. King, W. A., and J. P. LaMaster. 1950. Kudzu and fescue-ladino clover silages for dairy cows. J. Dairy Sci. 33:389.
40. ––––, ––––, and J. H. Mitchell, 1953. Tall fescue and ladino clover pasture for dairy cattle. South Carolina Agric. Exp. Stn. Bull. 410.
41. Lassiter, C. A., D. M. Seath, J. W. Woodruff, J. A. Taylor, and J. W. Rust. 1956. Comparative value of Kentucky bluegrass, Kentucky 31 fescue, orchardgrass, and bromegrass as pasture for milk cows. II. Effect of kind of grass on the dry matter and crude protein content and digestibility and intake of dry matter. J. Dairy Sci. 39:581–588.
42. Lawton, G. W. 1974. Tall fescue. J. Agric. South Aust. 77:26–29.
43. Lechtenberg, V. L., W. H. Smith, S. D. Parsons, and D. C. Petritz. 1974. Storage and feeding of large hay packages for beef cows. J. Anim. Sci. 39:1011–1015.
44. ––––, ––––, D. C. Petritz, and K. G. Hawkins. 1975. Performance of cows and calves grazing orchardgrass, tall fescue, and tall fescue-legume pastures. Indiana Agric. Exp. Stn. 1975 Indiana Beef-Forage Research Day. p. 3–7.
45. Levy, E. B. 1970. Grasslands of New Zealand, 3rd Ed. A. R. Shearer, Gov. Printer, Wellington, New Zealand. 374 p.
46. McClure, W. C., and C. V. Ross. 1974. Comparison of two methods of feeding round baled fescue on the performance of pregnant ewes grazing winter fescue pasture. Missouri Agric. Exp. Stn. Sheep Day 1974. p. 21–29.
47. McCullough, M. E., W. E. Neville, and P. E. Sell. 1952. The suitability and utilization of winter forage for dairy cattle. Georgia Agric. Exp. Stn. Mimeo Ser. 26.
48. ––––, O. E. Sell, and J. H. Shands. 1953. Forage intake and grazing performance of dairy cows. J. Range Manage. 6:25.
49. Morrison, R. L., J. D. Rhoades, and C. V. Ross. 1974. Effect of method of feeding fescue hay on performance of pregnant ewes wintered on fescue pasture plus supplement. Missouri Agric. Exp. Stn. Sheep Day 1974. p. 30–37.
50. Morrison, S. H., J. J. Sheuring, R. A. Marden, and J. F. Deal. 1952. Different grazing intervals on ladino clover-fescue pasture as affecting milk production and flavor of milk. J. Dairy Sci. 35:502.
51. Mostert, J. W. C. 1956. Grass-clover pastures for sheep. Farming in South Africa 32(1):57.
52. Mott, G. O., and H. L. Lucas. 1952. The design, conduct, and interpretation of grazing trials on cultivated and improved pastures. p. 1380–1385. *In* R. E. Wagner, W. M. Myers, and S. H. Gaines (ed.) Int. Grassl. Congr., Proc. 6th, State College, Pennsylvania. 17–23 Aug. 1952.
53. Parker, C. F., E. L. Potter, and R. W. Van Keuren. 1971. Winter nutrition and management of the ewe during the gestation and lactation periods. Ohio Agric. Res. Dev. Ctr. Res. Summary 53:22–26.
54. ––––, and R. W. Van Keuren. 1970. An evaluation of treated corn silage, legume silage, and deferred pasture for wintering ewes during gestation and lactation periods. Ohio Agric. Res. Dev. Ctr. Res. Summary 42:10–19.
55. ––––, and ––––. 1973. Comparison of mechanically harvested field-stored stacks

and round bales for outdoor wintering of ewes during gestation. Ohio Agric. Res. Dev. Ctr. Res. Summary 67:7–11.

56. ————, and ————. 1975. Fall calving valuable for Ohio beef herds. Ohio Rep. 60(6):96–98.

57. Peterson, R. C., W. M. Beeson, M. E. Heath, G. O. Mott, and T. W. Perry. 1964. Forage utilization studies with the beef cow herd in southern Indiana. Indiana Agric. Exp. Stn. Res. Prog. Rep. 126.

58. ————, and R. Garrigus. 1964. Supplementing winter pasture and round bales of tall fescue for the Southern Indiana Forage Farm beef cow herd. Indiana Agric. Exp. Stn. Res. Prog. Rep. 125.

59. Pratt, A. D., and J. L. Haynes. 1950. Herd performance on Kentucy 31 fescue. Ohio Farm Home Res. 35(262):10–11.

60. ————, and R. R. Davis. 1954. Kentucky 31 fescue. Ohio Farm Home Res. 38(291):93–94.

61. Ray, M. L., A. E. Spooner, C. J. Brown, R. M. Smith, Jr., and R. D. Child. 1968. Weight, grade, and value of beef calves as affected by cow quality, forage quality, season of birth, creep feeding, and sex. Arkansas Agric. Exp. Stn. Bull. 729.

62. ————, ————, and R. W. Parham. 1969. Cow and calf nutrition and management under different grazing pressures. Arkansas Agric. Exp. Stn. Bull. 749.

63. ————, ————, and ————. 1972. Feeding management of cows and calves grazing heavily-stocked, year-round pastures. Arkansas Agric. Exp. Stn. Bull. 775.

64. Redmon, G. 1974. Results of breeding cows grazed year around for fescue-ladino clover pastures without supplementation. Missouri Agric. Exp. Stn. Forage Systems Res. Ctr. Field Day Rep.

65. Renoll, E. S., W. B. Anthony, L. A. Smith, and J. L. Stallings. 1971. Comparison of baled and stacked systems for handling and feeding hay. Alabama Agric. Exp. Stn. Prog. Rep. Ser. No. 97.

66. Rollins, G. H., C. S. Hoveland, W. R. Langford, and R. A. Burdett. 1960. Lactation performance of dariy cows grazing certain annual and perennial grasses in Alabama. II. Tall fescue, oats, and wheat. J. Dairy Sci. 43:446.

67. Rommann, L. M., W. E. McMurphy, and B. D. Boyer. 1973. Tall fescue in bermudagrass. Oklahoma Agric. Ext. Serv., Oklahoma State Univ. Ext. Facts No. 2564.

68. Ross, C. V., A. J. Dyer, and K. L. Krieg. 1967. Rations for wintering ewes. Missouri Agric. Stn. Bull. 861.

69. Seath, D. M. 1952. Effect of kind of pasture and the feeding of supplements on persistency of milk production in summer. J. Dairy Sci. 35:502.

70. ————, C. A. Lassiter, and G. M. Bastin. 1954. How kind of pasture affected the yield of TDN and persistency of milk production when grazed by milk cow. Kentucky Agric. Exp. Stn. Bull. 609.

71. ————, ————, J. W. Rust, M. Cole, and G. M. Bastin. 1956. Comparative value of Kentucky bluegrass, Kentucky 31 fescue, orchardgrass, and bromegrass as pastures for milk cows. I. How kind of grass affected persistence of milk production, TDN yield, and body weight. J. Dairy Sci. 39:574–580.

72. Smith, W. H., V. L. Lechtenberg, S. D. Parsons, and D. G. Petritz. 1974. Suggestions for the storage and feeding of big-package hay. Indiana Coop. Ext. Ser. ID-97.

73. Spooner, A. E., and W. R. Jeffery. 1964. Tall fescue for use in Arkansas permanent pastures. Arkansas Agric. Exp. Stn. Farm Res. 13(2):9.

74. Stricker, J. A., V. E. Jacobs, G. B. Thompson, F. A. Martz, A. G. Matches, H. N. Wheaton, H. D. Currence, and C. L. Mottesheard. 1976. Missouri Agric. Exp. Stn. Forage Systems Res. Ctr. Field Day Rep.

75. Stuedemann, J. A., S. R. Wilkinson, D. J. Williams, H. Ciordia, J. V. Ernst, W. A. Jackson, and J. B. Jones, Jr. 1975. Long-term broiler litter fertilization of tall fescue pastures and health and performance of beef cows. p. 264–268. *In* Managing livestock wastes. Proc. 3rd Int. Symp. Livestock Wastes. Am. Soc. Agric. Eng., St. Joseph, Mich.

76. Thompson, G. B. 1974. Performance of cows and calves of fescue-ladino clover pastures with different levels of nitrogen and creep vs. no creep. Missouri Agric. Exp. Stn. Forage Systems Res. Ctr. Field Day Rep.

77. ————, A. J. Byer, and P. Q. Guyer. 1955. The value of Alta fescue pasture for wintering pregnant ewes. J. Anim. Sci. 14:1241.

78. Thompson, N. R., and C. W. Holdaway. 1954. Alfalfa-orchardgrass, ladino clover-orchardgrass, and ladino clover-Kentucky 31 fescue pasture for milk production. J. Dairy Sci. 37:666.

79. Van Keuren, R. W. 1970a. Symposium on pasture methods for maximum production in beef cattle: Pasture methods for maximizing beef production in Ohio. J. Anim. Sci. 30: 138-142.
80. ―――. 1970b. Summer pasture for beef cows. Ohio Rep. 55(3):43-45.
81. ―――. 1972. All-season forage systems for beef cow herds. p. 39-44. *In* The earth around us. Proc. 27th Annu. Meet. Soil Cons. Soc. Am., Portland, Oreg. 6-9 Aug. 1972.
82. ―――, and C. F. Parker. 1971. Forage systems for sheep. Ohio Agric. Res and Dev. Ctr. Res. Summary 53:8-15.
83. ―――, and ―――. 1973. Comparison of round bales and hay rolls for wintering ewes during gestation. Ohio Agric. Res. Dev. Ctr. Res. Summary 67:1-6.
84. ―――, and ―――. 1974a. In-field conservation for wintering sheep in midwestern United States. p. 542-549. *In* V. G. Iglovikov and A. P. Movsisyants (eds.) Proc. XII Int. Grassland Congr., Moscow, U.S.S.R. 11-20 June 1974.
85. ―――, and ―――. 1974b. All-season forage systems for beef cow herds in the temperate zone of the United States. p. 534-541. *In* V. G. Iglovikov and A. P. Movsisyants (eds.) Proc. XII Int. Grassland Congr., Moscow, U.S.S.R. 11-20 June 1974.
86. ―――, ―――, and W. E. Gill. 1973. Big-package forage handling for beef cows. Ohio Agric. Res. Dev. Ctr. Res. Summary 68:55-63.
87. Vartha, E. W., and P. T. Clifford. 1971. Grasses in oversowing. Tussock Grassland and Mountain Lands Institute Review (22):55-62. Grassl. Div. DSIR, Lincoln, New Zealand.
88. Wheaton, H. N. 1974. The carrying capacity of fescue pastures. Missouri Agric. Exp. Stn. Forage Systems. Res. Ctr. Field Day Rep.
89. Wilson, D. B., and R. D. Clark. 1961. Performance of four irrigated pasture mixtures under grazing by sheep. Can. J. Plant Sci. 41:533-543.
90. Wilson, L. L., R. Peterson, Jr., M. E. Heath, C. J. Callahan, C. J. Kaiser, and K. Hawkins. 1967. Grain supplementation of spring-calving beef cows on tall fescue winter pasture and round bales. Indiana Agric. Exp. Stn. Res. Prog. Rep. 291.
91. ―――, ―――, ―――, and R. E. Erb. 1965. Restricted versus unrestricted winter grazing of round fescue bales and aftermath for the beef cow herd on the forage farm. Indiana Agric. Exp. Stn. Res. Prog. Rep. 189.

Chapter 12 Tall Fescue Pasture for Growing and Finishing Animals

A. E. SPOONER

Agronomy Department
University of Arkansas
Fayetteville, Arkansas

W. S. MC GUIRE

Crop Science Department
Oregon State University
Corvallis, Oregon

Tall fescue (*Festuca arundinacea* Schreb.) is used throughout most of the United States as a source of feed for growing animals. Two systems of feeding are being used, backgrounding and finishing. Backgrounding, as used in this chapter, is defined as the period from weaning (around 200 kg) to the time the animal has reached an acceptable weight (around 340 kg) for finishing either in the feedlot or continued on pasture. Finishing, as used in this chapter, is defined as the period from backgrounding to the time the animal has reached an acceptable weight (410 to 475 kg) and grade (high good to low choice) for slaughter.

The discussion in this chapter will be limited to experiments where performance of the animal is the criteria for measuring production. Yields of herbage as affected by different cultural and management practices are discussed in detail in Chapters 4, 9, and 10. The two systems of management on tall fescue (backgrounding and finishing) will be discussed separately.

BACKGROUNDING

A. Tall Fescue

Tall fescue has been evaluated to a limited extent for growing or backgrounding beef animals. A large portion of the research work has compared tall fescue with other species.

The effects of N rates applied to tall fescue and orchardgrass (*Dactylis glomerata* L.) on the average daily gains, animals/ha, and beef produced/ha were evaluated by Peterson et al. (1962) in Indiana (Table 12-1). The highest average daily gain was obtained from the orchardgrass. Daily gains decreased as N rates were increased. The number of animals/ha was greatest for the tall fescue and animals/ha increased with each increase

Copyright © 1979 ASA-CSSA-SSSA, 677 South Segoe Road, Madison, WI 53711 USA. *Tall Fescue.*

Table 12-1—Effect of N fertilizer upon the production of tall fescue and orchardgrass. Average of 4 years. (From Peterson et al., 1962).

Grass	N rates (kg/ha)			
	0	84	168	Average
Average daily gans (g)				
Tall fescue	463	409	436	436
Orchardgrass	568	522	518	536
Animals/ha (no.)				
Tall fescue	3.31	4.77	5.58	4.54
Orchardgrass	3.04	3.90	4.50	3.80
Beef produced (kg/ha)				
Tall fescue	271	340	427	346
Orchardgrass	284	348	401	345

Table 12-2—Effects of dates of N applications on animal performance and tall fescue stands. Average of 4 years. (From Spooner and Ray, 1974).

N treatment†	Average	Animal	Animal	Fescue plants (end of 4 years)
	g	days	kg/ha	
A	381	1,354	487	10
B	345	1,603	554	22
C	372	1,613	588	18
L.S.D. at 5%	ns	ns	58	

† Details of N treatments:

Treatment	Dates of application (kg N/ha)			
	October	February	June	August
A	0	0	134	134
B	134	134	0	0
C	67	67	67	67

Table 12-3—Comparison of fall saved forage from different grasses on animal performance. Average of 3 years. (From Wedin et al., 1970).

Kind of pasture	Average daily gain	Animal days/ha	Animal gains/ha
	g	days	kg
Tall fescue	610	1,205	708
Reed canarygrass	600	1,052	611
Smooth bromegrass	800	908	679
Orchardgrass	650	1,050	640

in rate of N. Beef produced/ha increased with each increase in N rate for both grasses. Beef produced/ha was the same on the two grasses when averaged over all N rates.

The amount of fertilizer, especially N, applied to tall fescue is extremely important with respect to the amount of forage produced/ha. The timing of the applied N is also very important. Spooner and Ray (1974) studied the effects of three timings of N applications on a bermudagrass (*Cynodon dactylon* (L.) Pers.)-tall fescue pasture on animal performance and tall

GROWING & FINISHING ANIMALS

Table 12-4—Stocking rates, grazing pressures, and performance of cattle grazing Kentucky 31 and Kenwell tall fescue. Average of 4 years. (From Carlson et al., 1973).

Measurement	Cultivar	Data
Days grazed	Both	145
Animals days/ha	Ky 31	909 a*
	Kenwell	794 b
Animals/ha	Ky 31	6.54 a
	Kenwell	5.73 b
Average daily gain (g)	Ky 31	449 a
	Kenwell	310 b
Gain/ha (kg)	Ky 31	411 a
	Kenwell	248 b

* Varietal means are significantly different (5% level of probability).

fescue stands (Table 12-2). They found that splitting 269 kg/ha of N into four equal applications gave the highest gains/ha and maintained the most desirable stand of tall fescue over a 4-year period. Timing of N applications did not affect average daily gains.

Tall fescue was compared with three other cool-season grasses in Iowa with respect to average daily gains, animal days/ha, and animal gains/ha by Wedin et al. (1970). Average daily gains from tall fescue compared favorably with those obtained from reed canarygrass (*Phalaris arundinacea* L.) and orchardgrass, but was significantly lower than smooth bromegrass (*Bromus inermis* Leyss.) (Table 12-3). Tall fescue produced more animal grazing days and more gain/ha than any of the other grasses.

There is an apparent difference among cultivars of tall fescue on animal performance. Carlson et al. (1973) evaluated the animal response from grazing 'Kentucky 31' and 'Kenwell' tall fescue (Table 12-4). Steers were backgrounded for an average of 145 days each year over a 4-year period. The Kentucky 31 cultivar produced significantly more animal days/ha, animals/ha, higher daily gains, and more gains/ha. These data would indicate that Kentucky 31 produced more forage of a higher quality than 'Kenwell'; however, Buckner and Burrus (1965) and Berg (1971) have reported that the two cultivars yielded about the same in clipping trials. Possibly the high alkaloid content of Kenwell may have depressed animal gains even though none of the cattle on this test showed symptoms of fescue toxicity. Gentry et al. (1969) showed that Kenwell contains more perloline and alkaloids than Kentucky 31, and they suggested that the Kentucky 31 may have been more digestible than Kenwell, although Buckner et al. (1967) found no differences among cultivars. Bush et al. (1970) have shown, however, that perloline inhibited cellulose digestion, in vitro, and this inhibition might have been greater in Kenwell because of the higher alkaloid content. A detailed discussion of animal disorders on tall fescue is presented in Chapter 13.

Different grazing systems using tall fescue have been evaluated by Matches et al. (1974) in southwest Missouri (Table 12-5). The pasture systems are listed in the table. The forages from each of these systems were grazed throughout the year as the forage was produced. Note that the tall fescue all-season system produced the highest average daily gain and gave

Table 12-5—Grazing results as affected by different pasture systems. (From Matches et al., 1974).

Pasture systems	Animal gains/ha	Average daily gain
	kg	g
Tall fescue all season†	382	708
Tall fescue-pearl millet‡	352	604
Tall fescue-caucasian bluestem‡	382	604
Tall fescue-switchgrass‡	304	595

† Calendar days of grazing—232 days.   ‡ Calendar days of grazing—225 days.

Table 12-6—Stocking rate, grazing pressure, and performance of cattle grazing tall fescue without supplement (o) or with molasses (M) or molasses + urea (M+U) fed free choice. Average of 2 years. (From Hart et al., 1973).

Measurement	Treatment	Data
Days grazed	All	200
Animals/ha	0	5.83 b*
	M	7.44 a
	M+U	7.02 ab
Animal days/ha	0	1,173 b
	M	1,490 a
	M+U	1,407 ab
Supplement/animal (kg/day)	M	3.18
	M+U	3.16
Average daily gain (g)	0	228
	M	232
	M+U	234
Gain/ha (kg)	0	274
	M	358
	M+U	332

* Treatment means followed by different letters are significantly different (5% level, Duncan's multiple range test).

the most calendar days of grazing. The system was equal to the tall fescue-caucasian bluestem (*Bothriochloa caucasica* (Trin.) C. E. Hubb) system in producing gains/ha. The tall fescue-pearl millet (*Pennisetum americanum* (L.) Leeke) system was next in production/ha and the tall fescue-switchgrass (*Panicum virgatum* L.) system produced the least daily gains and gains/ha. The major conclusion drawn from this study was that the performance of growing animals could be best measured by providing grazing on a season-long basis.

Many researchers have felt that tall fescue is too low in energy to obtain acceptable gains with growing animals. Hart et al. (1973) conducted an experiment to determine the effects of feeding molasses and molasses plus urea to steers grazing tall fescue pasture (Table 12-6). Neither supplement significantly increased average daily gain or gain/ha but the supplements did significantly increase the animal days/ha over the no supplement treatment. The addition of urea to the molasses did not have any effect on any measurement taken on the steers. Neither feeding molasses nor molasses plus urea to animals on tall fescue pastures had an advantage, they concluded.

One of the major problems with tall fescue is its uneven seasonal

GROWING & FINISHING ANIMALS

Table 12-7—Seasonal distribution of gains by steers on tall fescue. (From Spooner and Jeffery, 1964).

Season	Average daily gain	Steer days/ha	Steer gains/ha
	g	days	kg
Fall 13 Oct. to 17 Dec.	472	346	169
Spring 15 Mar. to 18 July	508	709	298
Total season	490	1,055	467

Table 12-8—Continuous vs. rotational grazing on tall fescue-bermudagrass pastures. Average of 4 years. (From Spooner and Ray, 1974).

Pasture species and method of grazing	Average daily gain	Animal days/ha	Animal gains/ha
	g	days	kg
Bermuda			
Continuous	390	1,272	435
Rotational	390	1,326	503
Bermuda-tall fescue			
Continuous	372	1,549	558
Rotational	359	1,489	530
L.S.D. at 5%	ns	ns	64

production. This factor affects the farmer who keeps his calves after weaning or who purchases calves in the fall feeder calf sales. Fall growth, in the southern area of the transition zone, is dependent on rainfall in late summer and early fall. Many years the fall growth will be one-third or less than that received during the spring and early summer. Spooner and Jeffery (1964) conducted an experiment in Arkansas to measure the response of growing beef animals to this seasonal production (Table 12-7). These data verify the fact that about one-third of the tall fescue production occurs in the fall and the other two-thirds in the spring and early summer. Other researchers have reported similar results from clipping trials. These data are discussed in Chapter 10.

The type of grazing management used (continuous vs. rotational) is extremely important when grazing with growing animals. Spooner and Ray (1974) conducted an experiment to determine the best grazing method to use (Table 12-8). They found that continuous grazing of a bermudagrass-tall fescue pasture produced significantly more gains/ha than rotational grazing. Higher daily gains and animal grazing days/ha occurred even though these were not significant at the 5% level. The major conclusion drawn from this study was that tall fescue forage gets more fibrous if allowed to reach a height of 20 to 25 cm and probably limits intake and/or digestibility of the forage.

Stockpiling of or deferring tall fescue for winter grazing is a common practice for beef cow herds. Details of this system are discussed in Chapter 11. The use of stockpiled tall fescue for growing calves is used to a great extent due primarily to the fact that the nutritive value and intake increases after cold weather. Wedin et al. (1966, 1967) reported that fall-saved tall

Table 12-9—Chemical composition of tall fescue forage from pastures being grazed with light† and heavy grazing pressures‡. (From Ray and Spooner, 1969, 1972).

Year	N		P		Ca	
	Light	Heavy	Light	Heavy	Light	Heavy
			%			
1963	1.42	1.66	0.34	0.34	0.58	0.62
1965	1.71	2.06	0.28	0.30	0.26	0.25
1966	1.68	2.12	0.26	0.27	0.21	0.22
1967	1.91	2.39	0.30	0.35	0.27	0.30
1968	2.21	3.12	0.21	0.23	0.37	0.37
1969	2.78	3.36 3.36	0.29	0.28	0.34	0.36

† Light grazing pressure—grass kept grazed to 15 to 20 cm.
‡ Heavy grazing pressure—grass kept grazed to less than 5 cm.

Table 12-10—Dry matter digestibility of tall fescue cultivars and other cool-season perennial grasses. (From Hoveland et al., 1970).

Grass	Dry matter digestibility of forage†		
	23 Feb.	10 Mar.	25 Apr.
		%	
Kolea	78	74	67
Hardinggrass	77	74	67
Aubrun reed canarygrass	79	76	63
Kentucky 31 tall fescue	73	74	58
Goar tall fescue	74	71	49

† In vivo determinations were made by placing nylon bag of coursely ground forage in fistulated steers for 24 hours. Dry matter digestibility of standard comparison forages made at the same time was coastla bermuda 58% and alfalfa hay 74%.

fescue increased in nutritive value in early November and that voluntary intake also increased in late season. Buckner et al. (1967) reported that the palatability of tall fescue increases in the fall following cold weather. There is apparently a change in the types of carbohydrates in the tall fescue plant following cold weather that makes the plant more palatable and digestible.

The chemical composition of the tall fescue forage is affected to a large extent by the grazing pressure used. Ray et al. (1969, 1972) measured the effects of light and heavy grazing pressures on the percent N, P, and Ca in the forage (Table 12-9). The forage was sampled each 28 days and composited over the growing season for analyses. The percent N was higher each year in the forage obtained from the heavy grazing pressures. Percent P and Ca in the forage from the two grazing pressures was not markedly different.

Digestible dry matter (DDM) is considered to be a good measure of forage quality (Reid and Jung, 1965). In studies conducted by Hoveland et al. (1970) the in vivo DDM of two cultivars of tall fescue and other pasture forages were compared (Table 12-10). The DDM in Kentucky 31 was higher than 'Goar' in each of these studies. The tall fescue cultivars compared favorably with alfalfa (*Medicago sativa* L.) hay in early spring but lower in DDM in late spring. Bryan et al. (1970) found that the DDM of tall fescue was lowest in June and increased each month until late November. The percent DDM in tall fescue was about the same in Iowa as in Alabama.

GROWING & FINISHING ANIMALS

Table 12-11—Performance of steers grazing tall fescue and tall fescue-ladino clover. Average of 5 years. (From Blaser et al., 1956).

	Average daily gain	Steer days/ha	Steer gains/ha
	g	days	kg
Tall fescue†	418	1,015	440
Fescue-ladino clover‡	468	773	373

† Maintenance fertilizer 242-35-65 kg/ha.   ‡ Maintenance fertilizer 0-35-65 kg/ha.

Table 12-12—Effect of different pasture species mixtures on the performance and finish of beef steers. Average of 4 years. (From Gross et al., 1966).

Pastures	Length of grazing	Average daily	Animal days/ha	Animal gains/ha	Slaughter grade†
	days	g	no.	kg	
Ladino-orchardgrass	121	835	568	384	9.0
Ladino-tall fescue	113	922	363	394	8.2
Ladino-orchardgrass & alfalfa	132	276	536	338	8.8
Tall fescue & coastal bermudagrass	178	486	1,005	486	7.9

† 7 = standard; 8 = high standard; 9 = low good.

B. Tall Fescue-Legume Mixtures

Legumes are used in pastures for two reasons: 1) to provide atmospheric N for growth and 2) to improve forage quality for higher daily gains and finish on cattle. The research on the fertilization practices for tall fescue-legume mixtures have been conducted on clipping plots. The data from the clipping trials are discussed in detail in Chapter 10. Grazing data from trials designed to specifically measure animal response to different fertilizer levels, ratios, and timing of applications are limited.

White clover (*Trifolium repens* L.) is the predominant legume seeded with tall fescue for backgrounding calves. Alfalfa and red clover (*T. pratense* L.) are used in some areas of the United States. Korean lespedeza (*Lespedeza stipulcea* Maxim.) is used in areas where summer grazing is preferred (Blaser et al., 1956; Gross et al., 1966; High et al., 1965; Ray et al., 1969, 1972). Tall fescue legume mixtures require an intensive overall management program to obtain maximum gains and to maintain stands of the legume. In most cases, it is desirable to manage the pasture (fertilization and grazing) to favor the legume.

Limited research has directly compared tall fescue with tall fescue-legume pastures on animal performance. Blaser et al. (1956) compared tall fescue with tall fescue-ladino clover (*T. repens* L.) by grazing with steers (Table 12-11). The only advantage of the ladino clover was an increase in daily gain. Steer days/ha and steer gains/ha were higher from the tall fescue. Notably, the tall fescue received 242 kg/ha of N, whereas the tall fescue-ladino clover did not receive any applied N.

Gross et al. (1966) compared tall fescue-ladino clover with orchardgrass-ladino clover for backgrounding and finishing steers in the Piedmont

Table 12-13—Effect of different species mixtures on the performance and finish of beef steers. Average of 4 years. (From High et al., 1965).

	Pasture treatment†			
	O–C	O–F–C	F–C	F–C–N
Average weight and gain per head				
Initial weight (kg)	224	221	221	223
Final weight (kg)	388	372	374	368
Total gain (kg)	164	151	153	145
Daily gain (g)	600	550	560	530
Average animal grades‡				
Final condition	8.3	8.1	7.9	6.3
Productiivty of pastures:				
Grazing days/ha	284	312	268	339
Estimated steer grans/ha (kg)	414	388	333	415

† O—Orchardgrass; C—Ladino clover; F—Tall fescue; N—23 kg ammonium nitrate each 56 days beginning in February and ending in September. No additional N was applied to the other treatments.
‡ 7.0 = Average standard; 8.0 = High standard.

Table 12-14—Animal performance on tall fescue interseeded into a bermudagrass sod. Average of 4 years. (From Spooner and Ray, 1974).

	Average daily gain	Animal days/ha	Animal gains/ha
	g	days	kg
Bermudagrass	390	1,300	469
Bermudagrass-tall fescue	368	1,519	543

area of North Carolina (Table 12-12). They found that the tall fescue-ladino clover produced higher daily gains and gains/ha, but less grazing days than orchardgrass-ladino clover. There was a slight advantage in slaughter grade from the steers grazed on the orchardgrass-ladino clover pasture. High et al. (1965) obtained higher daily gains, gains/ha, and final condition from ladino-orchardgrass pasture (Table 12-13). The final condition of the backgrounded steers was similar to those in North Carolina. The data show that animal performance varies with the species of grasses and grass-legume mixtures. Most researchers agree that forage quality is usually higher when a legume is seeded with the grass.

C. Tall Fescue Interseeded Into a Warm-Season Grass Sod

Data are extremely limited on the interseeding of tall fescue into warm-season grass sods for backgrounding beef animals. In Arkansas Spooner and Ray (1974) seeded tall fecue in 40-cm rows into a bermudagrass sod and grazed it year-round with beef steers. Excellent gains were obtained throughout the year (Table 12-14) over a period of 4 years. Excellent stands of bermudagrass and tall fescue were maintained over the 4-year period. The steers were backgrounded on these pastures for 11 months and were put into a feedlot for a short grain feeding period before being slaughtered.

Table 12-15—Performance of yearling steers grazing different grasses and grass-legume mixtures under irrigation. Average of 3 years. (From Van Keuren and Heinemann, 1958).

	Average daily gain	Animal days/ha	Animal gains/ha
	g	days	kg
Orchardgrass-alfalfa	958	1,104	1,049
Tall fescue-alfalfa	922	1,035	952
Orchardgrass-ladino†	1,090	973	1,064
Tall fescue-ladino†	940	1,082	1,020
Orchardgrass	795	758	595
Tall fescue	790	879	719

† Average of 2 years, 1954–55. No ladino the third year.

Table 12-16—Effect of different pasture systems on slaughter steer production. Average of 3 years. (Adapted from High et al., 1965).

	Treatments			
	Tall fescue‡ Clover‡ Normal clip†	Tall fescue‡ Clover‡ Close clip‡	Tall fescue‡ Clover‡ Lespedeza	Orchardgrass‡ Clover
Weight gain (kg/head)§	111	115	123	147
Daily gain (g)	345	363	390	447
Slaughter grade¶	6.9	6.7	7.1	8.1
Grazing days/ha	613	588	556	652
Estimated beef gains/ha (kg)	234	227	228	254
Carrying capacity, ha/steer	0.51	0.54	0.57	0.49

† Clipped to a level of 10.16 to 15.24 cm (4 to 6 in.) when needed to remove seed heads. ‡ Clipped to a level of 10.16 to 15.24 cm when needed to control excessive growth and seed heads.
§ 316 days. ¶ Standard = 7; Good = 10.

D. Irrigation

The use of irrigation for producing pasture for backgrounding animals has been primarily limited to the arid northwestern U.S. Since the value of land and the cost of fertilizer have increased markedly in the past decade, irrigation of pastures in the more humid regions, especially in the southeast, has become more feasible. The amount of water needed to produce large amounts of forage varies tremendously from area to area. The amount of water applied in the arid region to produce high yields of tall fescue may be as much as 100 to 130 cm/growing season. This amount may be 15 to 25 cm in the humid area of the southeast to get the same animal gains/ha. Van Keuren and Heinemann (1958) conducted a study in Washington to determine the response of tall fescue and orchardgrass seeded alone and with ladino clover and alfalfa to irrigation (Table 12-15). They found that a legume grown with the grasses gave higher average daily gains, animal days/ha, and animal gains/ha. They also found that the alfalfa persisted over the 3 years of the study, but the ladino clover stand had disappeared at the end of 2 years. One of the problems that has received much attention on irrigated grass-legume pastures is bloat. These researchers did not experi-

Table 12-17—Effect of supplemental feeding and nitrogen fertilization upon animal production on tall fescue pasture. Average of 4 years. (From Mott et al., 1971).

	N levels (kg/ha)			
	0	84	168	Average
Average daily gain/steer (g)				
without grain	252	210	218	227
with grain†	630	548	548	576
average for N levels	442	379	383	
Animals/ha (no.)				
without grain	3.59	5.97	7.15	5.57
with grain†	4.10	7.22	8.01	6.44
average for N level	3.85	6.60	7.58	
Gain/ha (kg)				
without grain	165	246	290	234
with grain†	499	739	836	691
average for N level	332	293	563	

† Crimped corn fed at 1 kg/100 kg of body weight.

ence any difficulty with bloat. They attributed this experience to the balance of the grass-legume mixture that they maintained by rotationally grazing the pastures.

It has been shown that by supplying a minimum of 2.54 cm of water/week, either as rainfall or irrigation throughout the year, that tall fescue-white clover can be kept producing all year except in extremely cold weather in northern Arkansas (A. E. Spooner and J. L. Ray, unpublished data). One of the major problems encountered in this study was the problem of proper internal and external drainage to remove the excess water following a heavy rain or during a long period of excessive rainfall. They have found that under these conditions both the grass and legume will become non-productive and may die.

FINISHING

There are two systems used to finish animals on tall fescue pasture: without grain and with grain. The first system normally makes use of a legume with tall fescue, and the animals are marketed at a lower weight and grade than animals that have been fed grain. In most experiments on finishing animals on tall fescue, a legume has been used to improve the quality of forage produced; therefore, experiments conducted on stands of tall fescue will be discussed along with those containing a legume.

Data are very limited on finishing animals on tall fescue without the feeding of grain because of the low carcass grades of the animals obtained from these experiments, a limitation shown in the data presented in Table 12-16. In an experiment conducted in Tennessee by High et al. (1965) it was evident that tall fescue and orchardgrass with a legume did not produce the degree of finish desired for slaughter animals during the grazing season. All of the animals from this experiment graded low to high standard and had to be fed for approximately 100 days in drylot to reach an acceptable slaughter

Table 12-18—Effect of irrigated orchardgrass and tall fescue with and without grain on the performance and finish of beef steers. Average of 3 years. (From Heinemann and Van Keuren, 1958).

	Average daily gain	Carrying capacity	Animal gains/ha	No. of cattle grading†		
				C	G	S
	g	steers/ha	kg			
Orchardgrass + grain	949	5.19	768	4	2	0
Orchardgrass	772	3.95	444	0	0	6
Tall fescue + grain	758	5.93	722	3	3	0
Tall fescue	622	4.20	383	0	0	6

† C = Choice; G = Good; S = Standard.

grade. All pastures containing tall fescue-legumes were inferior in all measurements taken to the orchardgrass-clover pasture.

The effects of supplemental feeding of grain and different N fertilization levels upon animal production on tall fescue pastures were studied by Mott et al. (1971). The data from this study are presented in Table 12-17. Average daily gain/steer was significantly increased by feeding grain (227 g vs. 576 g). The average daily gains were significantly decreased by the addition of N; however, there was no difference in the 84 and 168 kg/ha of N levels. There was an increase of approximately one animal/ha where grain was fed. The addition of N to the tall fescue significantly increased the number of animals/ha. The greatest increase was obtained from the first increment of N (84 kg/ha) which almost doubled the number of animals/ha. The addition of the second increment of 84 kg/ha of N increased the number of animals/ha by approximately one. Gains per hectare were significantly increased by both grain feeding and N levels. Finish and carcass grades were not available for this study.

Heinemann and Van Keuren (1958) conducted an experiment in Washington to evaluate the effects of supplemental grain fed on tall fescue and orchardgrass pastures (Table 12-18). These pastures were irrigated to provide adequate water for high production. Orchardgrass produced higher daily gains and gains/ha than tall fescue with and without grain supplementation. Tall fescue gave a higher carrying capacity than the orchardgrass. The grades of the animals between the two species differed little; however, the grade of the animals that were fed grain increased markedly. The grades of the animals that were not fed grain in this experiment agree with those presented in Table 12-16.

Edwards et al. (1968) conducted an experiment in South Carolina to measure average daily gains, feed consumption, and estimated slaughter grades of cattle finished on various pasture feeding programs (Table 12-19). Tall fescue was compared with annual ryegrass (*Lolium multiflorum* Lam.)-clover in this experiment. Full grain and limited grain were fed to the animals on each pasture species. The animals were fed on each species until all animals graded approximately the same, a result that accounts for the differences in the number of days on test. The animals on tall fescue were fed approximately 50 days longer than those on the annual ryegrass-clover

Table 12-19—Average gains, feed consumption, and estimated slaughter grades of cattle finished on various pasture feeding programs. Average of 3 years. (From Edwards et al., 1968).

	Annual ryegrass and clover		Tall fescue	
	Full grain	Limited grain	Full grain	Limited grain
Days on test	198	197	248	247
Initial weight (kg)	221	223	210	207
Final weight (kg)	401	393	395	369
Total gain (kg)	180	170	185	162
Daily gain (g)	913	863	749	654
Feed/animal concentrates (kg)	216	121	291	188
Roughage (non-pasture) (kg)	14	15	0	0
Estimated slaughter grade†	11.7	11.2	11.8	10.8

† 9 = Low good; 10 = Average good; 11 = High good; 12 = Low choice.

Table 12-20—Performance of steers fed different levels of grain on pasture. (Adapted from Wise et al., 1965).

	Average daily gain	Corn consumed	Carcass grade (no.)		
			C	G	S
	g	kg/head			
Ladino clover pasture:					
No corn†	550	0	0	5	0
Limited corn‡	780	245	4	1	0
Medium level corn§	910	441	4	1	0
Liberal corn¶	900	507	4	1	0
Deferred corn#	570	166	2	3	0
Ladino-tall fescue pasture:					
No corn†	560	0	2	1	2
Limited corn‡	570	245	2	3	0
Medium level corn§	840	490	4	1	0
Liberal corn¶	840	651	4	1	0
Deferred corn#	570	207	2	3	0

† No corn during the grazing period. ‡ 0.5% of body weight/day. § 1.0% of body weight/day. ¶ 1.5% of body weight/day. # 1.5% of body weight/day for the last 60 days only.

pasture. Note in Table 12-19 that the animals fed on the annual ryegrass-clover pasture reached the same grade in the same number of days for the two levels of grain, whereas the animals fed tall fescue achieved a difference of 1.0 grade between the two levels of grain.

In North Carolina, Wise et al. (1965) evaluated the performance of grain on a ladino clover pasture and a ladino-tall fescue pasture (Table 12-20). Higher average daily gains were obtained from the steers fed limited corn on the ladino clover pasture, but no differences were obtained between no-corn and limited-corn feedings on the ladino-tall fescue pasture. The differences in grain consumption by the steers on the two pastures were attributed to the fact that the steers would not eat all of their grain allowance when the pasture was good. Ladino clover pasture maintained high quality forage for a longer period of time than the ladino-tall fescue pasture. The carcass grades of the animals were in the choice and good grades for all treatments except the no-corn on ladino-tall fescue.

Table 12-21—Performance of steers when fed a full feed of grain in the drylot and on a tall fescue-white clover pasture. Average of 5 years. (From Spooner and Ray, 1972).

	Drylot	Pasture
Days fed	66	66
Initial weight (kg)	339	339
Final weight (kg)	435	433
Total gain (kg)	96	94
Average daily gain (kg)	1,450	1,430
Initial grade†	12.2	12.2
Final live grade†	11.7	11.4
Carcass grade†	11.3	11.4
Total feed/head (kg)		
Grain	658	607
Cottonseed hulls	254	0
Feed/head daily (kg)		
Grain	10.0	9.2
Cottonseed hulls	3.9	0

† Scale: Choice = 13; Good = 10.

The performance of steers on a full feed of grain in the drylot and on a tall fescue-white clover pasture was evaluated by Spooner and Ray (1972) in Arkansas (Table 12-21). Both groups of steers were fed for an average of 66 days before being sent to slaughter. The performance of the steers was the same. The carcass grades for the two groups of steers were in the high good range. The buyer at the slaughterhouse did not object to the carcasses of the steers that were fed on pasture. The steers that were fed in drylot consumed more grain/head/day than those on pasture. The conclusions drawn from this study were: 1) weaned calves can be backgrounded on good quality tall fescue-white clover pasture to a weight of around 340 kg and 2) they can be fed grain on pasture for a relatively short period and finished to an acceptable slaughter weight of around 450 kg with a carcass grade of high good to low choice.

LITERATURE CITED

1. Berg, C. G. 1971. Forage yield of switchgrass (*Panicum Virgatum*) in Pennsylvania. Agron. J. 63:785–786.
2. Blaser, R. E., R. C. Hammes, Jr., H. T. Bryant, C. M. Kincaid, W. H. Skrdla, T. H. Taylor, and W. L. Griffeth. 1956. The value of forage species and mixtures for fattening steers. Agron. J. 48:508–513.
3. Bryan, W. B., W. F. Wedin, and R. L. Vetter. 1970. Evaluation of reed canarygrass and tall fescue as spring-summer and fall-saved pasture. Agron. J. 62:75–80.
4. Buckner, R. C., and Paul B. Burrus, II. 1965. Kenwell tall fescue. Characteristics and management. Kentucky Agric. Exp. Stn. Circ. 601.
5. ————, J. R. Todd, P. B. Burrus, II, and R. F. Barnes. 1967. Chemical composition, palatability, and digestibility of ryegrass-tall fescue hybrids, 'Kenwell' and 'Kentucky 31' tall fescue varieties. Agron. J. 59:345–349.
6. Bush, L. P., C. Streeter, and R. C. Buckner. 1970. Perloline inhibition of in vitro ruminal cellulose digestion. Crop Sci. 10:108–109.
7. Carlson, G. E., James Bond, and R. H. Hart. 1973. 'Kenwell' vs. 'Kentucky 31' tall fescue under grazing. Agron. J. 65:130–132.
8. Edwards, R. L., G. C. Skelley, Jr., D. W. Eady, W. C. Godley, R. F. Wheeler, J. W. Hubbard, and H. C. Gilliam, Jr. 1968. A comparison of drylot and supplemental pasture systems for finishing beef cattle. South Carolina Agric. Exp. Stn. Bull. 537.

9. Gentry, C. E., R. A. Chapman, L. Henson, and R. C. Buckner. 1969. Factors affecting the alkaloid content of tall fescue (*Festuca arundinancea* Schreb.). Agron. J. 61:313-316.
10. Gross, H. D., Lemuel Goode, W. B. Gilbert, and G. L. Ellis. 1966. Beef grazing systems in Piedmont, North Carolina. Agron. J. 58:307-310.
11. Hart, Richard H., James Bond, T. S. Rumsey, and G. E. Carlson. 1973. Gains and ruminal pH, NH_3, and VFA of beef steers fed molasses or molasses-urea on tall fescue pasture. Agron. J. 65:99-100.
12. Heinemann, W. W., and R. W. Van Keuren. 1958. Fattening steers on irrigated pastures. Washington Agric. Exp. Stn. Bull. 585.
13. High, Joe W., Jr., L. M. Safley, O. H. Long, H. R. Duncan, and T. W. High, Jr. 1965. Combinations of orchardgrass, fescue, and ladino clover pastures for producing yearling steers. Tennessee Agric. Exp. Stn. Bull. 388.
14. High, T. W., Jr., E. J. Chapman, B. L. Whittenberg, and J. W. High, Jr. 1965. Fescue pastures, under different management systems, and orchardgrass-clover for yearling slaughter steer production. Tennessee Agric. Exp. Stn. Bull. 385.
15. Hoveland, C. S., E. M. Evans, and D. A. Mays. 1970. Cool-season perennial grass species for forage in Alabama. Alabama Agric. Exp. Stn. Bull. 397.
16. Matches, A. G., F. A. Martz, and G. B. Thompson. 1974. Multiple assignment tester animals for pasture-animals systems. Agron. J. 66:719-722.
17. Mott, G. O., C. J. Kaiser, R. C. Peterson, Randall Peterson, Jr., and C. L. Rhykerd. 1971. Supplemental feeding of steers on *Festuca arundinacea* Schreb. pastures fertilized at three levels of nitrogen. Agron. J. 63:751-754.
18. Peterson, R. C., G. O. Mott, M. E. Heath, and W. M. Beeson. 1962. Comparison of tall fescue and orchardgrass for grazing in southern Indiana. Purdue Univ. Res. Prog. Rep. 26:34-44.
19. Ray, Maurice L., A. E. Spooner, and R. W. Parham. 1969. Cow and calf nutrition and management under different grazing pressures. Arkansas Agric. Exp. Stn. Bull. 749.
20. ———, ———, and ———. 1972. Feeding management of cows and calves grazing heavily-stocked year-round pastures. Arkansas Agric. Exp. Stn. Bull. 775.
21. Reid, R. L., and G. A. Jung. 1965. Influence of fertilizer treatment on the intake of digestibility, and palatability of tall fescue hay. J. Anim. Sci. 24:615-625.
22. Spooner, A. E., and Will R. Jeffery. 1964. Tall fescue for use in Arkansas permanent pastures. Arkansas Farm Res. Vol. 13:9.
23. ———, and Maurice L. Ray. 1972. Finishing yearling steers on pasture with grain. Arkansas Agric. Exp. Stn. Bull. 772.
24. ———, and ———. 1974. Influence of timing of nitrogen application on pastures and on performance of beef cattle. Arkansas Agric. Exp. Stn. Bull. 791.
25. Van Keuren, R. W., and W. W. Heinemann. 1958. A comparison of grass-legume mixtures and grass under irrigation as pastures for yearling steers. Agron. J. 50:85-88.
26. Wedin, W. F., I. T. Carlson, and R. L. Vetter. 1966. Studies on nutritive value of fall-saved forage, using rumen fermentation and chemical analyses. p. 424-428. *In* A. G. G. Hill (ed.) Proc. 10th Int. Grassl. Congr. Helsinki, Finland. 7-16 July 1966. Valtioneuvoston Kirjapaino, Helsinki, Finland.
27. ———, ———, and ———. 1967. Fall-saved forage. Crops Soils 19(9):17-18.
28. ———, R. L. Vetter, and D. T. Carlson. 1970. The potential of tall grasses as autumn-saved forage under heavy nitrogen fertilization and intensive grazing management. N.Z. Grassl. Assoc. Proc. 32:160-167.
29. Wise, M. B., E. R. Barrick, and T. N. Blumer. 1965. Finishing steers with grain on pasture. North Carolina Agric. Exp. Stn. Bull. 425.

Chapter 13

Animal Disorders

LOWELL BUSH
Agronomy Department
University of Kentucky
Lexington, Kentucky

JAMES BOLING
Animal Science Department
University of Kentucky
Lexington, Kentucky

SHELLY YATES
SEA-USDA
Peoria, Illinois

INTRODUCTION

Tall fescue (*Festuca arundinacea* Schreb.) is an excellent agronomic forage crop, particularly for much of the eastern U.S. When relating chemical composition to forage quality, tall fescue is a high quality forage. However, cattlemen occasionally see disorders in animals grazing tall fescue at different seasons of the year and under many different environmental conditions.

Fescue toxicosis has been a term used for many symptoms observed in animals grazing tall fescue. For the most part, and correctly so, grass tetany and nitrate toxicity have not been included in fescue toxicosis. The early symptoms of a staggering gait and muscle tremors of grass tetany may be confused with fescue toxicosis, but the development of other symptoms and a history of the animal (calving time, available feed, and season of the year) should help one avoid confusing grass tetany with fescue toxicosis.

Animal symptoms associated with fescue toxicosis have been described by many authors and most always include reduced rate of gain and/or milk production and a rough hair coat. The principal syndromes and terms associated with fescue toxicosis are fescue foot, poor performance, summer syndrome, and fat necrosis. The extreme symptoms of fescue foot and fat necrosis are easily distinguished, but it is not known if they are different animal responses caused by the same agent(s) or are an extension of the less acute response of poor performance and the summer syndrome. The interrelationships of the causative agents and the many syndromes of fescue toxicosis are not known. For this discussion fescue foot will be considered that syndrome most commonly associated with soreness and lameness in one or more feet, rough hair coat, dry gangrene in the extremities (tail, legs, and ears), and mainly occurring at times other than the summer. Fat

Copyright © 1979 ASA-CSSA-SSSA, 677 South Segoe Road, Madison, WI 53711 USA. *Tall Fescue.*

necrosis will be considered as that syndrome associated with digestive disturbance, bloating, reduced passage of digesta, kidney dysfunction, or other physiological dysfunctions resulting from the occurrence of the necrotic or hard fat lesions. Fat necrosis can be confirmed by post-mortem examination or in many cases by rectal palpation. Summer syndrome will refer to poor performance and is associated with the symptoms of rough hair coat, loss of weight, rapid breathing, increased body temperature, and a general unthrifty condition during the warmest grazing season. Summer syndrome is not usually fatal; however, the poor animal performance is a much greater cost to animal production than the acute syndromes covered in this chapter. If summer syndrome is different from fescue foot and fat necrosis, one can conclude from present data that the most likely causative agent is the nonprotein fraction, especially the alkaloids. The following discussion considers fescue foot, grass tetany, and fat necrosis as separate entities, and the nonprotein N section is presented with the summer syndrome being considered the principal animal response.

FESCUE FOOT

A. Description of the Syndrome

The diagnosis of fescue foot in cattle by a veterinarian is complicated by a host of clinical signs which range from loss of weight and a rough hair coat to necrotic feet, tail, and ears. The early clinical signs of fescue foot in cattle—loss of weight or a reduced rate of gain, rough hair coat, and an arched back (Fig. 13-1)—are characteristic only of a sick animal (Merriman, 1955; Price and Miller, 1959; Jacobson et al., 1963; Jacobson and Hatton, 1973; Garner, 1973). However, if cattle are grazing tall fescue, especially in cool weather, these signs may herald fescue foot. Other gross clinical signs may include a dull hair coat, scouring, lameness, and in warm weather, a tendency for the animal to try to keep cool (Cunningham, 1949; Pulsford, 1950; Merriman, 1955; Jensen et al., 1956; Jacobson et al., 1963; Jacobson and Hatton, 1973; Garner, 1973; Williams et al., 1975). As these clinical signs progress, the cattle may appear tranquilized (Garner, 1973). If the animals are immobilized and examined more closely, the following clinical signs may be noted: trembling in cold weather, elevated temperatures, elevated respiration rate, elevated pulse rate, and occasionally, the apparent absence of rumen motility (Merriman, 1955; Jacobson et al., 1963; Williams et al., 1975). A tremor may be apparent which starts in the animal's hindquarters and moves to the forequarters. A cough may be produced as this tremor passes the chest cavity (Garner, 1973). Walking may cause elevation of body temperature (Jacobson et al., 1963).

Early clinical signs may be apparent 3 to 7 days after cattle graze toxic forage. These signs may include changes in the extremities such as swelling and skin discoloration. A characteristic sign is a red line forming at the coronary band of the hind foot (Fig. 13-2) (Price and Miller, 1959; Jacob-

ANIMAL DISORDERS

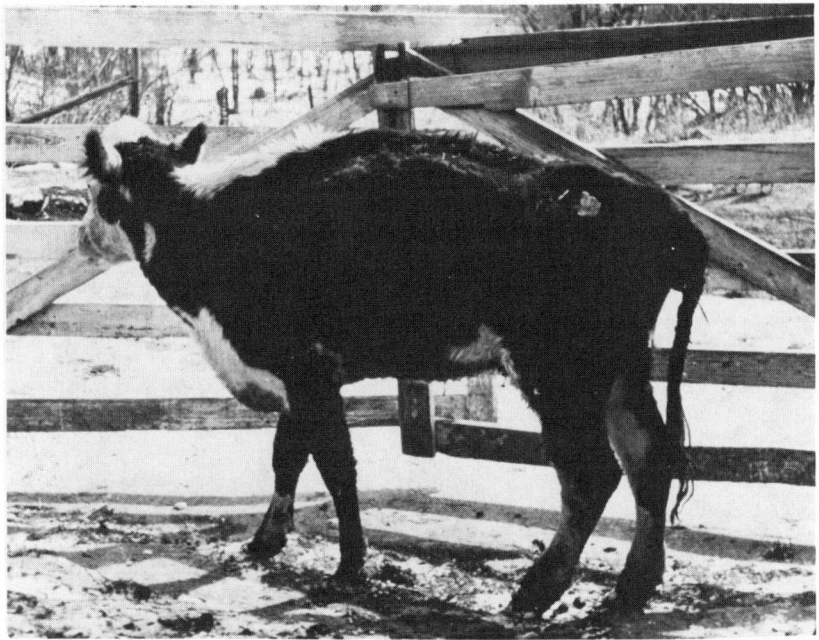

Fig. 13-1—Typical appearance of a cow in advanced stage of fescue foot. Reprinted from Yates et al. (1969), by permission of the American Chemical Society.

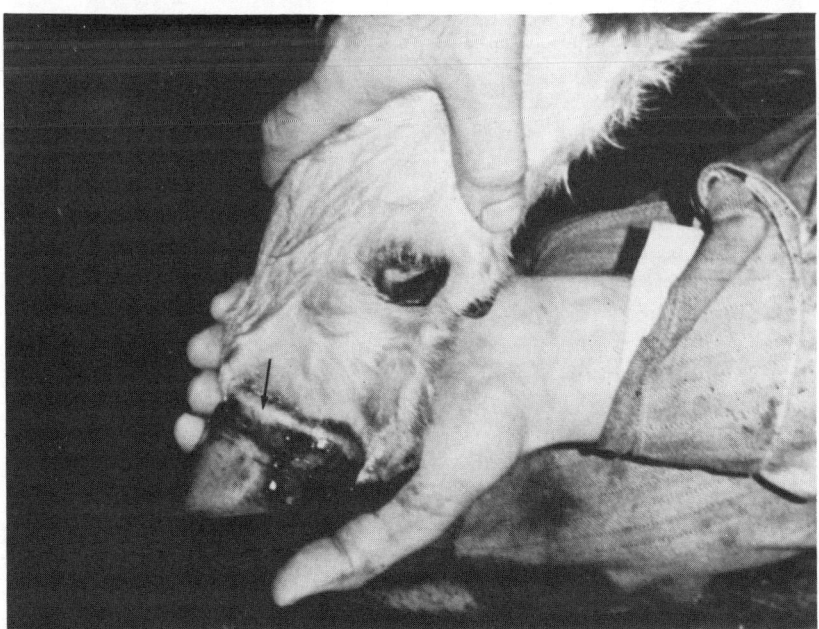

Fig. 13-2—Necrosis of coronary band: Pronounced red line at coronary band.

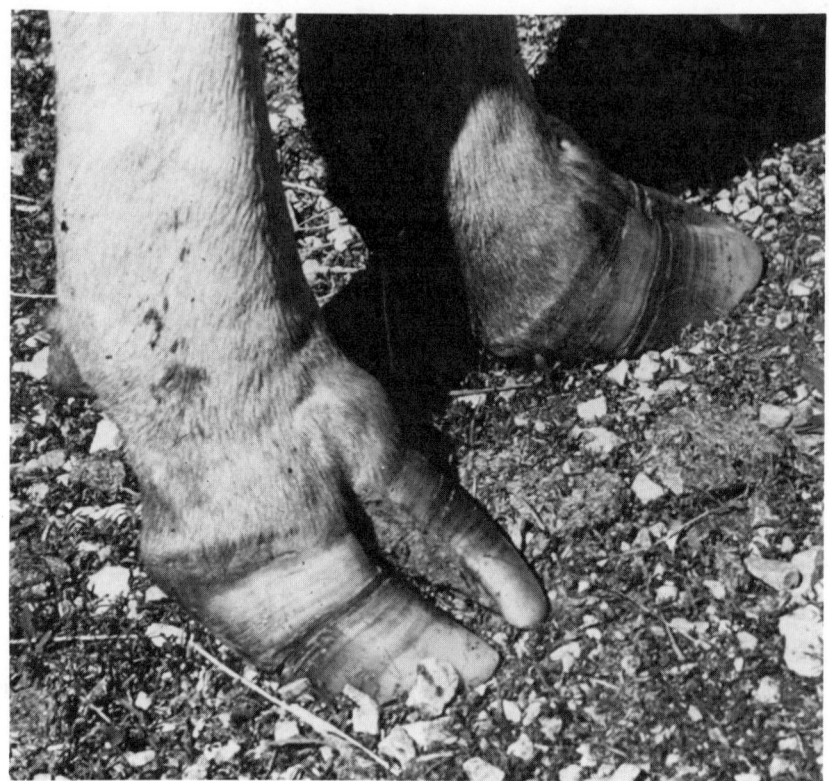

Fig. 13-3—Altered hoof growth. Reprinted from Jacobson et al. (1963) by permission of Dairy Science Association.

son et al., 1963; Jacobson and Hatton, 1973; Garner, 1973; Williams et al., 1975). The fetlock of the hind foot may swell and portions of the tail may become enlarged and pulpy. Signs of advanced fescue foot may be noticed as soon as 2 weeks after initial grazing of toxic forage, although many weeks may pass before they appear. For the feet and legs, these signs may include soreness, swelling, knuckling, reddening between the dewclaw and hoofs, altered hoof growth (Fig. 13-3), and loss of hair from coronary band. Eventually, a line of demarcation may form between healthy tissue and a gangrenous hoof—from which fescue foot gets its name—and the gangrenous portion is sloughed (Fig. 13-4) (Cunningham, 1949; Pulsford, 1950; Jensen et al., 1956; Price and Miller, 1959; Jacobson et al., 1963; Jacobson and Hatton, 1973; Garner, 1973; Williams et al., 1975). In some instances, instead of the characteristic fescue foot, lesions appear on the side of the legs and around the coronary band (Fig. 13-5) (Watson, 1957; Jacobson et al., 1963; Jacobson and Hatton, 1973). It is not clear whether these lesions on the feet and legs are due to different toxins or to the differences in animal response. Fescue foot may occasionally involve the fore-

ANIMAL DISORDERS

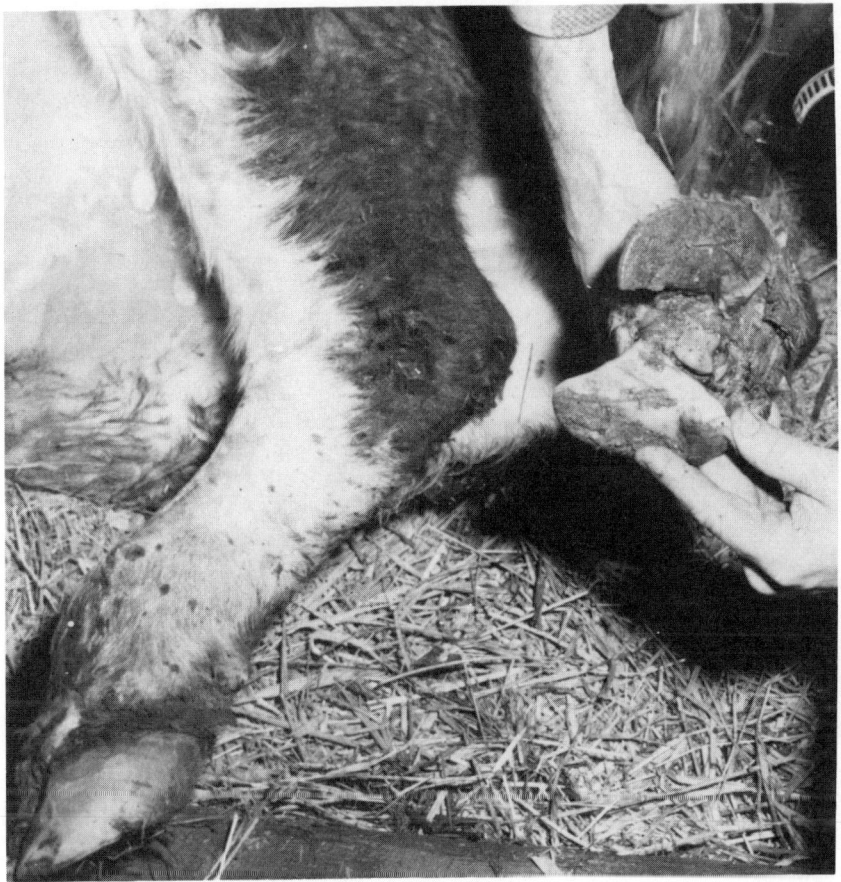

Fig. 13-4—Sloughed hoof of a cow grazing a toxic tall fescue pasture. Reprinted from Yates et al. (1969), by permission of the American Chemical Society. Photo courtesy of Dixon Springs Experiment Station, Robbs, Ill.

limbs in the same manner (Jensen et al., 1956; Watson et al., 1957; Jacobson et al., 1963; Garner, 1973).

Advanced clinical signs in the tail do not necessarily parallel those in the feet and legs. In one instance, seven to 57 cattle on a tall fescue pasture were lame. Upon close observation of these cattle, 52 of the 57 had gangrenous tails (Tookey et al., 1972). Advanced clinical signs at the tip of the tail were discoloration, loss of hair, necrosis, and sloughing of the gangrenous portion of the tail (Goodman, 1952; Jensen et al., 1956; Jacobson et al., 1963; Jacobson and Hatton, 1973; Garner, 1973; Williams et al., 1975). Under extreme conditions, the ears may show necrosis (Goodman, 1952; Garner, 1973; Williams et al., 1975).

Post-mortem findings and changes in the blood of cattle having fescue

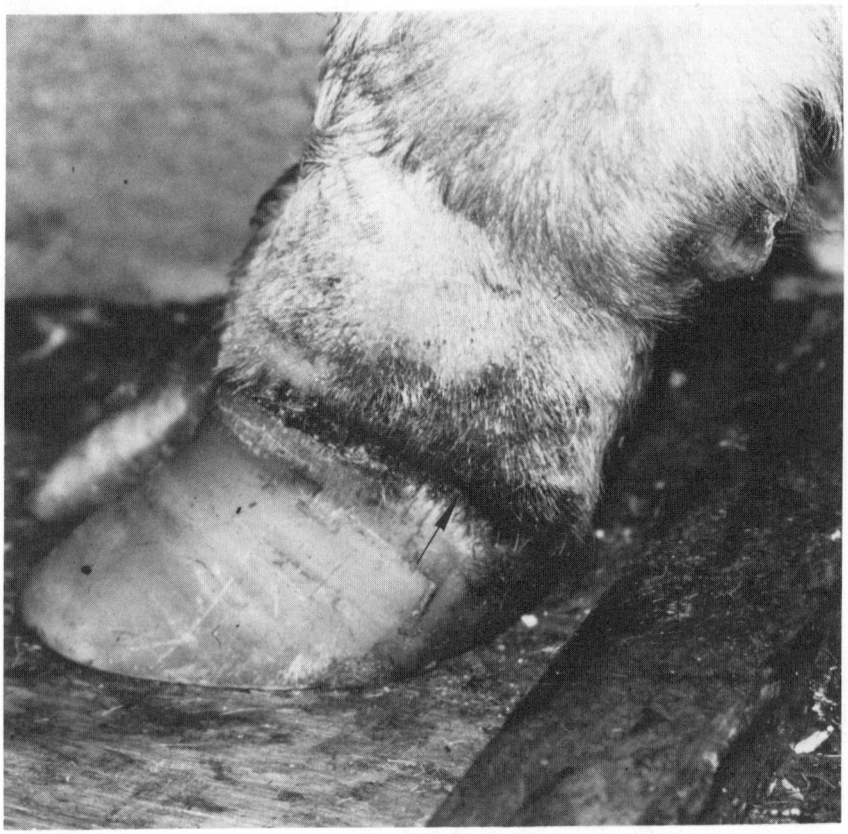

Fig. 13-5—Necrotic lesions on side of foot.

foot are inconsistent (Jacobson et al., 1963; Williams et al., 1975). Characteristic lesions in the visceral organs of cattle which have had fescue foot have not been reported. However, some important internal lesions have been noted. Jensen et al. (1956) reported thrombosis of the bulbar and proper digital arteries of affected feet. Congested blood vessels in necrotic extremities were noted by Jensen et al.(1956) and Williams et al. (1975). The most striking characteristic—blood vessels of affected feet with thickened or swollen walls and small lumens (Fig. 13-6)—was mentioned by Corley et al. (1973) and Williams et al. (1975). Arteries with small lumens, attributed to swollen arterial walls, were found next to the necrotic areas of 50% of the affected limbs of cattle showing advanced clinical signs of fescue foot (D. M. Johnson as cited by Corley et al., 1973). There was no evidence of hyperplasia; whether this swelling was a primary or secondary lesion was unanswered. That no definitive histological evidence was found in postmortem examinations indicated to Corley et al. (1973) that the primary lesion is transitory. Corley et al. (1973) subsequently showed that arterial changes do occur in the extremities of cattle affected with fescue foot and

ANIMAL DISORDERS

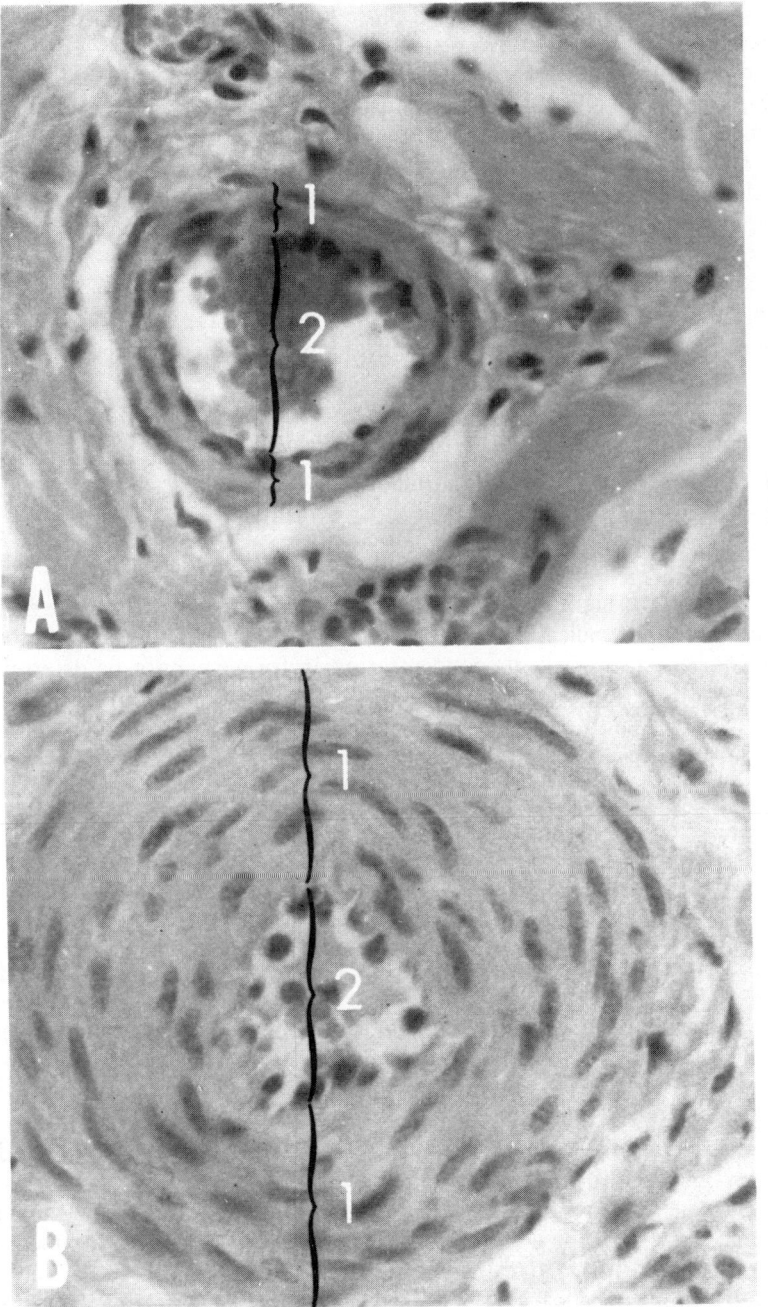

Fig. 13-6—Cross sections of arterioles of rear coronary bands from normal calf (A) and fescue foot-affected calf (B). Note the thick walls (1) and small lumens (2) of the blood vessels of the fescue foot-affected calf. Reprinted from Williams et al. (1975) by permission of American Veterinary Society.

that new circulation can develop in cattle severely affected with fescue foot. This finding agrees with observations that clinical signs regress or disappear when cattle are removed from the toxic source (Pulsford, 1950; Merriman, 1955; Watson et al., 1957; Price and Miller, 1959; Jacobson et al., 1963; Garner, 1973; Williams et al., 1975). Williams et al. (1975) noted that three of four cattle showing clinical signs of fescue foot (produced experimentally by dosing with fractions from a toxic hay source) had normal digital arteries, but the coronary band vessels had small lumens because of thickened arterial walls (Fig. 13-6).

Abortion in cattle showing clinical signs of fescue foot is rare (Pulsford, 1950; Garner, 1973). Garner (1973) stated that farmers in Missouri often nurse a severely crippled cow through parturition to get the calf.

B. Alteration of Clinical Signs

The Pasture

The clinical signs of fescue foot may be altered or made more severe by various factors, such as the concentration of the toxin. Because the toxin has not been identified, its concentration in forage cannot be determined before cattle are placed on the pasture. Environmental temperature, pasture management, supplemental rations, animal variation, and animal health also influence clinical signs. Usually only 10 to 30% of the herd show clinical signs of fescue foot on a toxic pasture (Garner and Harmon, 1973). If the cattle are removed from toxic pastures and maintained in a protected environment (shed or deep bedding), they usually recover within a few days (Garner, 1973). If, however, the concentration of toxin is high (a large percent of the herd is affected and/or hoofs are sloughed above the dewclaw), clinical signs of fescue foot may continue to worsen even though the cattle are removed from toxic pasture (cf. Case No. 1 and No. 3; Garner and Harmon, 1973).

Fescue foot is a seasonal problem, occurring more often in the winter months than at any other time (Cunningham, 1949; Pulsford, 1950; Goodman, 1952; Jensen et al., 1956; Watson et al., 1957; Price and Miller, 1959; Garner, 1973; Garner and Harmon, 1973). In Missouri, clinical signs of fescue foot are often related to the first snow or ice storm, or to a sudden drop in temperature (Garner, 1973).

Garner (1973) reported that in Missouri two methods of pasture management are used in low-cost cow-calf operations: 1) grazing summer-fall accumulated growth and 2) grazing fall regrowth after removal of a seed crop and baling the stubble. Apparently neither of these systems reduces the toxin concentration as measured by fescue foot in cattle in Missouri. Legumes are sometimes seeded into tall fescue pastures. This practice may help reduce the incidence of fescue foot because cattle consume forage other than tall fescue, but it does not eliminate the problem. Garner and Harmon (1973) found cattle grazing tall fescue supplemented with corn (*Zea mays*

L.) and protein showed less severe signs of fescue foot than cattle grazing tall fescue without supplements.

Garner and Harmon (1973) observed that many of the tall fescue pastures which produced clinical signs of fescue foot in cattle in Missouri had been heavily fertilized with N. Therefore, they applied 337 kg N/ha (part in mid-August and the remainder in mid-September) to experimental plots of pure 'Kentucky 31' tall fescue. Clinical signs of fescue foot were produced in cattle on these experimental plots 3 out of 4 years.

In southern Illinois, tall fescue pastures and year-round grazing are common, yet fescue foot is almost never reported (P. Trovillion, personal communication). Apparently in this area, tall fescue pastures are not heavily fertilized and some farmers feed baled tall fescue hay during the winter months.

The Animal

Animal variation may explain why only 10 to 30% of a cattle herd show clinical signs of fescue foot; some animals are more susceptible to fescue foot than others. When feeding three cattle toxic tall fescue (Cunningham, 1949), one animal showed lameness in both rear feet after 17 days; another became lame after 21 days but recovered while eating the same hay; the third animal had no clinical signs of fescue foot.

Cattle in good physical condition are less likely to show clinical signs of fescue foot. Ashley (1958) reported on three groups of cattle (321 head) in the same cross-fenced pasture. The group (171 head) that were thin at the time they were placed on pasture had severe clinical signs of fescue foot; 18 died, 15 had slight to severe lameness, and all lost the switch of their tail. The other two groups, which were in better physical condition, showed mild signs of fescue foot; 15 developed a slight limp and some lost their tail switch.

Some early researchers thought that cattle on tall fescue all year were more resistant to fescue foot than those unaccustomed to tall fescue (Cunningham, 1949; Pulsford, 1950); however, the possibility of induced resistance remains an unknown.

C. Theories on the Cause of Fescue Foot in Cattle

Etiological Agents

There are numerous kinds of lameness in cattle that are well defined and of known etiology (Prentice and Neal, 1972). Fescue foot is well defined, but its etiology is obscure. The theories concerning the cause of fescue foot in cattle include:
  a. Tall fescue alkaloids.
  b. Vasoconstrictor alkaloids of ergot produced in the infected ovary of tall fescue and other grasses by the fungus *Claviceps purpurea.*

c. Ergot-like vasoconstrictor alkaloids.
   d. Mycotoxins produced by one or more field fungi on grasses, or a toxic phytoalexin produced by tall fescue in response to fungal infection or nematode invasion.
   e. A toxin (not present in or on the plant) which is produced in the animal's rumen by microbial action.
   f. An anion, a negatively charged compound that can be adsorbed onto an anion exchange resin, present in the plant which under stress (heat, cold, drought, N content of the soil too high, or too low, etc.) increases to a level which is toxic to cattle.

a. Tall Fescue Alkaloids. Usually alkaloids produced by a plant are closely related structurally; however, tall fescue produces two types of alkaloids. One type, the loline family, contains a pyrrolizidine nucleus; the other type, contains a diazaphenanthrene nucleus (perloline and perlolidine). Unsaturated pyrrolizidine alkaloids are known as hepatotoxins, the most toxic being esters of amino-alcohols and branched chain hydroxy-acids (Mattocks, 1971). Loline has no reported hepatotoxic activity (Tookey and Yates, 1972) because it contains a saturated ring which is not dehydrogenated in the liver to a pyrrole, the true hepatotoxin (Mattocks, 1968). The hydrochloride salt of loline has a low oral toxicity; 1,000 mg/kg given as a single oral dose caused no effect in mice (Yates and Tookey, 1965). Neither loline nor N-acetyl loline cause contraction of rat duodenal muscle (Bruce et al., 1971). However, either of these alkaloids when added to the smooth muscle preparation, before addition of acetylcholine, potentiated its ability to produce contraction. Perloline is mildly toxic when given to animals by intravenous or intraperitoneal routes, but it is much less toxic when taken orally, probably because it is destroyed in the liver (Cunningham and Clare, 1943). Compared with eosin or rose bengal, perloline is a mild photosensitization agent (Cunningham and Clare, 1943; Reifer and Bathurst, 1943).

An alkaloidal subfraction prepared from toxic tall fescue hay was given intraruminally to a cow at a dosage equivalent to 21 lb of hay/day for 11 consecutive days (Yates et al., 1973). No signs of fescue foot occurred, although clinical signs of fescue foot were produced by an aqueous residue subfraction of the same hay (Jacobson et al., 1963). The daily dose of the alkaloidal subfraction would have contained about 8 g of loline dihydrochloride and an undetermined amount of perloline. Later tests with an alkaloidal subfraction referred to as "the chloroform phase," obtained by ion exchange chromatography, have shown that the alkaloids were nontoxic in two cattle; a non-alkaloidal subfraction from this same hay produced clinical signs of fescue foot (Williams et al., 1975).

Rifas et al. (1973) suggested that an atypical alkaloid might be present in tall fescue which remains in the aqueous residue subfraction. Paper chromatography revealed an alkaloid-like compound in the aqueous phase after the alkaloids were removed by chloroform extraction at pH 11. Further purification yielded betaine, which was present at 0.29% in toxic tall fescue. However, rations containing large amounts of betaine in sugar beet (*Beta vulgaris* L.) pulp and molasses from beets did not produce fescue

ANIMAL DISORDERS

foot when fed to cattle (Davies, 1936) (Linders, J. D. 1967. Liquid protein concentrate for cattle. M.S. thesis. Colorado State Univ., Fort Collins).

b. Vasoconstrictor Alkaloids of Ergot. Ergot was an early suspect as the causative agent of fescue foot because of the close relationship of the clinical signs of gangrenous ergotism to fescue foot (Burfening, 1973). The vasoconstrictor action of ergot is attributed to various indole alkaloids; more specifically, peptide derivatives of lysergic acid (Kingsbury, 1964). These alkaloids are produced in the sclerotia of the fungus. Therefore, a grass containing no sclerotia or seed head should not contain ergot alkaloids. Cunningham (1949) and Jacobson et al. (1963) stated that tall fescue hay containing no seed heads or ergot sclerotia had produced fescue foot in cattle.

c. Ergot-Like Vasoconstrictor Alkaloids. Maag and Tobiska (1956) did not find ergot sclerotia in tall fescue hay that produced lameness in cattle. They postulated the presence of other ergot-like alkaloids in this hay on the basis of the following tests:
  i. Tall fescue hay, mountain grass hay, or ergot sclerotia were steeped with ether-acidified with tartaric acid. The ether solutions were extracted with 10% sodium carbonate and these basic solutions allowed to stand. The fescue solution turned from orange to rose to violet, the mountain grass solution was light yellow, and the ergot solution was rose colored.
  ii. Alkaline ether extracts of the fescue hay and the controls were purified by solvent partitioning to give a final 1% tartaric acid extract. Upon treatment with p-dimethylaminobenzaldehyde, the fescue extract was light blue, the mountain grass extract was colorless, and the ergot extract was deep blue.
  iii. The ultraviolet absorption spectra of the tall fescue and ergot extracts were quite similar.

When investigating causes of fescue foot in Georgia, Bacon et al. (1975) and Porter et al. (1975) discovered additional evidence of ergot-like alkaloids. The fungi *Balansia epichloe* and *B. henningsiana,* were found on grasses in pastures where cattle had fescue foot. *B. epichloe* was detected on *Festuca* sp. in these toxic pastures. This fungus was detected as a black mass (pseudomorphic ascostromata) on the leaf surface. Studies indicate that the fungus will not sporulate at temperatures lower than 20 C, therefore, in cooler climates this fungus may parasitize grass tissue and not be readily detected (Bacon et al., 1975). *Balansia* sp. may have the capability, on specific hosts, under certain climatic conditions to produce ergot-like alkaloids. Bacon et al. (1975) showed that *B. epichloe* will produce substances toxic to the chick embryo which appear to contain an indole nucleus as is present in ergot alkaloids.

In an effort to find a field indicator of toxicity, Porter et al. (1975) examined fungal cultures of *Claviceps* and *Balansia* isolated from grasses in toxic fields and several grass samples from both toxic and supposedly nontoxic fields. Two ergostaenones produced by the fungi in laboratory cultures were detected in chloroform extracts of the grass samples.

Although these compounds were present in both toxic and nontoxic samples, the amounts were small in the nontoxic samples. Ergostatetraenone is not toxic to the chick embryo. Ergostatetraenone is produced by other fungi (*Candida utilis, Formes officinalis, Lampter omyces japonicus, Penicillium rubrum,* and possibly several species of *Aspergillus*); therefore, its presence does not confirm the presence of *Claviceps* or *Balansia.*

   d. *Mycotoxins, Toxic Phytoalexins.* It is possible that fungi produce toxins (mycotoxins) or cause the plant to produce toxins (phytoalexins) which, when ingested, lead to fescue foot in cattle. The sporadic, seasonal, and regional nature of fescue foot suggests a fungal cause (Yates et al., 1969). The fact that other field diseases in livestock are caused by fungi gives credence to this theory; i.e., *Rhizoctonia leguminacola* on red clover (*Trifolium pratense* L.) produces slaframine, the slobber factor (Aust et al., 1968), and spores from *Pithomyces chartarum* Brook and White, 1966) cause facial eczema in sheep. Yates (1963) reported that the loline family of tall fescue alkaloids was modified, presumably due to an infection of tall fescue by the fungus *Stemphylium.* If this fungus can influence the secondary metabolism of tall fescue, perhaps a fungus can cause the production and storage of a toxic phytoalexin. Unfortunately, the phytoalexin concept has received little research attention.

   The mycotoxin hypothesis has been pursued vigorously in recent years. Aside from research at the Russell Research Center on *Balansia,* which might also be classified as mycotoxin research, researchers at the Northern Regional Research Center have considered *Fusarium* fungi as the potential causative organism of fescue foot and the group at Mississippi State University has implicated *Aspergillus terreus.* Possibly, in different regions different fungi cause fescue foot.

   Examination of the fungi present on samples of toxic tall fescue from Missouri revealed that the fungal population was about the same year after year. In order of decreasing abundance, genera classified on a typical sample were: *Alternaria, Cladosporium, Fusarium, Epicoccum, Collectotrichum, Phoma, Streptomyces, Helminthosporium,* plus 15 other genera of fungi, some slime molds, bacteria, and nematodes (Yates, 1971).

   In an early study on fungal involvement, grass samples from a toxic tall fescue pasture near Sturgeon, Mo., were gathered biweekly, and unusual or infrequent fungi isolated, cultured, and classified. None of the fungi isolated were conspicuous by their continued presence on toxic tall fescue and absence on nontoxic fescue (Yates et al., 1969). Toxic pastures and hay from toxic pastures appeared the same as that from nontoxic areas. Keyl et al. (1967), however, found that extracts of some tall fescue hay samples from toxic fields produced a necrotic rabbit skin test. Certain of the fungi (*Epicoccum nigrum, Cladosporium cladosporioides,* and *Fusarium nivale*) isolated from a toxic hay produced compounds which caused a positive rabbit skin test and also produced visceral hemorrhage in mice. The *F. nivale* was later found to fit the Snyder and Hansen concept of *F. tricinctum* (Ellis and Yates, 1971).

   Nontoxic tall fescue hay enriched with glucose, peptone, and salts served as a substrate for the three fungi Keyl et al. (1967) indicated were most toxic. *E. nigrum* failed to grow and was not tested in cattle. An extract of the *Cladosporium cladosporioides* cultured on hay was nontoxic to cattle. An extract of the *F. tricinctum* molded hay killed one of two cattle when given a single dose orally at a rate equivalent to 0.80 kg of moldy hay. The second cow died when given an additional dose equivalent to 0.45 kg of hay. The extract produced no clinical signs when dosed at a rate equivalent to 0.18 kg of hay/day for 15 days (Yates et al., 1969).

Fungi were isolated from hay samples taken from a toxic tall fescue pasture where 11 of 100 head of cows were severely affected with fescue foot (rough hair coat, arched back, emaciated, lame, sloughing of hoofs and tails) (Yates et al., 1969). Two hundred fungal isolates from this toxic pasture and nearby pastures were grown, extracted, and tested in mice. One of 24 *Alternaria* was toxic, one of 35 *Epicoccum* was toxic, and 23 of 50 *Fusarium* were toxic. The toxic *Fusarium* cultures contained either 4-acetamido-4-hydroxy-2-butenoic acid γ-lactone, $4\beta,15$-diacetoxy-8α-(4-methylbutyryloxy)-12,13-epoxytrichothec-9-en-3α-ol, or both (Yates et al., 1970). The γ-lactone has a LD_{50} (intraperitoneal in mice) of 43.6 mg/kg, and the epoxytrichothecenol, 3.04 mg/kg (Yates et al., 1968). The γ-lactone contains a butenolide ring structure, and has been referred to as the butenolide. The epoxytrichothecenol isolated from the T-2 strain of *F. tricinctum* (Bamburg et al., 1967) was characterized by Bamburg et al. (1968) and is referred to as T-2 toxin. These two pure crystalline mycotoxins produced by *F. tricinctum* were also tested in cattle (Grove et al., 1970, 1973; Kosuri et al., 1970; Tookey et al., 1972).

T-2 toxin was given intramuscularly to a 295-kg steer at 0.1 mg/kg (Grove et al., 1970, 1973) for 65 days. During this test the steer lost 17% of its initial weight, showed signs of internal bleeding, and died on day 65. Post-mortem examination revealed numerous hemorrhages in the digestive tract and a general loss of body fat. These clinical signs were not suggestive of fescue foot but suggested the hemorrhagic syndrome seen in cattle consuming moldy corn. T-2 toxin has since been isolated from field corn which caused moldy corn toxicosis in dairy cattle (Hsu et al., 1972).

The butenolide was tested during the winter months (cold weather contributes to the development of gangrene in fescue foot) in a 3-year study at a University of Wisconsin experimental farm (Grove et al., 1970, 1973; Kosuri et al., 1970; Tookey et al., 1972). In the first test the butenolide was given intramuscularly to a 286-kg heifer on a tall fescue ration at 3.8 mg/kg for 90 days. Gross clinical signs were loss of weight, arched back, and gangrene of the tail tip. The next year the butenolide was given orally to six cattle maintained on a tall fescue ration. Oral doses of butenolide (in gelatin capsules) again produced tail necrosis, however, the control animal also had a gangrenous tail (2 cm of tail tip) which complicated interpretation of the results (Tookey et al., 1972). The final experiment was designed to eliminate (1) the possibility that the fescue ration contained toxins (though it was taken from fields with no history of toxicity), (2) the danger of frost bite, and (3) the corrosive action of the butenolide in the rumen. Cattle were housed in an unheated barn and fed tall fescue or timothy (*Phleum pratense* L.). The butenolide was given as a solution directly into the rumen through a small rumen fistula. The control cattle fed timothy gained weight, whereas those fed tall fescue remained at about the same weight. Cattle on the timothy or tall fescue ration plus butenolide lost weight, but the animals on the timothy ration died; whereas, those on the tall fescue ration survived. Relative to timothy, tall fescue did not enhance the lethal effect of butenolide. The cattle on both timothy and tall fescue showed the beginning of necrosis in the end of the tail. These results indicate that the clinical signs in the tail in the first two tests were due to the butenolide and not to frost bite. Although the butenolide produced some of the clinical signs of fescue foot in cattle (loss of weight, arched back, and gangrene in the distal portion of the tail), it did not produce a characteristic fescue foot or lameness in any of the cattle. Also, the butenolide affected several visceral organs and was very corrosive to the digestive tract, signs not characteristic of fescue foot. Furthermore, no butenolide has been detected (Yates, unpublished data) in ca. 100 samples of tall fescue taken from pastures which produced fescue foot in cattle. Results from cattle tests thus far indicate that neither T-2 toxin nor the butenolide is the primary cause of fescue foot. It remains to be seen whether precursors or derivatives of these mycotoxins are involved in the production of the fescue foot syndrome.

e. Toxins Produced in the Rumen. Research workers at Mississippi State University developed a theory that fescue foot was caused by the fungus, *Aspergillus terreus,* in any one of the following ways:
  i. Toxins could be produced and stored by the plant.
  ii. Toxins may be produced in association with the growth of *A. terreus* within the rumen.
  iii. Antibiotics may be produced in the rumen as a result of the growth of *A. terreus,* and these antibiotics kill beneficial bacteria in the rumen which normally detoxify tall fescue.
  iv. *A. terreus* acts as a mammalian pathogen.

Futrell et al. (1973) reported that in fescue pastures, propagules of *A. terreus* increased with increasing amounts of N applied. (A propagule was defined as "any fungal body that reproduces itself.") They also stated that the number of propagules of *A. terreus* per gram of grass was higher on fescue pastures that produce clinical signs of fescue foot in cattle than for nontoxic pastures. Pastures that produced fescue foot in grazing cattle in 1972 contained *A. terreus*; in these same pastures in 1973 no *A. terreus* was detected and little or no fescue foot was observed. The number of propagules of *F. tricinctum* in these pastures was similar during 1972 and 1973, and was considerably lower than *A. terreus* in 1972.

Farnell et al. (1973) examined a fescue pasture which had produced fescue foot (limping, edema, and reddening of the soft tissues adjacent to the hoof) in 40 of 65 cattle in December 1971. The pasture had been fertilized with chicken litter and, in October, 337 kg of ammonium nitrate/ha had been applied. After removing the cattle, two fistulated steers were placed on this pasture, and periodic rumen content was determined (Table

ANIMAL DISORDERS

Table 13-1—The number of propagules of two fungal species/g of rumen content of two steers grazing toxic fescue pastures at Bassfield (Jefferson Davis County), Miss. in 1972.†

	No. of propagules			
	Steer I		Steer II	
Days on fescue	F‡	A‡	F‡	A‡
0	18	0	12	0
15	27	100	16	75
22	33	600	18	500
34	40	1,200	18	500
41	42	1,500	16	800
49	48	1,450	17	950
55	45	1,400	17	1,000

† From Futrell et al., 1973, 1974. ‡ F = *Fusarium tricinctum* and A = *Aspergillus terreus*.
§ Fescue toxicosis apparent in both steers on day 29.

13-1) (Futrell et al., 1973, 1974; Farnell et al., 1973). Several genera of fungi were detected in the rumen fluid, but only *Fusarium* and *Aspergillus* were consistently present in the amounts shown in Table 13-1. Clinical signs of fescue foot (diarrhea, general inactivity, lameness) appeared in these steers by day 29, when the count of *A. terreus* was between 500 and 1,200 propagules/g of rumen fluid.

These workers sought to protect cattle against fescue foot with thiabendazole (TBZ), an antihelmintic and fungicide [cf. facial eczema of sheep controlled by application of TBZ to pastures (Sinclare and Howe, 1968)]. Six Hereford cows were placed on the previously mentioned toxic pasture; three were treated orally at a rate of 5 g of TBZ/45.4 kg of body weight at the beginning of the experiment and, thereafter, at intervals of 7 days or less. Clinical signs of fescue foot (stiff gait, sore feet, limping, loss of condition, and diarrhea) were observed in all the untreated animals after 3 weeks; treated animals were normal at 4 weeks when the experiment was terminated (Farnell et al., 1973). Cattle treated with 1 g of TBZ/45.4 kg of body weight and placed on this toxic pasture showed clinical signs of fescue foot.

Clinical signs of fescue foot (Table 13-2) were produced in three fistulated steers (two control steers remained normal) on a tall fescue pasture by intraruminal administration of *A. terreus* petri cultures (Futrell et al., 1973, 1974). The steers were dosed intraruminally with 20 petri cultures three times each week for 2 months. *A. terreus* remained viable throughout the digestive tract. In this series of experiments, diarrhea, loss of condition, sore feet, swelling, lameness, and open sores on the feet were listed as clinical signs of fescue foot; gangrene was not observed. The condition of the internal organs of experimental animals was not reported.

F. *Anions*. Scientists at the University of Missouri and Northern Regional Research Center are investigating the cause of fescue foot in cattle. They hope to isolate and characterize the toxin from botanically pure tall fescue, and then determine its source. Garner et al. (1972) and Williams et al. (1975) developed a cattle assay to guide their fractionation studies. The bioassay consists of dosing 180-kg calves intraperitoneally with subfractions

Table 13-2—Clinical signs in three steers given Petri cultures of *Aspergillus terreus* Intraruminally three times weekly for 2 months.†

Days on test	Clinical signs		
	Steer No. 003	Steer No. 009	Steer No. 008
0	No clinical signs	No clinical signs	No clinical signs
6	Lameness in right rear leg	No clinical signs	No clinical signs
7	Lameness and open sore in right rear leg	Lameness in right rear leg	No clinical signs
9	Lameness in both rear feet and right front foot	Lameness in right rear leg with swelling, bleeding skin break at hairline of hoof	No clinical signs
18	Lameness more severe, reluctant to walk	Right rear foot tender with broken skin	No clinical signs
19	Same as above	Right rear foot swollen at hairline of hoof, left swollen and tender	No clinical signs
23	Lameness more severe, open sore larger	Right rear foot, open sore spread to area between digits	Slight lameness in right rear foot, lost large patch of hair on neck
37	Signs more severe	Signs more severe	No clinical signs

† From Futrell et al. (1973, 1974).

of a concentrate obtained from an 80% ethanol extract of tall fescue hay which is harvested from a plot maintained to produce toxic forage. The experimental plots are pure stands of Kentucky 31 tall fescue, which are fertilized both spring and fall (Garner et al., 1972). The hay is generally harvested in December, January, or early February, about the time cattle on these plots show clinical signs of fescue foot. Each subfraction of the concentrate is tested in two cattle (at least one of which is an offspring of a cow which has shown clinical signs of fescue foot), at a maximum rate of 2.7 kg of hay equivalent/day for 14 days. Observations are made daily for gross clinical signs such as soreness, swelling, lameness, coronary band erosion, and gangrene. Usually, the earlier clinical signs are seen after 5 to 7 days. Post-mortem examinations are made on cattle which have died during the experiment and on cattle showing clinical signs of fescue foot. Control calves given saline are included in the bioassay.

Toxic tall fescue hay was extracted with 80% ethanol, and the solvent removed to give a concentrate similar in nature to Concentrate I assayed in cattle by Jacobson et al. (1963). This concentrate was clarified by centrifugation to give an aqueous solution similar to Aqueous II (Jacobson et al., 1963). This solution, upon daily intraperitoneal infusion, produced clinical signs of fescue foot in two cattle (necrotic lesions of the coronary band and tail) (Garner et al., 1972), showing that the rumen is not necessary to the production of clinical signs of fescue foot. A similar aqueous solution was further fractionated to give chloroform soluble cations (includes alkaloids), residual cations, neutrals, and anions. The anion fraction was shown to produce clinical signs of fescue foot in two calves (discoloration of one or both rear coronary bands, discoloration of tail tip, lameness). Post-mortem examination showed hemorrhage of hair follicles in tail tip, vessels congested with sludged blood, and, in one animal, thickened walls and small lumens of the vessels of the coronary band (Fig. 13-6) (Williams et al., 1975). This anion, once characterized, may be a specific compound such as a mycotoxin or toxic phytoalexin, or it may be any of a series of compounds, such as unusual organic acids.

Physiological Mechanisms

There is little evidence on the mechanism by which the fescue toxin(s) produces the gross clinical signs known as fescue foot. A few researchers have speculated that this malady is a contact dermatitis and that it is not systemic. Since the butenolide and T-2 toxin isolated by Yates et al. (1968) from *F. tricinctum* are both skin irritants, Kosuri et al. (1970) endeavored to test the contact dermatitis theory by applying these compounds topically to the rear feet of heifers. Both the butenolide and T-2 toxins, at 1.5, 3.0, or 6.0 mg in 0.2 ml of dimethylsulfoxide were applied to the withers of six heifers. Concurrently, these compounds, at 3 mg in 0.2 ml acetone, were applied to the coronary regions and interdigital spaces of the right hind feet of these heifers. Suitable controls were maintained. Neither the butenolide nor T-2 toxin produced the clinical signs of fescue foot but caused inflamed encrusted lesions that gradually healed.

The physiological mechanism most accepted is that fescue foot is caused by vasoconstriction. The oldest concept is that a vasoconstrictor acts directly, like ergotamine (Mueller-Schweinitzer and Sturmer, 1974). Jensen et al. (1956), who favor this interpretation, relate that the toxic agent in tall fescue causes vasocontriction which leads to thrombosis. The result is ischemia in the extremities. This condition, plus cold weather, can lead to dry gangrene in the extremities. It appears that gangrene occurs most often in those extremities farthest from the heart, i.e. the rear feet or tail.

Alternatively, vasoconstriction may be brought about by thickening of the vessel walls by hyperplasia or hypertrophy. Johnson (as cited by Corley et al., 1973) said that 50% of the animals with fescue foot examined at the University of Missouri School of Veterinary Medicine had swollen vessel walls in the areas adjacent to the affected feet (no evidence of hyperplasia). Since most animals were showing advanced clinical signs, this lesion could be secondary. Corley et al. (1973) suspected that the reason they did not identify a primary lesion was because it was transitory. They report thrombi caused by in vivo platelet aggregation, or by erythrocyte aggregation, could initiate ischemia and then be lysed before any of the thrombi were detected. Williams et al. (1975) reported that three of four cattle showing clinical signs of fescue foot in their fractionation-bioassay experiment had blood vessels from the coronary band area with thickened walls and small lumens.

Whittow (1962) found evidence that variations in skin temperatures of the extremities of the ox are the result of changes in blood flow. When the animal is in a cold environment, it helps to maintain its body heat by restricting blood flow to the extremities. Cattle in the field (Garner, 1973) and cattle in experiments (Jacobson et al., 1963; Tookey et al., 1972; Williams et al., 1975) show a temperature drop in the skin of the extremities, indicating a decrease in blood flow in the extremities. Corley et al. (1973) reported that a temperature decrease would result in an increase in blood viscosity; therefore, stasis of the blood could occur with a minimum vasoconstriction unless blood pressure increased. Corley et al. (1973) also relate that in animals showing clinical signs of fescue foot, new circulation can develop. This finding is consistent with the fact that lame cattle appear to recover, yet remain sensitive to toxic tall fescue. Some cattle also appear to recover while continuing to graze the same pasture which produced clinical signs of the disease.

When one considers the variations in clinical signs in cattle in various regions on toxic fescue—a distinct necrotic foot in Missouri, or swollen, bleeding feet with necrotic patches of skin sloughing off observed farther south—it is difficult to conceive that they are caused by the same mechanism. The Shwartzman reaction or its systemic equivalent, the Shwartzman-Sanarelli reaction (Barrett, 1974) may help to reconcile these apparent differences. In the past, this phenomenon has been considered to be closely related to skin hypersensitivity. However, currently being considered are hypotheses that this phenomenon is a reaction at the blood vessel wall (Barrett, 1974; Gaynor et al., 1970). The Shwartzman reaction can be elicited by injfection of bacterial endotoxin. A second intravenous injection of a bacterial endotoxin 24 hours later produces a hemorrhagic necrosis at the initial injection site. The second challenge may be nonantigenic materials such as starch or agar (Barrett, 1974). Selye (1965) produced acral necrotic lesions in rats by a single challenge of carrageenin (25 to 50 mg carrageenin by subcutaneous infiltration) followed 3 days later by exposing the treated rats to cold. Necrosis of the tail and feet can be produced by forcing the rats to walk on ice for 5 hours, at room temperature. Perhaps in Missouri where temperatures near 32 C (0 F), snow, and ice may occur during the winter grazing season, fescue foot in cattle is produced in the same manner as the necrotic tails and feet in these rats. Although no evidence definitely links fescue foot with the Shwartzman reaction, likewise no evidence eliminates this phenomenon.

A final mechanism which may be considered is that the toxin(s) acts as an antimetabolite. However, since the toxin(s) itself is not known, no research has examined the merits of this theory.

GRASS TETANY

A. Hypomagnesemia in Cattle

The condition referred to as grass tetany in cattle has often been associated with their grazing cool-season grasses. The level of Mg in the blood of cattle is lower than normal when the animals exhibit grass tetany, consequently hypomagnesemia is one of the most common clinical characteristics of true grass tetany. Tetany usually occurs in the spring when pasture growth is very lush; however, cases have been reported in fall-calving cows.

Hypomagnesemic tetany has been described by Sjollema (1932), Dozsa (1959), Curtis and MacMaster (1966), Crookshank and Sims (1955), Custer (1959), and others. The condition is most prevalent in females and usually occurs more frequently subsequent to the second lactation. Cows are also more susceptible to tetany within a few weeks after calving. The initial physical or outward symptoms of grass tetany may begin with the animal having a decreased appetite, a dull appearance, and she may occasionally isolate herself from the remainder of the herd. The animal may also develop a staggering gait and walk in a stiff manner. As the condition progresses, the animal becomes highly excitable, nervous, exhibits muscular tremors, and may even show signs of an instinct to fight or charge surrounding persons. Very shortly after these symptoms become apparent, the animal usually falls to the ground and the tetanic muscular spasms continue, with the legs thrashing the ground. Death will usually follow soon after collapse if the animal does not receive medical treatment.

The lowered blood levels of Mg observed in cattle exhibiting tetany are usually related to a low intake of Mg but also may be influenced by other dietary, hormonal, and environmental factors. Blood levels of Mg can be variable, and values in grazing beef cows average approximately 2 mg/100 ml serum. Allcroft and Green (1934) observed normal blood serum Mg values to range from 1.85 to 3.17 mg/100 ml. Crookshank and Sims (1955) observed a range in serum Mg of 1.0 to 2.8 mg/100 ml, with a mean value of 2.05 mg/100 ml serum. Halse (1970) observed plasma Mg values in cows to range from 1.5 to 3.0 mg/100 ml. Concern is usually expressed when blood Mg falls below 1.5 mg/100 ml in beef cattle. However, Todd and Thompson (1960) reported that blood serum values fell from 2.19 to 1.29 mg/100 ml in lactating beef cows within a few days after they were turned out to graze and had no clinical signs of tetany.

Plasma Mg levels are usually below 1.0 mg/100 ml at the time of tetany (Dozsa, 1959; Stewart, 1954; Kemp and Geurink, 1967). Plasma Mg averaged 0.58 mg/100 ml in five cows while they were exhibiting tetany (T. O. Okolo, J. A. Boling, N. Gay, and N. W. Bradley, unpublished data).

B. Occurrence in Cool-Season Grasses

Grass tetany has been recognized for many years, but an increased incidence of tetany has been suggested to accompany current intensive cow-calf management practices. It is most prevalent in the spring when pasture is growing rapidly. Grass tetany has been reported in many states of the United States and in many temperate countries of the world (Reid and Jung, 1973). It is most likely to occur in areas of the world where spring temperatures range from 5 to 15 C (Church, 1972). Grass tetany cases may be rather sporadic within a specific geographical location and quite variable from year to year. McLaren et al. (1975) recently referred to the observed variation in the incidence of tetany with respect to location and year. They indicated that tetany was one of the greatest causes of death in adult beef cows.

Spring grasses usually contain less Mg than legumes. Therefore, cattle grazing pastures of pure grass stands are likely to be more susceptible to tetany. Also, the first pasture growth in the very early spring is likely to be from grasses rather than legumes, which further accentuates the problem. Excellent reviews of grass tetany, including a summary of incidence of the disorder were compiled by Grunes et al. (1970) and Grunes (1973).

Results of a 3-year survey (1971–73) of grass tetany in beef cows in Kentucky (Murdock et al., 1975) indicated that approximately 1.1% of the beef cow population was affected with the disorder in 1973. Of the cattle diagnosed as having grass tetany during the 3-year study, 37.7% of the cases resulted in death. Underwood (1966) reported that grass tetany usually occurred in older lactating animals within approximately 10 weeks post-calving and that it could occur in animals of diverse ages and stages of lactation. He also pointed out that the economic importance of the disorder was related to the high death rate, and that death was usually sudden and unexpected. However, one must recognize the economic value of the decreased production of those cattle which are included in the incidence statistics, but recover from the syndrome.

C. Factors Affecting Plant Composition of Magnesium and Potassium

Magnesium and potassium were implicated as factors involved in the hypomagnesemic tetany syndrome in the early reports of Sjollema (1932) and Dryerre (1932). Since that time experiments have attempted to elucidate the role of these two minerals in the occurrence of grass tetany. Studies have been oriented towards obtaining information on factors affecting the plant content of Mg and K, and their subsequent influence on animal metabolism. Kemp (1960) compared the serum Mg, herbage Mg, and incidence of tetany in 822 cows. He suggested that 0.20% Mg in the herbage dry matter should be a fairly safe level to prevent the likelihood of grass tetany.

Todd (1961) and Stuedemann et al. (1975) pointed to the seasonal variation in Mg content of grasses. Todd (1961) observed that the Mg content of several grasses was lower (on a dry matter basis) in April than when sampled later in the year. The composition of cool-season grasses also can be changed considerably by fertilization, management, and environmental influences. Wolton (1960) pointed to the antagonism between Mg and K, and also noted that K applications resulted in increased K and decreased Mg contents in plants. Stillings et al. (1964) fertilized S-37 orchardgrass (*Dactylis glomerata*) forages with low or high rates of ammonium nitrate. The high rate of N fertilization resulted in increased total N, NO_3-N, Mg, and K in the plant. Lowrey and Grunes (1968) fertilized rye (*Secale cereale* L.) with N, and P; N, P, and K; N, P, and K plus Mg. Nitrogen, P, and K fertilization resulted in decreased Mg and increased K concentrations in the plant (January and February samplings) as compared with the N and P fertilization regime. Nitrogen, P, K, and Mg fertilization resulted in higher plant Mg concentrations but concentrations of K similar to those observed in the N, P, K fertilized forage.

Kemp (1960) administered different fertilizer applications to permanent pasture grassland. Potassium concentration in the forage was increased 40 to 60% with high rates of K fertilization. When the high rate of K was combined with a high rate of N, a further increase in K concentration was observed in the forage. Magnesium content of the herbage was decreased 15 to 20% by a heavy application of K. However, Mg content increased with increasing levels of N application. Heavy application of N also increased the herbage crude protein.

Hannaway and Reynolds (1976) in Tennessee fertilized Kentucky 31 tall fescue with 0 and 112 kg N/ha as ammonium nitrate. The split applications were applied in October and March. Plant K was higher in March and April because of N fertilization. Magnesium content of the herbage was similar at both levels of N during the months of February through May. Magnesium concentrations averaged approximately 0.30% during February, March, and April and increased during the summer months. Forage dry matter yields and concentration of crude protein in the plant were increased owing to N fertilization. Hannaway and Reynolds (1976), in a subsequent experiment compared three levels of N and two levels of K fertilization. The N levels were 56, 112, and 168 kg/ha in split applications in October, March, and June. The K levels were 0 and 168 kg/ha and were applied in October. At the low level of N, K fertilization resulted in increased K and decreased Mg in the forage at the January through May samplings. Magnesium content in the forage was below 0.20% for both K fertilization levels in April and May. The second application of N in March resulted in an increased K level in April and May. All Mg values were below 0.20% in April and May but increased to approximately 0.30% in June. The third application of N was made in June which is past the time the spring tetany syndrome usually occurs in the southeastern U.S. and the Mg content of the forage reamined at approximately 0.30% from July through October.

Kentucky 31 tall fescue was unfertilized or fertilized with 75 kg N/ha in late winter in Kentucky (W. C. Templeton, Jr., T. H. Taylor, J. A. Boling, and R. E. Tucker, unpublished data). First-cut plots were mowed at weekly intervals from 27 March through mid-July. Nitrogen fertilization resulted in increased total N in the forage. During this same time period, N fertilization resulted in increased K and Mg concentration at most sampling times. Magnesium levels in forage from unfertilized plots averaged 0.22% and from the fertilized plots 0.29% for the 27 March through April samplings. In May, Mg levels averaged 0.18% in the unfertilized forage and 0.20% in the fertilized forage. Reynolds (1975) stockpiled Kentucky 31 tall fescue from September to May and from May to September. The forage was harvested at the end of the stockpiling period. Fertilized at 112 kg N/ha (split applications in mid-to-late March and late May or early June), the forage harvested in May had lower Mg concentrations than that harvested in September. The May harvested forage contained slightly less than 0.20% Mg in 2 of the 3 years of the study. September harvested herbage averaged approximately 0.30% Mg. Potassium was higher in the May harvested herbage than was observed for the September harvest. Fertilization of Kentucky 31 or 'Kenwell' tall fescue in north Georgia with NH_4NO_3 or broiler litter increased N concentrations in the plant. Magnesium and Ca levels in the herbage increased with 448 kg N/ha. Broiler litter fertilization increased plant K and Mg content and the K/(Ca + Mg) ratio (D. L. Grunes, S. R. Wilkinson, V. A. Lazar, P. K. Joo, W. A. Jackson. 1974. Effect of broiler litter and nitrogen fertilizer on the grass tetany potential of tall fescue. Agron. Abstr. p. 139).

As indicated in the review by Grunes (1973), several investigators have reported increased incidence of tetany following cool periods. Potassium is more readily absorbed by the plant at low temperatures than any other cation (Barnett and Reid, 1961). Therefore, during cool periods grasses have a higher concentration of K and a reduced level of Mg. Utilizing growth chambers, Grunes (1967) grew perennial ryegrass (*Lolium perenne*) at temperatures of (20 C day and 14 C night) and (26 C day and 23.3 C night). He observed decreased Mg levels in the herbage at the cooler temperatures. Also, the values for K/(Ca + Mg) were higher at the cooler temperatures. Voisin (1963) reported that a rapid rise in temperature after a cool period resulted in an immediate increase in plant uptake of K. Calcium and Mg content of the plant did not change significantly; therefore, the ratio of K/(Ca + Mg) was increased.

Certain plant organic acids have been implicated in the induction of the tetany syndrome. Stout et al. (1967) collected plants of 94 forage species for aconitate determination. Of the two aconitate stereoisomers, the *trans* form accumulated to higher levels in most species. Tall fescue was classed as a "medium" accumulator of *trans*-aconitate (medium = 0.2 to 1%). The possible role of such organic acids may be that of complexing Ca and Mg in a manner that would make these two cations unavailable to the animal. Malic acid content increased in Kentucky 31 and Kenwell tall fescues fertilized with either NH_4NO_3 or broiler litter. Citric acid was also increased, but the extent was not so great as was observed for malic acid (D. L. Grunes, S. R. Wilkinson, V. A. Lazar, P. K. Joo, and W. A. Jackson. 1974. Effect of broiler litter and nitrogen fertilizer on the grass tetany potential of tall fescue. Am. Soc. Agron. Abstr., p. 139). Teel (1966) observed increased succinate, malate, and total organic acids owing to fertilization of tall fescue with ammonium nitrate. Hannaway (Hannaway, D. B. 1975. Chemical composition and yield of tall fescue as influenced by N and K fertilization. M.S. Thesis. Univ. of Tennessee, Knoxville) observed increased malic acid concentrations in tall fescue in the spring months owing to fertilization with 112 kg N/ha.

D. Magnesium Metabolism in Animals

Magnesium represents approximately 0.04% of the animal body (Maynard and Loosli, 1969). The skeleton contains about 70% of the total body Mg, and the remainder is distributed throughout the body fluids and soft tissues. It was suggested that about one-third of the bone Mg could be mobilized for soft tissue and body fluid needs if inadequate dietary intake persists. Magnesium is involved in many enzymatic reactions in metabolism. It is required for cellular oxidation, especially in those reactions leading to ATP formation. It activates the enzymatic reactions which require thiamine pyrophosphate. Magnesium is also involved in many reactions in protein and lipid metabolism (White et al., 1968; Pike and Brown, 1975). The biochemical involvement of Mg in metabolism has also been summarized by Wacker (1965), Rook and Storry (1962), and Wacker and Parisi (1968).

In addition to tissue requirements, Martin et al. (1964) demonstrated ruminal requirements for Mg. They observed decreased ruminal cellulolytic activity both in vitro and in vivo in steers fed a Mg-deficient diet. The decreased cellulose digestion was accompanied by a depression in feed intake. Ammerman et al. (1971) observed that feeding sheep a diet almost devoid of Mg for 4 days resulted in a 32% reduction in voluntary feed intake over that of control animals. Magnesium added to the diet restored normal feed intake. Insufficient Mg also resulted in decreased cellulose digestion.

O'Kelley and Fontenot (1969) calculated dietary Mg levels necessary to maintain blood levels of 2.0 mg/100 ml serum for beef cows in early, mid, and late lactation. They determined that 20.9, 22.1, and 18.0 g dietary Mg was required daily to maintain blood serum levels of 2.0 mg/100 ml at the three stages of lactation, respectively. Based on an intake of 10.2 kg of dry matter/day, these values represented 0.18, 0.19, and 0.16% of the diet, respectively. Kemp (1960) suggested that 0.20% Mg, on a dry weight basis in herbage, should be sufficient to meet the cow's requirement and minimize possible occurrence of tetany. The Mg requirement of the gestating cow is lower than that of the lactating cow. O'Kelley and Fontenot (1973) determined that the requirement of dietary Mg for cows at 145, 200, and 255 days of gestation to maintain blood serum levels of 2.0 mg/100 ml was about 0.12, 0.10, and 0.13% of the diet, respectively.

Magnesium availability in forages has been reported to be rather low. Kemp et al. (1961) reported that 17% of the herbage Mg was available to cows. The range in availability was from 7 to 33%. They observed increased Mg availability as the herbage matured. Stillings et al. (1964) also observed rather low values for apparent availability of Mg from forage. Apparent Mg availability ranged from 11 to 16% in high N containing forages and from 18 to 24% in low N containing forages. Notably the high N forages were also higher in Mg and K. Butler (1962) found that pastures associated with the tetany syndrome had lower Mg and higher K, and a higher ratio of K/(Ca + Mg) than was observed for less tetany prone pastures.

Gerken and Fontenot (1967) observed that Mg availability for steers was higher from magnesium oxide than from dolomitic limestone. Ammerman et al. (1972) compared the availability of Mg from magnesite, reagent-grade magnesium carbonate, magnesium oxide, and magnesium sulfate. Magnesium was well-utilized from magnesium carbonate, magmesium oxide, and magnesium sulfate; however, magnesite magnesium was not well utilized. Moore et al. (1971) compared Mg absorption in steers fed supplemental Mg as magnesium oxide, magnesium carbonate, and dolomitic limestone. Magnesium absorption was greater from magnesium oxide than from dolomitic limestone, with magnesium carbonate having an intermediate value. However, Mg retention was lowest for magnesium carbonate.

In addition to magnesium, two of the major factors implicated in the development of hypomagnesemic tetany have been the N and K content of herbage. Kemp (1960) noted that high rates of N or potash fertilizer applied to forages resulted in decreased serum Mg levels.

Suttle and Field (1969) fed ewes different levels of Mg and K. Plasma Mg concentrations were reduced by reduced dietary Mg or increased K intake. They suggested that the two effects were additive. Newton et al. (1972) fed lambs diets supplemented with 0.6 or 4.9% K but containing the same amounts of Mg. The apparent absorption of Mg was decreased by 46% by the high level of K intake. The average absorption values for the low K and high K diets were 49.1 and 26.4%, respectively, when expressed as a percent of Mg intake. Utilizing ^{28}Mg in a subsequent experiment, they concluded that the high dietary K reduced absorption from the intestine rather than increasing its excretion into the lumen of the gut.

Since high dietary N has been implicated as a possible factor in the occurrence of tetany, Moore et al. (1972) observed the influence of form and level of N on Mg metabolism. They fed lambs 10 and 32% crude protein on a dry matter basis. Within each protein level, 0 and 35% of the crude protein was fed as urea (nonprotein N). Ruminal fluid pH was significantly increased in lambs fed the urea containing diets. However, apparent absorption of Mg was not significantly different among the groups of lambs fed the four diets. These data suggest that the high levels of ammonia in ruminal fluid of the lambs on the urea-supplemented diet did not interfere with Mg absorption. Urinary Mg excretion was greater in lambs on the high N diets, regardless of form of N. The increased urinary Mg excretion resulted in a significant decrease in Mg retention in lambs on the high N diets.

Fontenot et al. (1973) conducted studies designed to separate the effects of N and K on Mg absorption and retention. In summary, high dietary N does not seem to affect absorption of Mg, but does result in increased urinary excretion of Mg. The decrease in Mg absorption owing to high K levels in diets appears to be rather conclusive.

Several plant organic acids have been implicated in influencing the onset of tetany. Bohman et al. (1969) experimentally induced tetany resembling grass tetany in cattle by oral administration of KCl plus citric acid or *trans*-aconitic acid. The administration of KCl or the acids alone did not produce tetany symptoms. They observed that administration of KCl over long periods lowered blood Mg, but a combination of KCl and citric acid lowered blood Mg more rapidly. House and Van Campen (1971) fed lambs different levels of K and citric acid and observed that high levels of K decreased absorption of Mg, resulting in an increased fecal Mg output. They suggested that the level of dietary citric acid used in their study did not significantly affect Mg metabolism. Even though the values were not significantly different, lambs fed KCl plus citrate had the lowest plasma Mg concentrations. Burt and Thomas (1961) fed citric acid (as sodium citrate) to calves at 1% of the dietary dry matter. They observed a significant depression in serum Mg in calves fed the citric acid-containing diet.

Other factors influence Mg metabolism in ruminants. Kemp et al. (1966) reported that additions of animal fat to rations of milking cows resulted in decreased Mg availability, as indicated by increased fecal excretion of Mg. This finding suggests the formation of Mg-containing soaps from which the Mg is unavailable. They also observed that increased higher fatty acid content of the plant was associated with an increased crude protein content. They suggested that the influence of N fertilization on the Mg availability to the animal could be at least partially explained by the fat content in the rations.

Since many tetany cases occur after cool periods, the thyroid status of the animal has been suggested to play a role in the occurrence of tetany. Madsen et al. (1975) studied ^{28}Mg metabolism in sheep using normal and hypothyroid sheep. Magnesium absorption and retention was higher in the hypothyroid sheep. The decreased rate of passage of feed residues through the gastrointestinal trace of the hypothyroid sheep suggested that retention time in the gut influenced Mg availability to the animal.

E. Corrective Measures for Grass Tetany

Several systems have been devised in an attempt to provide Mg to ruminants on a daily basis. One of the most critical problems that has faced cow-calf producers in supplementing Mg to cattle has been the palatability of the Mg source. Since Mg is not stored readily in the ruminant body, a continuous supply is needed. The level of supplemental Mg needed for a particular herd may vary. Usually, 57 g/head/day of magnesium oxide is recommended to meet the requirements of lactating cows. However, in areas where forage is extremely high in K, higher levels of Mg may be needed to overcome the antagonistic effects of K on Mg absorption. It has also been suggested that one of the major reasons that the incidence of tetany is higher in older cows is that the skeletal Mg is not mobilized as readily in the older animal as it is in the young.

Kemp and Todd (1970) used magnesium alloy bullets which were placed in the rumen of cows. The cattle which had bullets placed in the rumen had a rapid decline in plasma Mg. Also, they observed bullets in the pasture which came from both regurgitation and fecal excretion. Consequently, this method of supplementation is not adequate for cows on an extremely low Mg diet.

Pasture dusting procedures were discussed by Young (1973). He suggested that 12 to 15 kg finely ground calcined magnesite or 55 kg dolomite/ ha was successful in preventing tetany. The material should be applied at 7- to 10-day intervals. The level of rainfall is the major factor affecting the quantity of Mg adhering to the herbage over a period of time. It also appears that the quantity of standing herbage is extremely critical in any practice requiring application of material to the pasture. Stuedemann et al. (1973a) reported that adding magnesium oxide to a mineral mix or to forage by foliar dusting of tall fescue did not consistently elevate blood Mg concentrations. They observed that magnesium oxide applied as a dust was easily blown or washed from the herbage. However, Stuedemann et al. (1974) applied Mg to tall fescue pastures as a magnesium oxide-bentonite-water slurry. Blood Mg levels of beef cows grazing these pastures increased within 48 hours after slurry application to the herbage. Cases of tetany were evident in the control group; however, tetany did not occur in cattle grazing the treated pastures. The bentonite also increased the persistence of Mg on the grass.

Hansard et al. (1975) discussed several supplements for cattle grazing pasture and noted the criticality of consumption of supplements by the cattle if they are to meet the supplemental Mg needs. Legumes are generally higher in Mg than grasses and supplementation of spring grass with legume hay is a practical means of Mg supplementation from a natural feedstuff. Magnesium for cows in high risk situations probably should be fed as an energy-containing supplement in order to insure adequate daily intake of Mg. Murdock et al. (1975) utilized a mixture of 20% steamed bone meal, 40% magnesium oxide, 20% plain salt, and 20% trace mineral salt for cows in "mild risk" situations. Cows consumed approximately 35 g of the mixture/day, resulting in a Mg intake of about 8 g. However, in low or mild risk situations, providing such a mineral supplement assumes that the cow will be obtaining a portion of her Mg requirements from the herbage being grazed. It should be emphasized that cattle receiving all of their supplemental Mg from a mineral mix should become accustomed to the mix several weeks before the initiation of the tetany season. Murdock et al. (1975) also developed supplements for high risk situations designed to supply 57 g of magnesium oxide/day. A free-choice supplement of 39% grain, 19% soybean meal, 10% dicalcium phosphate, 7% magnesium oxide, and 25% salt should give an intake of about 900 g/head/day. A hand-fed supplement at 0.9 to 1.4 kg/head/day consisting of 66% grain, 17% soybean meal, 10% dicalcium phosphate, and 7% magnesium oxide should provide adequate Mg for the cow. Feeding of these supplements should begin at least 2 to 3 weeks before the tetany season usually begins in a particular area.

NON-PROTEIN NITROGEN

The non-protein N (NPN) fraction of forage can be an important fraction and does vary with the physiological state of the plant. In general, the more favorable the growth conditions the higher the NPN fraction as well as the total N content. NPN content expressed as a percentage of forage dry matter decreases as growth progresses to maturity. The major portion of the NPN fraction consists of substances concerned with the synthesis of protein in plants. From the standpoint of animal response, the most important types of NPN substances are the alkaloids, NO_3-N, and nitrogenous compounds with a potential for rapid release of ammonia.

A. Alkaloids

Alkaloids of Tall Fescue

The alkaloid fraction of tall fescue has been separated into at least 9 components by paper chromatography (Yates, 1963) and 11 components by thin layer chromatography (Gentry et al., 1969). Perloline was first identified from perennial ryegrass (Melville and Grimmett, 1941). Grimmett and Waters (1943) identified perloline in tall fescue. Perlolidine was first isolated from perennial ryegrass (Grimmett and Waters, 1943). Perlolidine has not been positively identified in tall fescue, but by using two chromatography media and two solvent systems on each, Bush and Jeffreys (1975) reported the amounts of perlolidine in tall fescue forage. Perloline and perlolidine are novel because they are the only natural products with a diazaphenathrene (benzo[c][2,7] naphthyridine) ring structure.

Yates and Tookey (1965) described the pyrrolizidine alkaloid, loline, termed festucine in the original publication, from tall fescue. Robbins et al. (1972) identified the loline derivatives of N-formylloline, N-acetylloline and demethyl-N-acetylloline in tall fescue forage and seed. The alkaloids identified in tall fescue have been similar to those in ryegrass; therefore, in the *Festuca-Lolium* hybrids, one could reasonably expect to find many of the ryegrass alkaloids. Jeffreys (1970) reported the isolation of perlolyrine from ryegrass and discussed its relationship to other alkaloids in ryegrass annuloline, lolinine, and norloline. Culvenor (1973) reviewed the occurrence and toxicity of the alkaloids in Gramineae, including some of the simpler ones in ryegrass such as halostachine, octopamine, and histamine.

Alkaloids and Genotype Interaction

Genetic differences in perloline content of ryegrass forage were reported by Butler (1962). He did not find a significant correlation between perloline content and shoot growth. Buckner et al. (1973) found that *Lolium* sp. had less perloline than *Festuca* sp. *L. multiflorum* had the lowest concentration of perloline and *F. elator* had the greatest accumulation of perloline. Of the grasses with intermediate levels of perloline, *F. arundinacea* had higher levels than *F. gigantea* and *L. perenne*. Broad sense heritability estimated varied from 0.57 to 0.80 among polycross progenies of these groups of genetic materials. Perloline content in progeny of *L. multiflorum* × *F. arundinacea* analyzed by a modified diallel analysis suggested that perloline content is controlled primarily by a few major genes with a high degree of dominance for low perloline levels (Cornelius et al., 1974). Among tall fescue cultivars, the authors have observed 'Alta', 'Goar',

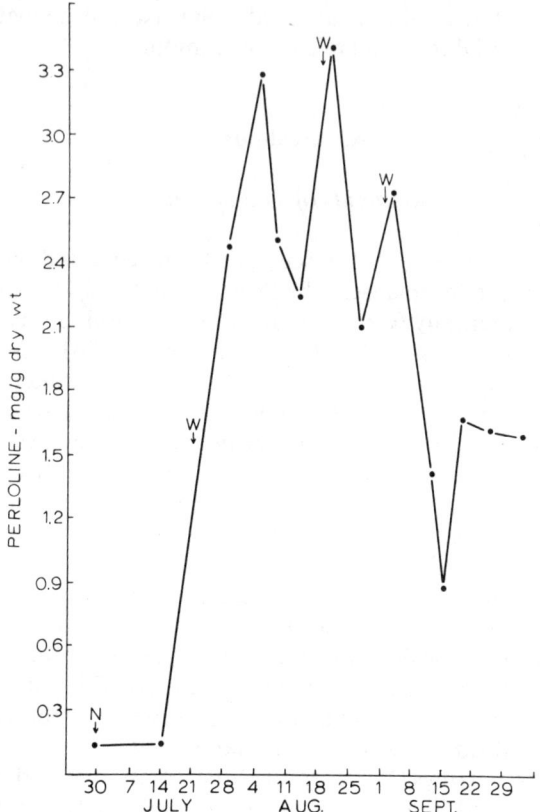

Fig. 13-7—Perloline content of Kenwell tall fescue sampled at weekly intervals during the summer and fall at Lexington, Ky. N designates application of 225 kg NH$_4$NO$_3$/ha. W designates the application of irrigation or rainfall in excess of 50 mm.

'Fawn', 'Manade', 'S-170' to be low; Kentucky 31, intermediate; Kenwell, high in perloline content. Kenhy, the ryegrass × tall fescue hybrid derivative, has perloline levels similar to those of Kentucky 31.

Influence of Environmental Factors on Alkaloid Composition

Bennett (1963) reported that increased fertilizer N increased the perloline content of perennial ryegrass forage, whereas increased soil phosphate levels had no effect on perloline levels in the forage. Gentry et al. (1969) reported a maximum of 6,700 µg perloline/g dry forage for tall fescue fertilized with 112 kg N/ha. More typical perloline levels for tall fescue forage in mid-summer are 1,000 to 3,000 µg/g. Many other reports in the literature confirm the observation of increased perloline with increased soil N. Bush and Buckner (1973) reported perloline accumulations in tall fescue increased as the amount of N in the nutrient solution increased. Wilkinson et al. (1971) found 1,500 µg perloline/g in Kentucky 31 pastures fertilized with 24 metric tons/year of broiler house litter (ca. 867 kg N/ha/year). Pastures fertilized with broiler litter or ammonium nitrate had low levels of perloline in spring and fall with higher levels in summer (Gentry et al., 1969; Robbins et al., 1973). Gentry et al. (1969) reported that perloline accumulated in response to hot, dry weather as well as N fertilizer. Figure 13-7 shows the fluctuation that can occur in the perloline accumulation pattern during late summer. Perloline levels increased shortly after the addition of N fertilizer as expected, but more important were the rapid fluctuations of perloline content. In this experiment perloline losses were associated with heavy rainfall or the application of irrigation water. The authors have observed the same seasonal accumulation pattern for perlolidine as perloline, as well as increased levels of perlolidine with added N fertilizer. Nitrogen fertilizer rates did not influence the levels of N-acetylloline and N-formylloline in tall fescue (Robbins et al., 1973), but concentrations of these two alkaloids were greatest in mid-summer and lower in spring and fall.

Perloline and total alkaloids exhibited a seasonal effect in perennial ryegrass grown in southeast Australia (Aasen et al., 1969). They also found the alkaloid levels to fluctuate widely, but observed a general inverse relationship between the number of hours of sunshine in the preceding 2 days and alkaloid levels. New seedlings, 8 to 15 days old, of perennial ryegrass contained the greatest concentration of alkaloids (1,400 to 3,200 µg/g) and alkaloid content decreased with successive harvests for the first 50 days. Perloline and total alkaloid concentrations were greatest in the stem, followed in decreasing order by the roots, leaves, heads, and seeds of Kentucky 31 tall fescue plants in the dough stage (Gentry et al., 1969). Vegetative regrowth of shoots contained 1,500 µg perloline and 1,800 µg total alkaloids/g, whereas the roots during vegetative regrowth had 1,900 µg perloline/g.

Relationship Between Alkaloids and Animal Performance

The most important animal response to grazing tall fescue is poor animal performance and lack of weight gain and/or milk production. This response commonly occurs during summer and is referred to as summer syndrome. Summer syndrome does not occur in cattle grazing tall fescue each grazing season or necessarily for a full grazing season when it is observed (Jacobson et al., 1970; Mott et al., 1971). (Readers are referred to Chapters 11 and 12 for a full discussion of animal performance on tall fescue.) Summer syndrome is most often noticed in summer, especially August and September. The greatest accumulation of alkaloids—perloline, perlolidine, and the lolines—in tall fescue corresponds in time to greatest occurrence of summer syndrome.

Early work on the biological activity of perloline indicated it might have considerable detrimental effects on grazing animals (Cunningham and Clare, 1943). In late summer, perloline levels of 2,000 to 3,000 µg/g dried forage have been observed frequently in tall fescue pastures (Gentry et al., 1969; Bush and Buckner, 1973; Buckner et al., 1973; Webb, 1972). Perlolidine levels of 800 to 1,300 µg/g and levels of loline derivatives of 1,000 to 2,000 µg/g have also been observed in tall fescue pastures during the summer (Bush, unpublished data; Robbins et al., 1973). Based on intake data for animals grazing tall fescue (Jacobson et al., 1970), a 250-kg animal would ingest 18 to 24 g of perloline and perlolidine/day. Studies by Bush et al. (1970, 1976) have demonstrated inhibition of in vitro ruminal functions by perloline and perlolidine. Cellulose digestion was significantly inhibited by 0.1 to 1.0 mM perloline. If one assumes a rumen volume of 25 liters and 18 g of alkaloid ingested, a rumen concentration of 2 mM perloline could easily occur. In vitro inhibition of cellulose digestion and volatile fatty acid production by perloline and perlolidine apparently were caused by the inhibition of growth of rumen bacteria (Bush et al., 1970; 1976). Rumen inoculum from animals on a high perloline tall fescue pasture digested powdered dried forage high in perloline content 27% better than rumen inoculum from animals on low perloline tall fescue pastures (Bush and Buckner, 1973). No difference was observed in the digestion of the low perloline forage with the two rumen inocula. A similar response was observed using the in vivo nylon bag technique. These observations suggest alkaloid components of the forage alter the microflora populations of the rumen which influence the rate of digestion.

Lambs fed a diet containing 0.5% perloline monohydrochloride had lower cellulose and crude protein digestion than lambs on a control diet (Boling et al., 1975). Apparent digestibility of crude fiber, N-free extract, ether extract, and ash also tended to be lower in the lambs on the perloline diet than the control group. Nitrogen retained per day was 0.94 g for the control group and 0.44 g for the perloline fed lambs. Body temperatures of the perloline fed lambs were higher than that of the control group on days 10 and 11, the last 2 days of the trial. This was a short-term experiment with perloline levels at the upper levels of those observed in pastures and the animal responses were similar to those often reported for summer syndrome (Jacobson and Hatton, 1973).

Aasen et al. (1969) produced symptoms of ryegrass staggers by parenteral administration of perloline to sheep. Dose levels used were amounts one would expect the animals to obtain by grazing ryegrass in Australia. However, the levels of perloline in the perennial ryegrass used were much lower than those typically observed in tall fescue. Ryegrass staggers or similar symptoms have not been considered a problem in animals grazing tall fescue. Unless other compounds are in ryegrass that are not in tall fescue which activate or are synergistic with perloline in causing ryegrass staggers, one must conclude that perloline is not involved in ryegrass staggers because, if it were, ryegrass staggers would be an important problem in animals grazing tall fescue.

The loline alkaloids have decreased feed intake and weight gains when fed to rats (Robbins et al., 1972). N-acetylloline and N-formylloline were not metabolized and did not inhibit cellulose digestion in in vitro rumen fermentation experiments. These results cannot be readily extrapolated to grazing ruminants, but the loline derivatives could be decreasing forage intake and not be inhibiting digestion. Because the seasonal accumulation pattern of the pyrrolizidine alkaloids coincides with summer syndrome, further investigations are needed for evaluating their effect on performance of ruminants. Additionally, many unidentified alkaloids with unknown biological activity in tall fescue may be pyrrolizidine-type alkaloids.

From the observations that tall fescue alkaloids have a seasonal accumulation maxima coinciding with summer syndrome and the observations reported from in vitro and in vivo experiments with tall fescue alkaloids in the diet, elevated alkaloid concentrations in the forage and poor animal performance appear to be associated. Bush and Buckner (1973) hypothesized that the in vivo effects of the alkaloid fraction in tall fescue is the inhibition of microflora activity, particularly cellulolytic activity, and subsequent decrease in the energy and nutrient availability to the animal. As the rate of digestion decreases, the rate of passage through the animal decreases, and the intake of the forage by the animal is reduced. The low palatability, low intake, and poor performance observed with animals grazing tall fescue (Cowan, 1956; Buckner et al., 1967; Webster and Buckner, 1971) could result from the action of alkaloids on the microflora in the rumen. The in vivo inhibition of crude protein digestion by addition of perloline to the diet (Boling et al., 1975) supports this hypothesis and more strongly warrants further research to test the hypothesis. To test the hypothesis adequately, isogenic lines for the diazaphenathrene alkaloids through plant breeding must be developed and these forages tested with grazing animals under many different stress conditions.

Relationship Between Alkaloids and Fungi

The alkaloid content of tall fescue was reduced in plants infected by *Rhizoctonia solani* and *Helminthosporium vagans* (Gentry, 1968). Lesions produced by *R. solani* and *H. vagans* contained less perloline than tissue surrounding the lesions and less perloline was in tissue surrounding the infection sites than in healthy tissue. A high positive correlation was obtained between perloline concentration and resistance to *R. solani* and *H. vagans*. No association between *Balansia* sp., *Fusarium* sp., or *Aspergillus* sp. infection of tall fescue and alkaloid concentration in the forage has been reported. (The reader is referred to section I.C. 1.) Insufficient data are available to determine if the alkaloid fraction in tall fescue is important to mycotoxin production and the fescue foot syndrome. As the causative agents of fescue foot are elucidated, the interaction with the alkaloid fraction of the forage will need to be considered.

Potential for the Future

The present association of high alkaloid concentrations with summer syndrome warrants work on development of genetically low alkaloid cultivars of tall fescue. The development of low alkaloid cultivars, especially low in the diazaphenanthrene alkaloids, is a real possibility because the low perloline genotype was dominant (Cornelius et al., 1974) and the genetic content depended on relatively few genes (Buckner et al., 1973). Ideally the low perloline materials would possess the vigor, disease resistance, and excellent agronomic characteristics of present cultivars. With such materials, one could test the effect of alkaloids on animal performance, and if alkaloids are detrimental to animal performance the plant breeder would have the genetic lines from which to develop new cultivars to solve the problem.

If high alkaloid (diazaphenanthrene and loline type) concentration is not a cause of disease resistance, but only associated with disease resistance, the high alkaloid materials would be a genetic source for resistance in breeding research. Conversely, if the alkaloids are involved with disease resistance, but are not involved in animal performance, they would be a useful tool for screening for sources of resistance as well as the genetic transfer of resistance. High and low alkaloid materials have tremendous potential use by plant breeders in development of new cultivars.

B. Ammonium Toxicity

Ammonium toxicity is usually associated with the use of urea as a feed additive. Ammonium toxicity from NPN compounds usually occurs when hungry animals are overfed NPN or when NPN-containing feeds are improperly mixed. Toxicity symptoms in cattle include uncoordination, tetany, slobbering, bloating, and labored breathing. Death is caused by acute circulatory collapse. The ruminal contents have a strong smell of ammonia. The general opinion is that the ammonium ion is the toxic factor, owing to the rapid increase in blood ammonium noted with toxicity of NPN compounds (Church, 1969). Many ammonium salts caused similar responses to those of urea and were about as toxic (Repp et al., 1955).

Toxicity symptoms to urea occur rapidly and are postulated to be due to the high urease activity in the rumen (Jones et al., 1964). The lack of toxicity of some NPN substances, especially the amides, is presumed to reflect a low amidase activity in the rumen (Church, 1969). This hypothesis is not unanimously accepted (Maddaloni, 1964) and needs to be substantiated by further research data. Ammoniated carbohydrates found in feeds apparently are toxic per se and are not toxic because of elevated blood ammonia levels (Church, 1969). These observations make it imperative that more components of the NPN fraction in tall fescue be identified.

Animals in poor condition could obtain lethal amounts of urea by ingesting 0.44 to 0.55 g urea/kg body weight. Assuming a 275-kg animal and a forage with 3% total N and 30% of the total N in NPN, the animal would have to ingest only 6 to 7.5 kg forage to obtain an equivalent amount of N to the lethal urea N dose.

Although data in the literature are scarce, 3% total N with one-third in the NPN fraction is not unrealistic. Ferguson and Terry (1954) reported 2.9% total N in ryegrass with 29.5% in the NPN fraction and of the NPN,

47% was amino N. Early work by Greenhill (1936) supports the foregoing calculations in that his observations showed that when the NPN exceeds 30% of the total N, digestive disturbances in the grazing animal may result. He reported NPN of ryegrass pastures causing digestive disturbance as a percentage of the total N in the spring to average 44%; however, during other times of the growing season the NPN averaged 17% of the total N fraction. Data on the NPN fraction in tall fescue or extensive data for any cool-season grass are not available because most reports of protein N in the literature are obtained by multiplying Kjeldahl N by 6.25 and it is impossible to calculate a NPN value. Data from tall fescue on maintenance fertility levels show 14 to 26% of the total N as NPN (Bush, unpublished data). From the little data available on tall fescue, we know protein levels to be 10 to 15% and total N levels to be 2.5 to 4%; therefore, the NPN levels would be sufficient to produce toxicity as described above. Jacobsen (1974) reported that as total N increased, the protein N percentage remained constant, indicating that the total amount of NPN would increase as total N increased at any stage of plant development. These data and the report by Maddaloni (1964) suggest investigations need to be directed at the NPN fraction of highly fertilized pastures.

C. Nitrate Toxicity

Description

Nitrate toxicity has been reviewed thoroughly by Wright and Davison (1964) and Singer (1972) and will be covered herein only to update the reader and as a preface to nitrate accumulation in tall fescue. The nitrate ion is relatively nontoxic to animals as it is readily absorbed and readily excreted. The term, nitrate toxicity, as commonly used, is actually nitrite toxicity. Ingested nitrates are metabolized (reduced) in the rumen in a series of steps from nitrate to ammonium; nitrate → nitrite → nitrate oxides → hydroxylamine → ammonia. Nitrates may not be reduced at all, or may be reduced to one of the intermediate forms and accumulate. The rate and degree of reduction depends on the type of microflora present in the rumen and the type of food ingested (Sapiro et al., 1949; Wang et al., 1961). If ingested nitrate is not rapidly reduced, it is absorbed from the gastrointestinal tract and apparently remains unchanged in the bloodstream. Excessive quantities have a strong diuretic effect and can bring about alkalosis, dehydration, and cause an excess excretion of certain halogen ions (Singer, 1972).

However, most often nitrate is reduced to nitrite in the rumen (Sapiro et al., 1949). If the nitrite is not reduced further, the nitrite can cause several conditions detrimental to animal health. These include methemoglobinemia, vasodilation leading to cardiovascular collapse, nitrite syncope, and abortion (Singer, 1972). The formation of methemoglobin is most often associated with nitrate toxicity. Nitrite oxidizes the ferrous iron of hemoglobin to ferric iron, producing the chocolate-brown methemoglobin, which is not capable of releasing oxygen to body tissues (Winter, 1962). When levels of methemoglobin become sufficiently high the first sign of nitrate toxicity appears as a brownish discoloration of white areas on the skin and the nonpigmented mucous membranes of the eyes, nose, mouth, and vulva. As the syndrome progresses, a staggering gait, rapid pulse, frequent urination, and labored breathing develop, followed by collapse, coma, and death.

However, Holtenius (1957) suggested that the conversion of hemoglobin to methemoglobin may not be the principal factor leading to death. He reported that up to 65% of the total hemoglobin could be in the form of methemoglobin and that his experimental animals could be lying in coma without any decrease in oxygen consumption. He suggested that the vasodilator effect of nitrate, with an associated drastic decrease in blood pressure, to be of primary importance. The lowering of the blood pressure is responsible for the cyanosis and the collapse of sheep that had been administered sodium nitrate before methemoglobinemia occurred.

Nitrite syncope is sudden, peripheral circulatory failure. A severe drop in blood pressure follows the collapse of the peripheral circulatory vessels. Nitrite collapse has not been studied extensively in ruminants, but syncope can occur with doses of nitrite that are insufficient to cause methemoglobinemia. Singer (1972) reported that animals with high nitrite levels in the blood and low methemoglobin levels developed syncope and died after exertion.

The effect of nitrates or nitrites on abortion has not been clearly elucidated. The effect of various levels of nitrate on cattle and sheep were variable (Setchell and Williams, 1962; Simon et al., 1958, 1959; Winter, 1962; Eppon et al., 1960), but poisoned pregnant cows often aborted soon after they had recovered from the poisoning. No other aspects of reproduction or growth were apparently affected (Davison et al., 1964; Jones et al., 1966).

Toxic Levels

Nitrates may not be reduced in the gastrointestinal tract or they may be reduced to one of the intermediates or ammonium, and accumulate. The extent and rate of reduction influence the toxicity of nitrate accumulation in feedstuffs. As little as 500 ppm NO_3-N in water has caused poisoning and death in sheep (Singer, 1972), whereas in other experiments 8,000 ppm NO_3-N in feed had no effects on sheep. The nitrate in the feed was converted to ammonia, and the sheep remained in excellent health. The rate and degree of reduction depends on the microflora present in the rumen and the diet (Holtenius, 1957; Sapiro et al., 1949). Sapiro et al. (1949) and Emerick et al. (1965) found that glucose increased nitrate and nitrite disappearance in the rumen. Sokolowski et al. (1960) showed that complete rations offered more protection against nitrates than did poor quality alfalfa hay.

In addition to acute nitrate toxicity, sublethal or chronic toxicities have been reported. These include impaired vitamin A, vitamin E, and iodine utilization or thyroid insufficiency. In dairy cows it appears that nitrate is unlikely to lower milk production unless it is fed at a level that keeps the cows on the threshold of collapse, or at a level that lowers feed consumption. Bradley et al. (1940) established the lower limit for toxic hay to be 2,000 ppm NO_3-N. Garner (1958) indicated that sublethal toxicity occurred and that the maximum safe level of NO_3-N in forage was 700 ppm. Davison et al. (1964) and Jainudeen et al. (1965) concluded from experiments with lactating dairy heifers that there is no sublethal toxicity of nitrates and that the lower toxic level was 4,600 ppm of NO_3-N in forage.

Factors Affecting Nitrate Accumulation in Forage

Nitrate accumulation in plants may result from plant characteristic and/or environmental factors. Stems usually contain more nitrate than leaves, and leaves contain more nitrate than floral parts (Wright and Davison, 1964). The most important factor in nitrate accumulation is the N availability in the soil. The higher the soil N the greater the nitrate accumulation (Wright and Davison, 1964). If high soil N is associated with adequate K greater nitrate accumulation will occur than if Na or some other cation must be absorbed with the nitrate. Phosphorus has had a variable influence on nitrate accumulation. High levels of nitrate accumulations are often associated with moisture stress.

From a practical point of view, moisture must be available long enough to permit suitable plant size for nitrate to be accumulated, but a moisture stress slows the rate of nitrate reduction while the soil is still wet enough to provide for nitrate uptake. Under these conditions, accumulation of nitrate may proceed rapidly. Also, reduction in light intensity is associated with an increase in nitrate content (Wright and Davison, 1964). Temperature is often confounded with moisture stress and light; when it is not, the influence of temperature will depend on type of plant and source of N.

Accumulation of Nitrates in Tall Fescue

Tall fescue accumulates significant amounts of NO_3-N and is grouped with other cool-season grasses such as orchardgrass, bromegrass, and timothy in potential to accumulate nitrate. Tall fescue generally accumulates nitrate to a lesser extent than sudangrass (*Sorghum sudanense* (Piper) Stapf.), sorghums (*Sorghum* spp.), corn, and small grain but to a greater extent than bluegrass (*Poa annua* L.), alfalfa, red clover, and ladino clover.

Hojjati et al. (1972, 1973) reported that the accumulation maxima of NO_3-N in tall fescue occurred within 14 days following fertilization, after which the values gradually declined. Length of time required for dissipation of the effect of N application on NO_3-N levels was directly related to the quantity of fertilizer used. At 200 kg N/ha the maximum NO_3-N found in tall fescue was 5,567 µg NO_3-N/g dry forage. Even at N fertilizer rates of 50 kg/ha, tall fescue accumulated more than 2,000 µg NO_3-N/g during certain growth periods. Kukulka (1970) reported a maximum NO_3-N accumulation of *F. pratensis* of 2,270 µg/g with orchardgrass always equal to the fescue in nitrate accumulation. Kukulka recommended a maximum application of 300 kg N/ha on fescue to avoid nitrate poisoning. Wilkinson et al. (1971) reported accumulation of 3,300 µg NO_3-N/g of tall fescue forage following fertilization with 24 metric tons of broiler litter/ha/year and 2,000 µg NO_3-N following N application of 135 kg N/ha/year from NH_4NO_3. George et al. (1972) fertilized tall fescue with 168 kg N/ha and obtained maximum NO_3-N accumulation of 1,800 µg/g which did not influence average daily gain compared with tall fescue pastures fertilized with 0 or 84 kg N/ha. Following 135 kg N/ha application in spring and early summer, Ryan et al. (1972) found NO_3-N accumulation of approximately 2,480 µg/g in early June. Maximum NO_3-N accumulation of 3,510 µg/g was observed following 540 kg N/ha application in July and harvesting the forage in November.

Early spring application of 224 to 896 kg N/ha resulted in NO_3-N accumulation in first growth tall fescue forage at 20 to 25 cm height of 4,400 to 8,200 µg/g (Murphy and Smith, 1967). Marty (1970) found 130 to 700 µg/g NO_3-N following application of 300 kg N/ha and found higher levels in grazed than ungrazed pastures. Only at extremely high levels of added N did the NO_3-N accumulation exceed the suggested safe level by Bradley et al. (1940) of 2,000 µg/g and only at the extremely high level was Davison's (1964) safe level of 4,600 µg/g exceeded. In the few instances where animal performance was evaluated, NO_3-N levels did not have an effect on animal performance.

FAT NECROSIS

A. Description

Bovine fat necrosis is characterized by the presence of hard masses of fat in the adipose tissue primarily of the abdominal cavity and is an animal condition given many different names in the literature (Papp and Williams, 1970). Necrotic fat lesions are most common in mesenteric fat surrounding the intestine and are found along the intestinal tract from the abomasum to the rectum (Fig. 13-8). Necrotic fat is often encapsulated or separated from normal fat by fibrous connective tissue. Normal carcass fat is bright yellow, whereas the fat lesions are a deeper yellow than the healthy adjacent fat and often contain chalky white or orange-colored areas (Papp and Williams, 1970). The lesions apparently are coagulative necrosis of adipose tissue which causes little acute inflammatory response. Necrotic cells are inter-

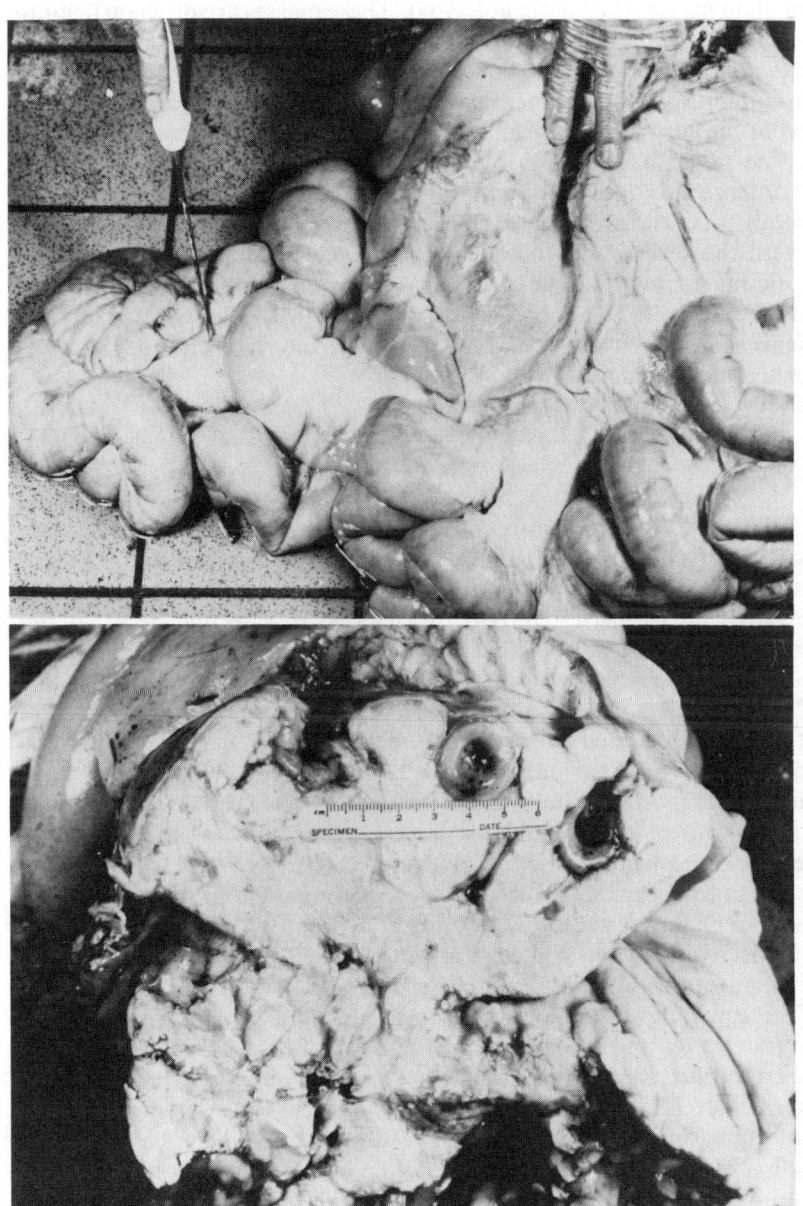

Fig. 13-8—Upper picture is an intact lesion of hard fat in a cow that had been grazing broiler-littered tall fescue. Bottom picture is the cross section of the lesion in the upper picture; the intestine is embedded and constricted in the hard fat mass. Photographs courtesy of J. A. Stuedemann.

spersed among apparently normal adipose cells. Frequently the necrotic cells contain a fine crystalline mass which causes the cells to be distended. The lesions are thus hard or firm and ranged in size from small loci to large 30-cm diam masses (Forney et al., 1969). These masses are often difficult to incise and occasionally grittiness can be felt (Williams et al., 1969). Necrotic fat has greater amounts of protein, cholesterol, and fatty acid salts but lesser amounts of ether extract than normal fat (Rumsey et al., 1976). The effects of the hard lesions are largely dependent upon their anatomical location. The larger hard masses often encase and cause constriction of the small or large intestine. Cattle with fat necrosis do not show symptoms until vital body processes are affected. Digestive upset is the most frequent symptom with the severity depending on the degree of intestinal constriction. Necrotic fat surrounding the birth canal has been associated with difficult births (Edgson, 1952; Papp and Williams, 1970). Edgson (1952) and Williams et al. (1969) also found fat necrosis to interfere with the functions of the heart and kidney.

B. Occurrence

Fat necrosis in cattle has been reported from several countries (Forney et al., 1969). Williams et al. (1969) first clinically diagnosed and observed fat necrosis as a herd problem in the United States in 1967. He found 67% of the cows in one herd and 31% of the cows in a second herd had abdominal fat necrosis. The clinical diagnosis followed signs of gastrointestinal disorder. A survey of six herds in north Georgia revealed that 23.5% of the cows had rectally palpable lesions of fat necrosis. Of 1,519 adult cattle slaughtered in northern Georgia, 2.5% had abdominal fat necrosis. Fat necrosis has been detected in most breeds, in cows of all ages, and has been reported in bulls. Herd problems of fat necrosis were initially associated with poultry production areas of the southeastern U.S. where cattle were clinically diagnosed to have fat necrosis when they had been continuously grazing fescue fertilized with 13 to 25 metric tons of broiler litter/ ha/year (Williams et al., 1969). However, subsequent research has shown that fat necrosis can occur in cows grazing fescue fertilized with either ammonium nitrate or broiler litter (Stuedemann et al., 1973b; 1975).

The apparent association of heavy fertilization of tall fescue pastures with broiler litter and fat necrosis was tested by Stuedemann et al. (1975). Three established Kentucky 31 tall fescue pastures were fertilized with three different N levels. The high level (789 kg N/ha/year) was obtained by addition of approximately 20 metric tons of broiler litter/ha/year. The intermediate (224 kg N/ha/year) and low (74 kg N/ha/year) N levels were obtained by addition of ammonium nitrate fertilizer. Cattle on all pastures were managed in a similar manner. The workers observed cow weights to be directly related to the level of N fertilization. The incidence of fat necrosis increased as N levels increased, and number of cows with fat necrosis increased with increased grazing time on highly fertilized pastures. During the 6 years of this experiment, six cows that grazed broiler litter-fertilized tall fescue died as a direct result of hard fat around the digestive tract. Because some cows from all N treatments, regardless of N source, developed palpable fat necrosis lesions, the etiology of fat necrosis apparently is associated with the level of plant nutrients supplied to tall fescue and is not associated with broiler litter other than the high nutrient composition of broiler litter. Williams et al. (1969) removed one cow with fat necrosis from a tall fescue pasture and fed only bermudagrass hay. Within 1 month the lesion began to get smaller and general body condition improved within 3 months, whereas no significant change was observed in the incidence or the character of the lesions in cows left on the tall fescue pasture. Stuedemann et al. (1973b) did not find necrotic fat lesions by post-mortem examinations or rectal palpations in cows not grazing predominantly tall fescue pastures. These observations support the theory of the dietary origin of fat necrosis in cows grazing tall fescue.

Dietary influences have been suggested by several workers and discussed by Williams et al. (1969) and Papp and Williams (1970). Hoflund (1953) theorized that during extended periods of elevated body temperatures, adipose tissue is modified. This modification could be similar to a modification caused by the diet in that the modification could be a result of a stimulus on the hepatic cells that altered fat metabolism. A dietary alteration of fat metabolism would suggest components of tall fescue pasture to be responsible.

C. Corrective Measures

As mentioned in the preceding section, no obvious external symptoms are associated with fat necrosis, but as interference with the digestive process occurred the animals began to lose weight, lose appetite, and the hair coat became rough. Cows grazing broiler litter-fertilized pastures had body temperatures above normal and sought ways of lowering heat stress by standing in water or shade and/or by lying in mud. These symptoms of weight loss, appetite loss, rough hair coat, and elevated body temperature are typical of observations of "fescue toxicity". The high incidence of fat necrosis in cows grazing tall fescue pastures receiving high levels of N/ha/year was also associated with potentially toxic levels of NO_3-N (3,300 $\mu g/g$) and accumulation of 1,500 μg perloline/g (Wilkinson et al., 1971).

Wilkinson et al. (1971) suggests that because of the potential health problems associated with high nutrient inputs from broiler litter, tall fescue pastures should not receive more than 9 metric tons/ha/year. The relative importance of perloline, NO_3-N, NPN, and fat necrosis in fescue toxicity is not understood, but the association between poor animal performance with high levels of perloline, NO_3-N, NPN, and fat necrosis needs further investigation.

LITERATURE CITED

1. Aasen, A. J., C. C. J. Culvenor, E. P. Finnie, A. W. Kellock, and L. W. Smith. 1969. Alkaloids as a possible cause of ryegrass staggers in grazing livestock. Aust. J. Agric. Res. 20:71–86.
2. Allcroft, W. M., and H. H. Green. 1934. Blood calcium and magnesium of the cow in health and disease. Biochem. J. 28:2220–2228.
3. Ammerman, C. B., C. F. Chicco, P. E. Loggins, and L. R. Arrington. 1972. Availability of different inorganic salts of magnesium to sheep. J. Anim. Sci. 34:122–126.
4. ———, C. F. Chicco, J. E. Moore, P. A. Van Walleghem, and L. R. Arrington. 1971. Effect of dietary magnesium on voluntary feed intake and rumen fermentations. J. Dairy Sci. 54:1288–1293.
5. Ashley, G. 1958. Fescue poisoning of cattle on Florida muck land. J. Am. Vet. Med. Assoc. 132, 493–494.
6. Aust, S. D., H. P. Broquist, and K. L. Rinehart, Jr. 1968. Slaframine: A parasympathomimetric from *Rhizoctonia leguminicola*. Biotechnol. Bioeng. 10, 403–412.
7. Bacon, C. W., J. K. Porter, and J. D. Robbins. 1975. Toxicity and occurrence of *Balansia* on grasses from toxic fescue pastures. Appl. Microbiol. 29, 553–556.
8. Bamburg, J. R., W. F. Marasas, N. V. Riggs, E. B. Smalley, and F. M. Strong. 1967. Toxic spiroepoxy compounds from fusaria and other hyphomycetes. Am. Chem. Soc. Abstr. Paper, No. Q-70.
9. ———, N. V. Riggs, and F. M. Strong. 1968. The structures of toxins from two strains of *Fusarium tricinctum*. Tetrahedron 24, 3329–3336.
10. Barnett, A. J. G., and R. L. Reid. 1961. Reactions in the rumen. Edward Arnold Ltd., London.
11. Barrett, J. T. 1974. Textbook of immunology, an introduction to immunochemistry and immunobiology, 2nd Ed. C. V. Mosby Co., St. Louis, Mo.
12. Bennett, W. D. 1963. A note on the effect of nitrate and phosphate on the perloline content of perennial ryegrass (*Lolium perenne* L.). N.Z. J. Agric. Res. 6:310–313.
13. Bohman, V. R., A. L. Lesperance, G. D. Harding, and D. L. Grunes. 1969. Induction of experimental tetany in cattle. J. Anim. Sci. 29:99–102.
14. Boling, J. A., L. P. Bush, R. C. Buckner, L. C. Pendlum, P. B. Burrus, S. G. Yates, S. P. Rogovin, and H. L. Tookey. 1975. Nutrient digestibility and metabolism in lambs fed added perloline. J. Anim. Sci. 40:972–976.
15. Bradley, W. B., H. F. Eppson, and O. A. Beath. 1940. Livestock poisoning by oat, hay and other plants containing nitrate. Wyoming Agric. Exp. Stn. Bull. 241.
16. Brook, P. J., and E. P. White. 1966. Fungus toxins affecting mammals. Annu. Rev. Phytopathol. 4:171–194.
17. Bruce, L. A., J. D. Robbins, and T. L. Huber. 1971. Smooth muscle response to fescue alkaloids. J. Anim. Sci. 32:373, Abstr. No. 18.
18. Buckner, R. C., L. P. Bush, and P. B. Burrus II. 1973. Variability and heritability of perloline in *Festuca* sp., *Lolium* sp., and *Lolium-Festuca* hybrids. Crop Sci. 13:666–669.
19. ———, J. R. Todd, P. B. Burrus, II, and R. F. Barnes. 1967. Chemical composition, palatability, and digestibility of ryegrass-tall fescue hybrids, 'Kenwell' and 'Kentucky 31' tall fescue varieties. Agron. J. 59:345–349.
20. Burfening, P. J. 1973. Ergotism. J. Am. Vet. Med. Assoc. 163:1288–1290.
21. Burt, A. W. A., and D. C. Thomas. 1961. Dietary citrate and hypomagnesaemia in the ruminant. Nature 192:1193.
22. Bush, L. P., and R. C. Buckner. 1973. Tall fescue toxicity. p. 99–112. *In* A. G. Matches (ed.) Antiquality components of forages. Crop Sci. Soc. Am., Madison, Wis.
23. ———, H. Burton, and J. A. Boling. 1976. Activity of tall fescue alkaloids and analogues in in vitro rumen fermentation. J. Agric. Food Chem. 24:869–872.
24. ———, and J. A. D. Jeffreys. 1975. Isolation and separation of tall fescue and ryegrass alkaloids. J. Chromatogr. 111:165–170.
25. ———, C. Streeter, and R. C. Buckner. 1970. Perloline inhibition of in vitro ruminal cellulose digestion. Crop Sci. 10:108–109.
26. Butler, E. J. 1962. The mineral element content of spring pasture in relation to the occurrence of grass tetany and hypomagnesaemia in dairy cows. J. Agric. Sci. 60:329–340.

27. Butler, G. W. 1962. Genetic differences in the perloline content of ryegrass (*Lolium*) herbage. N.Z. J. Agric. Res. 5:158-162.
28. Church, D. C. 1969. Digestive physiology and nutrition of ruminants. Vol. 1. Digestive physiology. Oregon State Univ. Book Stores, Inc., P.O. Box 489, Corvallis. p. 297-300.
29. ———. 1972. Digestive physiology and nutrition and ruminants. Vol. 2. Nutrition. Oregon State Univ. Book Stores, Inc., P.O. Box 489, Corvallis. p. 704.
30. Corley, E. A., C. E. Short, and G. B. Garner. 1973. Vascular changes in the rear limb of cattle with fescue foot visualized by a radiopaque material. p. 48-53. *In* Proc. Fescue Toxicity Conf., Lexington, Ky. 31 May-1 June 1973. Univ. of Missouri, Columbia.
31. Cornelius, P. L., R. C. Buckner, L. P. Bush, P. B. Burrus, II, and J. Byars. 1974. Inheritance of perloline content in annual ryegrass × tall fescue hybrids. Crop Sci. 14:896-898.
32. Cowan, J. Ritchie. 1956. Tall fescue. Adv. Agron. 8:283-320.
33. Crookshank, H. R., and F. H. Sims. 1955. Serum values in wheat pasture poisoning cases. J. Anim. Sci. 14:964-969.
34. Culvenor, C. C. 1973. Alkaloids. *In* G. W. Bulter, and R. W. Bailey (eds.) Chemistry and Biochemistry of Herbage, Vol. 1. Academic Press. p. 375-446.
35. Cunningham, I. J. 1949. A note on the cause of tall fescue lameness in cattle. Aust. Vet. J. 25:27-28.
36. ———, and E. M. Clare. 1943. A fluorescent alkaloid in ryegrass (*Lolium perenne* L.) V. Toxicity, photodynamic action, and metabolism of perloline. N.Z. J. Sci. Tech. 24:167B-178B.
37. Curtis, R. A., and D. MacMaster. 1966. Hypomagnesemic tetany. Can. Vet. J. 7:239.
38. Custer, F. D. 1959. Tetany in cattle on winter rations. p. 159-164. *In* Proc. Magnesium and Agric. Symp., West Virginia Univ., Morgantown, W. Va.
39. Davies, W. L. 1936. The metabolism of betaine and allied tertiary nitrogen bases in the ruminant. J. Dairy Res. 7:14-24 (Chem. Abstr. 30:4555, 1936).
40. Davison, K. L., W. Hansel, L. Krook, K. McEntee, and M. J. Wright. 1964. Nitrate toxicity in dairy heifers. I. Effects on reproduction, growth, lactation and vitamin A nutrition. J. Dairy Sci. 47:1065-1073.
41. Dozsa, L. 1959. Field case descriptions of spring tetany. p. 165-168. *In* Proc. Magnesium and Agric. Symp. West Virginia Univ., Morgantown, W. Va.
42. Dryerre, H. 1932. Lactation tetany. Vet. Rec. 12:1163-1168.
43. Edgson, F. A. 1952. Bovine lipomatosis. Vet. Rec. 64:449-454.
44. Ellis, J. J., and S. G. Yates. 1971. Mycotoxins of fungi from fescue. Econ. Bot. 25:1-5.
45. Emerick, R. J., L. B. Embry, and R. W. Seerley. 1965. Rate of formation and reduction of nitrite induced methemoglobin in vitro and in vivo as influenced by diet of sheep and age of swine. J. Anim. Sci. 24:221-230.
46. Eppon, H. F., M. W. Glenn, W. W. Ellis, and C. S. Gilbert. 1960. Nitrate in the diet of pregnant ewes. J. Am. Vet. Med. Assoc. 137:611-614.
47. Farnell, D. R., M. C. Futrell, V. H. Watson, W. E. Poe, and R. E. Coats. 1973. Field studies on etiology and control of fescue toxicosis. *In* Proc. Fescue Toxicity Conf. Lexington, Ky., Ext. Publ. Univ. of Missouri, Columbia, Mo.
48. Ferguson, W. S., and R. A. Terry. 1954. The fractionation of the nonprotein nitrogen of grassland herbage. J. Sci. Food Agric. 5:515-521.
49. Fontenot, J. P., M. B. Wise, and K. E. Webb, Jr. 1973. Interrelationships of potassium, nitrogen, and magnesium in ruminants. Fed. Proc. 32:1925-1928.
50. Forney, M. M., D. J. Williams, E. M. Papp, and D. E. Tyler. 1969. Limited survey of Georgia cattle for fat necrosis. J. Am. Vet. Med. Assoc. 155:1603-1604.
51. Futrell, M. C., D. R. Farnell, W. E. Poe, V. H. Watson, and R. E. Coats. 1973. Fungal populations in the rumen associated with fescue toxicosis. p. 131-139. *In* Proc. Fescue Toxicity Conf., Lexington, Ky., 31 May-1 June 1973. Univ. of Missouri, Columbia.
52. ———, ———, ———, ———, and ———. 1974. Fungal populations in the rumen associated with fescue toxicosis. J. Environ. Qual. 3:140-143.
53. Garner, G. B. 1958. Learn to live with nitrates. Missouri Agric. Exp. Stn. Bull. 708.
54. ———. 1973. Fescue foot syndrome in Missouri. p. 6-8. *In* Proc. Fescue Toxicity Conf., Lexington, Ky., 31 May-1 June 1973. Univ. of Missouri, Columbia.
55. ———, and B. W. Harmon. 1973. Experimental pastures and field case data. p. 42-47. *In* Proc. Fescue Toxicity Conf., Lexington, Ky., 31 May-1 June 1973. Univ. of Missouri, Columbia.

56. ─────, M. Williams, S. G. Yates, and H. L. Tookey. 1972. Fescue foot induction from experimental pastures. J. Anim. Sci. 35:228.
57. Gaynor, E., C. Bouvier, and T. H. Spaet. 1970. Vascular lesions: Possible pathogenetic basis of the generalized Shwartzman reaction. Science 170:986–988.
58. Gentry, E. G. 1968. Interrelationship of *Rhizoctonia solani* Kuhn, environment, and genotype on the alkaloid content of tall fescue, *Festuca arundinacea* Shreb. Ph.D. Thesis. Univ. of Kentucky. Univ. Microfilms. Ann Arbor, Mich. (Diss. Abstr. 30:1446B).
59. ─────, R. A. Chapman, L. Henson, and R. C. Buckner. 1969. Factors affecting the alkaloid content of tall fescue, *Festuca arundinacea* Schreb. Agron. J. 61:313–316.
60. George, J. R., C. L. Rhykerd, G. O. Mott, R. F. Barnes, and C. H. Noller. 1972. Effect of nitrogen fertilization of *Festuca arundinacea* Schreb. on nitrate nitrogen and protein content and the performance of grazing steers. Agron. J. 64:24–26.
61. Gerken, H. J., Jr., and J. P. Fontenot. 1967. Availability and utilization of magnesium from dolomitic limestone and magnesium oxide in steers. J. Anim. Sci. 26:1404–1408.
62. Goodman, A. A. 1952. Fescue foot in cattle in Colorado. J. Am. Vet. Med. Assoc. 121:289–290.
63. Greenhill, A. W. 1936. LXV. A study of the relative amounts of the protein and non-protein nitrogenous constituents occurring in pasture herbage, and their significance in the grazing of the herbage by stock. Biochem. J. 30:412–416.
64. Grimmett, R. E. R., and D. F. Waters. 1943. A fluorescent alkaloid in ryegrass (*Lolium perenne* L.). II. Extraction from fresh ryegrass and separation from other bases. N.Z. J. Sci. Technol. 24:151B–155B.
65. Grove, M. D., H. L. Tookey, and S. G. Yates. 1973. Relation of mycotoxins to fescue toxicity. p. 124–130. *In* Proc. Fescue Toxicity Conf., Lexington, Ky., 31 May-1 June 1973. Univ. of Missouri, Columbia.
66. ─────, S. G. Yates, W. H. Tallent, J. J. Ellis, I. A. Wolff, N. R. Kosuri, and R. E. Nichols. 1970. Mycotoxins produced by *Fusarium tricinctum* as possible causes of cattle disease. J. Agric. Food Chem. 18:734–736.
67. Grunes, D. L. 1967. Grass tetany of cattle as affected by plant composition and organic acids. Proc. Cornell Nutr. Conf. Cornell Univ., Ithaca, N.Y.
68. ─────. 1973. Grass tetany of cattle and sheep. p. 113–140. *In* A. G. Matches (ed.) Antiquality components of forages. Crop Sci. Soc. Am., Madison, Wis.
69. ─────, P. R. Stout, and J. R. Brownell. 1970. Gross tetany of ruminants. Adv. Agron. 22:331–374.
70. Halse, K. 1970. Individual variation in blood magnesium and susceptibility to hypomagnesaemia in cows. Acta Vet. Scand. 11:394–414.
71. Hannaway, David B., and John H. Reynolds. 1976. Seasonal changes in minerals, protein, and yield of tall fescue forage as influenced by nitrogen and potassium fertilization. Tennessee Farm Home Sci. Prog. Rep. 98:14–17.
72. Hannaway, David B., and John H. Reynolds. 1979. Seasonal changes in organic acids, water-soluble carbohydrates, and neutral detergent fiber in tall fescue forage as influenced by N and K fertilization. Agron. J. 71:493–496.
73. Hansard, S. L., F. C. Madsen, G. M. Merriman, and J. B. McLaren. 1975. Tennessee grass-fed cattle need supplementary magnesium. Tennessee Farm Home Sci. Prog. Rep. 93:36–38.
74. Hoflund, S., J. Holmberg, L. K. B. Appelviken, and H. Nihlen. 1953. On the aetiology pathogenesis and clinical picture of fat necrosis in cattle. p. 642–652. *In* Proc. Int. Vet. Congr. Stockholm, Sweden.
75. Hojjati, S. M., T. H. Taylor, and W. C. Templeton, Jr. 1972. Nitrate accumulation in rye, tall fescue, and bermudagrass as affected by nitrogen fertilization. Agron. J. 64:624–627.
76. ─────, W. C. Templeton, Jr., and T. H. Taylor. 1973. Post-fertilization changes in concentration of nitrate nitrogen in Kentucky bluegrass and tall fescue herbage. Agron. J. 65:880–883.
77. Holtenius, P. 1957. Nitrite poisoning in sheep, with special reference to the detoxification of nitrite in the rumen. Acta Agric. Scand. 7:113–163.
78. House, W. A., and D. Van Campen. 1971. Magnesium metabolism of sheep fed different levels of potassium and citric acid. J. Nutr. 101:1483–1492.
79. Hsu, Ih-Chang, E. B. Smalley, F. M. Strong, and W. E. Ribelin. 1972. Identification of T-2 toxin in moldy corn associated with a lethal toxicosis in dairy cattle. Appl. Microbiol. 24:684–690.

80. Jacobsen, A. 1974. Protein and protein value in grass and green fodder. Dan. Landbrug. 5:12-15.
81. Jacobson, D. R., S. B. Carr, R. H. Hatton, R. C. Buckner, A. P. Graden, D. R. Dowden, and W. M. Miller. 1970. Growth, physiological responses, and evidence of toxicity in yearling dairy cattle grazing different grasses. J. Dairy Sci. 53:575-587.
82. ———, and R. H. Hatton. 1973. The fescue toxicity syndrome. p. 55-73. *In* Proc. Fescue Toxicity Conf., Lexington, Ky., 31 May-1 June 1973. Univ. of Missouri, Columbia.
83. ———, W. M. Miller, D.M. Seath, S. G. Yates, H. L. Tookey, and I. A. Wolff. 1963. Nature of fescue toxicity and progress toward identification of the toxic entity. J. Dairy Sci. 46:416-422.
84. Jainudeen, M. R., W. Hansel, and K. L. Davison. 1965. Nitrite toxicity in dairy heifers. 3. Endocrine responses to nitrate ingestion during pregnancy. J. Dairy Sci. 48:217-221.
85. Jeffreys, J. A. D. 1970. The alkaloids of perennial ryegrass (*Lolium perenne* L.). Part IV. Isolation of a new base, perlolyrine; the crystal structure of its hydrobromide dihydrate, and the synthesis of the base. J. Chem. Soc., C, 1091-1103.
86. Jensen, R., A. W. Deem, and D. Knaus. 1956. Fescue lameness in cattle. I. Experimental production of the disease. Am. J. Vet. Res. 17:196-201.
87. Jones, G. A., R. A. MacLeod, and A. C. Blackwood. 1964. Ureolytic rumen bacteria. I. Characteristics of the microflora from a urea-fed sheep. Can. J. Microbiol. 10:371-378.
88. Jones, I. R., P. H. Weswig, J. F. Bone, M. A. Peters, and S. O. Alpen. 1966. Effect of high-nitrate consumption on lactation and vitamin A nutrition of dairy cows. J. Dairy Sci. 49:491-499.
89. Kemp, A. 1960. Hypomagnesaemia in milking cows. The response of serum magnesium to alterations in herbage composition resulting from potash and nitrogen dressings on pasture. Neth. J. Agric. Sci. 8:281-304.
90. ———, W. B. Deijs, O.J. Hemkes, and A. J. H. Van Es. 1961. Hypomagnesaemia in milking cows: intake and utilization of magnesium from herbage by lactating cows. Neth. J. Agric. Sci. 9:134-149.
91. ———, ———, and E. Kluvers. 1966. Influence of higher fatty acids on the availability of magnesium in milking cows. Neth. J. Agric. Sci. 14:290-295.
92. ———, and J. H. Geurink. 1967. Prevention of hypomagnesaemia in milking cows. Agric. Dig. 12:23-29.
93. ———, and J. R. Todd. 1970. Prevention of hypomagnesaemia in cows. The use of magnesium alloy bullets. Vet. Rec. 86:463-464.
94. Keyl, A. C., J. C. Lewis, J. J. Ellis, S. G. Yates, and H. L. Tookey. 1967. Toxic fungi isolated from tall fescue. Mycopathol. Mycol. Appl. 31:327-331.
95. Kingsbury, J. M. 1964. Poisonous plants of the United States and Canada. Prentice-Hall, Inc., Englewood Cliffs, N.J. p. 79-86.
96. Kosuri, N. R., M. D. Grove, S. G. Yates, W. H. Tallent, J. J. Ellis, I. A. Wolff, and R. E. Nichols. 1970. Response of cattle to mycotoxins of *Fusarium tricinctum* isolated from corn and *Fescue*. J. Am. Vet. Med. Assoc. 157:938-940.
97. Kukulka, I. 1970. Influence of nitrogen fertilization and environment conditions on content of nitrate in grasses. Poznan. Towarzy. Przyj. Nauk, Wydzial Nauk Roln. Lesn., Pr. Kom. Nauk Roln. Kom. Nauk Lesn. 29:219-256.
98. Lowrey, R. S., and D. L. Grunes. 1968. Magnesium metabolism in cattle as related to potassium and magnesium fertilization of rye forage. Proc. Ga. Nutr. Conf. Feed Manuf. p. 51-56.
99. Maag, D. D., and J. W. Tobiska. 1956. Fescue lameness in cattle. II. Ergot alkaloids in tall fescue grass. Am. J. Vet. Res. 17:202-204.
100. Maddaloni, J. 1964. Intoxicacion amoniacal en rumiantes. Estacion Experimental Agropecuoria Pergamino. Informes Technicos Publicados No. 30. Casilla de Correo 31, Argentina.
101. Madsen, F. C., G. E. Spalding, J. K. Miller, S. L. Hansard, and W. A. Lyke. 1975. Magnesium movement in hypothyroid sheep. Proc. Soc. Exp. Biol. Med. 149:207-214.
102. Martin, J. E., L. R. Arrington, J. E. Moore, C. B. Ammerman, G. K. Davis, and R. L. Shirley. 1964. Effect of magnesium and sulfur upon cellulose digestion of purified rations by cattle and sheep. J. Nutr. 83:60-64.
103. Marty, J. 1970. Observation on the nitrate nitrogen content of some forages. Fourrages 43:57-73.
104. Mattocks, A. R. 1968. Toxicity of pyrrolizidine alkaloids. Nature 217:723-728.

105. ————. 1971. Synthetic compounds with toxic properties similar to those of pyrrolizidine alkaloids and their pyrrolic metabolites. Nature 232:476.
106. Maynard, L. A., and J. K. Loosli. 1969. Animal nutrition. Sixth Ed. McGraw-Hill Book Co., New York. p. 14, 177.
107. McLaren, J. B., D. W. Saylor, S. L. Hansard, and L. M. Safley. 1975. Effects of supplemental magnesium on the incidence of grass tetany in beef cows. Tennessee Farm Home Sci. Prog. Rep. 94:21-23.
108. Melville, J., and R. E. R. Grimmett. 1941. Isolation of a new alkaloid from perennial ryegrass. Nature 148:782.
109. Merriman, G. M. 1955. Fescue poisoning. Tennessee Farm Home Sci. Prog. Rep. 16:8.
110. Moore, W. G., J. P. Fonenot, and R. E. Tucker. 1971. Relative effects of different supplemental magnesium sources on apparent digestibility. J. Anim. Sci. 33:502-506.
111. ————, ————, and K. E. Webb, Jr. 1972. Effect of form and level of nitrogen on magnesium utilization. J. Anim. Sci. 35:1046-1053.
112. Mott, G. O., C. J. Kaiser, R. C. Peterson, Randall Peterson, Jr., and C. L. Rhykerd. 1971. Supplemental feeding of steers on *Festuca arundinacea* Schreb. pastures fertilized at three levels of nitrogen. Agron. J. 63:751-754.
113. Mueller-Schweinitzer, E., and E. Stuermer. 1974. Mechanism of the venoconstrictor activity of ergotamine on isolated canine saphenous veins. Blood Vessels 11:183-190.
114. Murdock, L. W., C. W. Absher, N. Gay, and J. A. Boling. 1975. Beef: Grass tetany in beef cattle. Anim. Sci.-16. Kentucky Agric. Exp. Stn.
115. Murphy, L. S., and G. E. Smith. 1967. Nitrate accumulation in forage crops. Agron. J. 59:171-174.
116. Newton, G. L., J. P. Fontenot, R. E. Tucker, and C. E. Polan. 1972. Effects of high dietary potassium intake on the metabolism of magnesium by sheep. J. Anim. Sci. 35: 440-445.
117. O'Kelley, R. E., and J. P. Fontenot. 1969. Effects of feeding different magnesium levels to drylot-fed lactating beef cows. J. Anim. Sci. 29:959-966.
118. ————, and ————. 1973. Effects of feeding different magnesium levels to drylot-fed gestating beef cows. J. Anim. Sci. 36:994-1000.
119. Papp, E., and D. J. Williams. 1970. Bovine lipomatosis. Zentralbl. Veterinarmed. Reihe A. 17:735-742.
120. Pike, R. L., and M. L. Brown. 1975. Nutrition: An integrated approach, Second Ed. John Wiley and Sons, Inc., New York. p. 185-187.
121. Porter, J. K., C. W. Bacon, J. D. Robbins, and H. C. Higman. 1975. A field indicator in plants associated with ergot-type toxicities in cattle. J. Agric. Food Chem. 23:771-775.
122. Prentice, D. E., and P. A. Neal. 1972. Some observations on the incidence of lameness in dairy cattle in West Cheshire. Vet. Rec. 91:1-7.
123. Price, J. T., and R. B. Miller. 1959. Fescue grass poisoning in cattle. Jen-Sal J. 42:26-29.
124. Pulsford, M. F. 1950. A note on lameness in cattle grazing on tall meadow feasue (*Festuca arundinacea*) in South Australia. Aust. Vet. J. 26:87-88.
125. Reid, R. L., and G. A. Jung. 1973. Forage-animal stresses. p. 639-654. *In* M. E. Heath, D. S. Metcalfe, and R. E. Barnes (eds.) Forages. Iowa State Univ. Press, Ames.
126. Reifer, I., and N. O. Bathurst. 1943. A fluorescent alkaloid in ryegrass (*Lolium perenne* L.). III. Extraction and properties. N.Z. J. Sci. Technol. 24B:155-159.
127. Repp, W. W., W. H. Hale, E. W. Cheng, and W. Burroughs. 1955. The influence of oral administration of non-protein nitrogen feeding compounds upon blood ammonia and urea levels in lambs. J. Anim. Sci. 14:118-131.
128. Reynolds, John H. 1975. Yield and chemical composition of stockpiled tall fescue and orchardgrass forages. Tennessee Farm Home Sci. Prog. Rep. 94:27-29.
129. Rifas, L., S. Shaffer, M. Williams, B. Harmon, and G. B. Garner. 1973. Isolation and identification of an iodoplatinate positive spot from toxic tall fescue. p. 74-80. *In* Proc. Fescue Toxicity Conf., Lexington, Ky., 31 May-1 June 1973. Ext. Publ. Univ. of Missouri, Columbia.
130. Robbins, J. D., J. G. Sweeny, S. R. Wilkinson, and D. Burdick. 1972. Volatile alkaloids of Kentucky 31 tall fescue seed (*Festuca arundinacea* Schreb.). J. Agric. Food Chem. 20: 1040-1042.
131. ————, S. R. Wilkinson, and D. Burdick. 1973. Loline alkaloids of tall fescue seed and forage. p. 98-107. *In* Proc. Fescue Toxicity Conf., Lexington, Ky. 31 May-1 June 1973. Univ. of Missouri, Columbia.

132. Rook, J. A. F., and J. E. Storry. 1962. Magnesium in the nutrition of farm animals. Nutr. Abstr. Rev. 32:1055-1077.
133. Rumsey, T. S., J. A. Stuedemann, S. R. Wilkinson, and D. J. Williams. 1976. Composition of necrotic fat in beef cows grazing fertilized fescue. J. Anim. Sci. 43:332.
134. Ryan, M., W. F. Wedin, and W. B. Bryan. 1972. Nitrate-N levels of perennial grasses as affected by time and level of nitrogen application. Agron. J. 64:165-168.
135. Sapiro, M. L., S. Hofluend, R. Clark, and J. I. Quin. 1949. Studies on the alimentary tract of the merino sheep in South Africa. XVI. The fate of NO_3 in ruminal ingesta as studied in vitro. Onderstepoort J. Vet. Sci. Anim. Ind. 22:357-372.
136. Selye, H. 1965. Induced hypersensitivity to cold. Science 149:201-202.
137. Setchell, B. P., and A. J. Williams. 1962. Plasma nitrate and nitrite concentration in chronic and acute NO_3 poisoning in sheep. Aust. Vet. J. 38:58-62.
138. Simon, J., J. M. Sund, F. D. Douglas, M. J. Wright, and T. Kowalczyk. 1959. The effect of nitrate or nitrite when placed in the rumens of pregnant dairy cattle. J. Am. Vet. Med. Assoc. 135:311-314.
139. ———, ———, M. J. Wright, A. Winter, and F. D. Douglas. 1958. Pathological changes associated with the lowland abortion syndrome in Wisconsin. J. Am. Vet. Med. Assoc. 132:164-169.
140. Sinclare, D. P., and M. W. Howe. 1968. Effect of thiabendazole on *Pithomyces chartarum* (Berk. and Curt.) M. B. Ellis. N.Z. J. Agric. Res. 11:59-62.
141. Singer, R. H. 1972. The nitrate poisoning complex. U.S. Anim. Health Assoc. Proc. 310-322.
142. Sjollema, B. 1932. Nutritional and metabolic disorders in cattle. Nutr. Abstr. Rev. 1:621-632.
143. Sokolowski, J. H., U. S. Garrigus, and E. E. Hatfield. 1960. Some effects of varied levels of potassium nitrate ingestion by lambs. J. Anim. Sci. 19:1295.
144. Stewart, J. 1954. Hypomagnesemia and tetany of cattle and sheep. Scott. Agric. 34:68-73.
145. Stillings, B. R., J. W. Bratzler, L. F. Marriott, and R. C. Miller. 1964. Utilization of magnesium and other minerals by ruminants consuming low and high nitrogen-containing forages and vitamin D. J. Anim. Sci. 23:1148-1154.
146. Stout, P. R., J. Brownell, and R. G. Burau. 1967. Occurrences of *trans*-aconitate in range forage species. Agron. J. 59:21-24.
147. Stuedemann, J. A., S. R. Wilkinson, W. A. Jackson, and J. B. Jones, Jr. 1973a. Foliar dusting with MgO to prevent hypomagnesemia and/or grass tetany. J. Anim. Sci. 36:227.
148. ———, ———, ———, R. L. Wilson, Jr., and J. B. Jones, Jr. 1974. Controlling grass tetany with a foliar applied MgO-bentonite-water slurry. J. Anim. Sci. 39:254.
149. ———, ———, D. J. Williams, H. Ciordia, J. V. Ernst, W. A. Jackson, and J. B. Jones, Jr. 1975. Long-term broiler litter fertilization of tall fescue pastures and health and performance of beef cows. p. 264-268. *In* Managing livestock wastes. Proc. 3rd Int. Symp. on Livestock Wastes. Am. Soc. Agric. Eng., St. Joseph, Mich.
150. ———, ———, ———, W. A. Jackson, and J. B. Jones, Jr. 1973b. The association of fat necrosis in beef cattle with heavily fertilized fescue pastures. p. 9-22. *In* Proc. Fescue Toxicity Conf., Lexington, Ky. 31 May-1 June 1973. Univ. of Missouri, Columbia.
151. Suttle, N. F., and A. C. Field. 1969. Studies on magnesium in ruminant nutrition. 9. Effect of potassium and magnesium intakes on development of hypomagnesaemia in sheep. Br. J. Nutr. 23:81-90.
152. Teel, M. R. 1966. Nitrogen-potassium relationships and their influence on some biochemical intermediates and quality of crude protein in forages. p. 465-484. *In* Potassium Symp. Proc. 8th Congr. Int. Potash Inst., Brussels.
153. Todd, J. R. 1961. Magnesium in forage plants. I. Magnesium contents of different species and strains as affected by season and soil treatment. J. Agric. Sci. 56:411-415.
154. ———, and R. H. Thompson. 1960. Major blood electrolyte concentration in cattle during the development of hypomagnesaemia. Br. Vet. J. 116:437-442.
155. Tookey, H. L., and S. G. Yates. 1972. The alkaloids of tall fescue: Loline (festucine) and perloline. An. R. Soc. Esp. Fis. Quim. 68:921-935.
156. ———, ———, J. J. Ellis, M. D. Grove, and R. E. Nichols. 1972. Toxic effects of a butenolide mycotoxin and of *Fusarium tricinctum* cultures in cattle. J. Am. Vet. Med. Assoc. 160:1522-1526.
157. Underwood, E. J. 1966. The mineral nutrition of livestock. Commonwealth Agric.

Bureaux, Aberdeen, Scotland. p. 68-69.
158. Voisin, A. 1963. Grass tetany. Crosby Lockwood and Son, Ltd., London. p. 198-199.
159. Wacker, W. E. C. 1965. The biochemical functions of magnesium. p. 4-10. *In* Proc. Georgia Nutr. Conf. Feed Manuf.
160. ─────, and A. F. Parisi. 1968. Magnesium metabolism. New Engl. J. Med. 278:658-663.
161. Wang, L. C., J. Garcia-Rivera, and R. H. Burris. 1961. Metabolism of nitrate by cattle. Biochem. J. 81:237-242.
162. Watson, D. F., J. R. Rooney, and W. G. Hoag. 1957. Fescue foot lameness in cattle— some observations on the disease in Virginia. J. Am. Vet. Med. Assoc. 130:217-219.
163. Webb, P. J. 1972. Grasses. Alkaloid studies. Plant Breeding Inst. Cambridge, England. Annu. Rep.
164. Webster, G. T., and R. C. Buckner. 1971. Cytology and agronomic performance of *Lolium-Festuca* hybrid derivatives. Crop Sci. 11:109-112.
165. White, A., P. Handler, and E. L. Smith. 1968. Principles of biochemistry, Fourth Ed. McGraw-Hill Book Co., New York.
166. Whittow, G. C. 1962. The significance of the extremities of the ox (*Bos taurus*) in thermoregulation. J. Agric. Sci. 58:109-120.
167. Wilkinson, S. R., J. A. Stuedemann, D. J. Williams, J. B. Jones, Jr., R. N. Dawson, and W. A. Jackson. 1971. Recycling broiler house litter on tall fescue pastures at disposal rates and evidence of beef cow health problems. p. 321-324, 328. *In* Livestock waste management and pollution abatement. Columbus, Ohio. 19-22 Apr. 1971. Am. Soc. Agric. Eng. Publ. Proc. 271. Am. Soc. Agric. Eng. St. Joseph, Mich.
168. Williams, D. J., D. E. Tyler, and E. Papp. 1969. Abdominal fat necrosis as a herd problem in Georgia cattle. J. Am. Vet. Med. Assoc. 154:1017-1026.
169. Williams, M., S. R. Shaffer, G. B. Garner, S. G. Yates, H. L. Tookey, L. D. Kintner, S. L. Nelson, and J. T. McGinity. 1975. Induction of fescue foot syndrome in cattle by fractionated extracts of toxic fescue hay. Am. J. Vet. Res. 36:1353-1357.
170. Winter, A. J. 1962. Studies on nitrate metabolism in cattle. Am. J. Vet. Res. 23:500-505.
171. Wolton, K. 1960. Physiological disorders of grazing livestock. I. Some factors affecting herbage magnesium levels. p. 544-548. *In* C. L. Skidmore (ed.) Proc. Eighth Int. Grassl. Congr. Reading, England.
172. Wright, M. J., and K. L. Davison. 1964. Nitrate accumulation in crops and nitrate poisoning in animals. Adv. Agron. 16:197-248.
173. Yates, S. G. 1963. Paper chromatography of alkaloids of tall fescue hay. J. Chromatogr. 12:423-426.
174. ─────. 1971. Toxin-producing fungi from fescue pasture. p. 191-206. *In* S. J. Ajil, S. Kadis, and T. C. Montie (eds.) Microbial toxins, Vol. 7. Academic Press, New York.
175. ─────, M. D. Grove, and H. L. Tookey. 1973. Assay of toxic forage. p. 108-113. *In* Proc. Fescue Toxicity Conf., Lexington, Ky., 31 May-1 June 1973. Univ. of Missouri, Columbia.
176. ─────, and H. L. Tookey. 1965. Festucine, an alkaloid from tall fescue (*Festuca arundinacea* Schreb.): Chemistry of the functional groups. Aust. J. Chem. 18:53-60.
177. ─────, ─────, and J. J. Ellis. 1970. Survey of tall-fescue pasture: Correlation of toxicity of *Fusarium* isolates to known toxins. Appl. Microbiol. 19:103-105.
178. ─────, ─────, ─────, and H. J. Burkhardt. 1968. Mycotoxins produced by *Fusarium nivale* isolated from tall fescue (*Festuca arundinacea* Schreb.). Phytochemistry 7:139-146.
179. ─────, ─────, ─────, W. H. Tallent, and I. A. Wolff. 1969. Mycotoxins as a possible cause of fescue toxicity. J. Agric. Food Chem. 17:437-442.
180. Young, P. W. 1973. Grass staggers in beef cattle. p. 93-109. *In* Proc. Ruakura Farmers Conf. Hamilton, New Zealand.

Chapter 14 Turf

J. J. MURRAY
SEA-USDA
Beltsville Research Center
Beltsville, Maryland

JERREL B. POWELL
SEA-USDA
Beltsville Research Center
Beltsville, Maryland

INTRODUCTION

The use of tall fescue (*Festuca arundinacea* Schreb.) as a frequently mowed turfgrass is of relatively recent origin. Its use for conservation and erosion control on highway rights-of-way is considerably older (Chapter 16). Tall fescue was generally considered unsatisfactory for groomed turfgrass until the past 2 decades. A cultivar marketed for turf in 1923 called "Turfing Fescue" [actually meadow fescue (*F. pratensis* Huds.)] came under intense criticism from the USDA because "grossly erroneous statements have been made regarding suitability for this purpose" (Anon., 1923). The criticism was due mostly to its tall, erect, nonspreading growth habit, coarse texture, low density, and clumping in thin stands. Its appearance suggested little potential as a turfgrass. However, tall fescue's summer heat tolerance, its growth at low temperature, its good ground cover, and its persistence in spite of extreme neglect brought attention to its potential for turf. Seeding at high rates followed by regular, frequent mowing resulted in much finer leaf texture and the absence of clumping. This observation played a major role in the increased use of tall fescue for turf. Coupled with lower maintenance requirements and expressing deep-rooted, drought- and wear-resistant turf characteristics, tall fescue became a desirable turfgrass. The Fusarium blight problem on Kentucky bluegrasses (*Poa pratensis* L.) in the eastern U.S. has accelerated the shift to tall fescues by many homeowners. Some golf courses are now using tall fescue in the roughs and fairways.

Acceptance and use of tall fescue on commercial or residential lawns have been unusually rapid during the past 10 years (Fig. 14-1). A recent survey (J. J. Murray, unpublished data) revealed that landscape use in the transition zone of the United States doubled during this period. General purpose turf use increased approximately 25% during the same period. Tall fescue was reported being utilized for turfgrass in 42 of the 50 United States.

Copyright © 1979 ASA-CSSA-SSSA, 677 South Segoe Road, Madison, WI 53711 USA. *Tall Fescue.*

Fig. 14-1—Tall fescue used as a turfgrass on a home lawn.

HISTORY OF USE AND ATTRIBUTES AS A TURFGRASS

The use of tall fescue for turf developed slowly. Early investigations indicate that only the fine-leaved fescues were considered turf types. Even as late as 1950 Reed stated that tall fescue was unsuitable for turf. However, Musser and DeFrance (1948) had suggested that tall fescue be used for turf in parks, cemeteries, and home lawns; and Moorish and Harrison (1948) recommended tall fescue for rough-use turf areas because of its wear resistance and relatively high quality.

The Green Section of the U.S. Golf Association and USDA began testing tall fescue for turf in 1942. When used as a nurse grass in shady areas, tall fescue gave a quick cover while slower-growing species became established. In 1947, the front lawn of the Plant Industry Station at the Beltsville Agricultural Research Center, Beltsville, Md., was seeded to 'Alta' tall fescue (Grau, 1954) after Kentucky bluegrass and Chewings fescue failed. The excellent results stimulated the use of tall fescue for lawns (Fig. 14-2).

Harper and Hein (1954) recognized that tall fescue was adapted to use on play areas and athletic fields where a tough turf resistant to wear was more important than a fine texture. Grau (1954) rated tall fescue high when compared with other cool-season grasses because of its tolerance to widely varying soil textures and fertility. He postulated that tall fescue would extend further into the warm-season region of the U.S. than other cool-season grasses. Lantz (1955) listed tall fescue for turf with new and special

Fig. 14-2—Tall fescue used in landscaping institutional grounds. National Agricultural Library, Beltsville, Md.

purpose grasses; seeding of tall fescue alone and in mixtures with Kentucky bluegrass in Iowa remained green until November. He also observed that tall fescue did not turn brown during summer stress periods as contrasted with the Kentucky bluegrasses.

Tennessee was the first state to recommend tall fescue for turf (Hanson and Juska, 1969). Underwood (1964) stated that tall fescue had become a popular lawn grass in Tennessee because of its wide adaptation to variable soil pH and drainage, tolerance to sun, diseases, insects, and drought. In California, Youngner (1965) found that higher seeding rates than those recommended for pastures and roadsides gave nonclumping turf with finer leaf texture. He also found that tall fescue was the only cool-season grass that persisted year after year.

A. Adaptation

The adaptation range of tall fescue for turf is similar to that for conservation and forage (Chapters 2, 10, and 16). Once established, the strong vegetative bud regeneration perpetuates the sward virtually indefinitely. It is best adapted to the transition zone between the cool-humid and warm-humid regions where neither warm- nor cool-season species are well adapted (Juska et al., 1969; Beard, 1973). It is used in transition zones where other cool-season turfgrasses are less tolerant to high summer temperatures and relatively cool, but not severe, winter conditions. At colder latitudes, tall fescue is injured by low temperatures and stands gradually thin (Hanson and Juska, 1969; Beard, 1973). Adaptation outside of the transition zone is restricted to special situations where turf use dictates characteristics or performance not provided by other adapted species. Tall fescue is adapted to arid temperate regions when irrigated. It is intermediate in shade tolerance. In areas where leafspot (*Helminthosporium* sp.) disease occurs, it grows as well as red fescue (*F. rubra* var. *rubra* L.) in shade (Hanson and Juska, 1969).

Although adapted to a wide range of soil types, tall fescue is best adapted to fertile, well-drained, fine-textured soils. It tolerates low fertility but responds to fertilization. Growth on alkaline and saline soils is better than that of most other cool-season grasses. Optimum growth occurs in a pH range of 5.5 to 6.5 but it will tolerate pH ranges from 4.7 to 8.5 (Beard, 1973). Tall fescue is one of the most drought-tolerant, cool-season grasses used for turf (Wood and Buckland, 1966; Youngner, 1970; Butler, 1975). It tolerates wet soils during extended submersion periods, especially under cool temperatures and survives in compacted soils better than other cool-season turfgrasses (Youngner, 1965). Seedling vigor is one of the most important attributes of a turfgrass, and tall fescue ranks high.

B. Cultivars

Cultivars recommended for forages are also used for turf. Breeding of tall fescue to improve its texture, density, color, and mowability are major objectives of a number of institutions. Selection for finer leaf has often resulted in poor seed set and less persistence and vigor. Rhizomatous tall fescue types are also being evaluated for their potential utilization for turf. Presently, tall fescue variants with rhizomes are only weakly rhizomatous and much selection will be required to approach the accustomed characteristics of a rhizomatous grass such as Kentucky bluegrass.

Many tall fescue cultivars have been tested for turf. However, 'Kentucky 31' comprises about 97% of that seeded for turf in United States (J. J. Murray, 1977, unpublished survey). Alta, an Oregon selection, is commonly grown west of the Rocky Mountains. 'Kenhy', a recently released cultivar from Kentucky, is performing as well as or slightly better than Kentucky 31 in some tests located in the southern part of the transition zone.

ESTABLISHMENT

A. Seed, Seedbed, and Seeding

The basic principles of turf establishment, such as seedbed preparation, liming and fertilization, seed placement, and irrigation or moisture requirements are similar for all seeded turfgrass species. Tall fescue seed is relatively large compared with seed of most other turfgrass species. Often the large seed size leads to the misconception that a good seedbed is not necessary. However, since tall fescue is a bunch grass (nonspreading) and uniformity of stand is dependent on uniform seed distribution, the seedbed must be carefully prepared and free of clods and clumps.

Tall fescue seedings for high quality turf require a heavier seeding rate than seedings for forage or conservation. Heavy seeding rates cause seedling competition that encourages the development of a uniform stand with finer texture. Dense stands grow more erect and are easier to mow. Seeding rates of approximately 200 to 400 kg/ha have been most satisfactory (Youngner, 1965; J. J. Murray, unpublished data). This rate compensates to some extent for the lack of rhizomatous and stoloniferous growth. Lighter seeding rates can be used if a good cover is the primary objective and aesthetics is of secondary importance. Significantly lighter or heavier seeding rates often result in a coarse-textured, clumpy, open stand of turfgrass. Heavier seeding delays maturity and results in physiologically weak plants that are susceptible to injury from moisture stress, traffic, disease, and insect pests. This is especially important on sports turf areas where a strong perennial crown must be developed before tall fescue will withstand heavy traffic.

The intended use of the area and the expected intensity of maintenance are important factors to consider to determine the best seeding rate. On large expansive areas with light traffic and under a low-maintenance program, seeding rates of about 200 kg/ha have resulted in the highest turfgrass quality. Regardless of the intended use of the area and other maintenance practices, the lower seeding rate should be used where irrigation is not available.

The use of certified seed, free of weeds and other crop seeds, is very important when tall fescue is seeded for turf. In pasture seedings, a few weed seeds or other crop seed such as orchardgrass (*Dactylis glomerata* L.) can be tolerated; in turfgrass seedings, however, weeds or other crop seeds can be very objectionable and difficult to eradicate.

The time of seeding is not as critical for tall fescue as it is with other cool-season species such as Kentucky bluegrass. Successful seedings can be made almost anytime during the growing season if the area is mulched and irrigated as needed until the seedlings are well established. Germination takes a week to 10 days, and a complete stand is possible in 1 month with optimum conditions. As with other cool-season turfgrasses, late summer seeding is most desirable. Later seeding, especially in the northern areas, increases the chance for frost heaving or direct low temperature kill of young seedlings. Spring is the second best time for seeding.

A planting depth of 6 to 12 mm is most favorable for rapid germination. A poor stand is more likely to occur if the seed is planted on the soil surface rather than at a shallow depth. Successful establishment of surface-planted seed is possible if an adequate moisture level is maintained and the seed is not displaced by wind or water erosion. Firming the soil after seeding will provide good contact between the seed and soil. Rolling increases the rate and uniformity of seedling emergence by improving the seed-soil contact for imbibition of water.

Tall fescue may be seeded with several types of seeders including a) gravity, b) centrifugal, c) cultipacker, and d) hydroseeder. The calibrations for seeding rates given for most seeders are approximate and can vary considerably for tall fescue, depending on the seedlot. The seeder should be calibrated before use to insure desired rate of seed per unit area. The deep ridge cultipacker is the preferred type on extensive turfgrass areas because it provides uniform seed distribution at the proper depth and firms the soil in a single operation. With gravity or centrifugal seeders some type of shallow raking, aerating, harrowing, disking, or similar method of covering the seed should be performed as soon as possible after planting. A more uniform seed application is usually accomplished with a gravity-type seeder than with a centrifugal-type seeder. Hydroseeding has been most satisfactorily used on soils in areas of high rainfall or where irrigation is practiced. This method is also satisfactory on roadside slopes and similar areas where cultipacker seeders cannot be utilized or areas that cannot be easily cultivated.

B. Mulching

The use of mulching of new seedings is not as critical for germination and seedling establishment of tall fescue as for many other turfgrasses. Mulches are beneficial in controlling wind and water erosion and providing a favorable microenvironment for seed germination and seedling growth. It reduces the rate at which the soil dries out, prevents surface crusting, reduces the impact of falling raindrops, and modifies air temperature extremes. The application of a mulch is optional when tall fescue is seeded during the optimum seeding period, on level seedbed areas, and when adequate irrigation water is available. However, mulching is a good practice to insure successful stand establishment, and it is an absolute necessity on sloping areas. Straw and hay mulches at rates of 3.4 to 4.5 metric tons/ha have been effective in protecting late fall seedings from winter injury.

C. Monoculture vs. Mixtures

For lawns and general turf, tall fescue seeded alone provides a more attractive turf than a mixture of tall fescue and other turfgrass species (Juska et al., 1969; Youngner, 1965). Mixtures containing tall fescue and Kentucky bluegrass, fineleaf fescue, or perennial ryegrass (*Lolium perenne* L.) are frequently seeded and generally provide a medium quality turf for 3 to 5 years following establishment. In commercial sod production, 5 to 10% common Kentucky bluegrass is mixed with tall fescue to provide sufficient shear strength for handling the sod. However, it is difficult to maintain an adequate stand of tall fescue in mixtures. As the turf becomes old, the more competitive grasses crowd out tall fescue. The wide, coarse leaf blades of tall fescue provide a contrast in turf texture, and clumps are formed as the stand deteriorates (Lantz, 1955; Davis, 1958; Juska and Hanson, 1959; Juska, 1959; Miller, 1969). Mixed turf of this sort tends to have a ragged, uneven appearance which severely affects quality.

The rate of change in species composition of mixtures is influenced by management practices and the relative aggressiveness of species or cultivars included in the mixture (D. T. Duff. 1964. Tall fescue and tall fescue-Kentucky bluegrass turf as influenced by cutting height, nitrogen application, and irrigation. M.S. Thesis. Ohio State Univ.; Miller, 1969). In tall fescue-Kentucky bluegrass mixtures, mowing height has a marked influence on the ability of tall fescue to remain dominant in the stand. The shift to bluegrass is rapid as the mowing height is decreased below 7.5 cm. In Beltsville, Md. tall fescue seeded alone at 195 kg/ha and mowed at a height of 2.5 or 7.5 cm reverted to 86 and 23% Kentucky bluegrass, respectively, after 10 years. Davis (1958) compared a mixture of tall fescue and common Kentucky bluegrass in Ohio. Over a 5-year period tall fescue was reduced to 10 and 58% at 1.8- and 5.0-cm mowing heights, respectively. At cutting heights of 5.0 to 10.0 cm, tall fescue has generally been persistent in mixtures with Kentucky bluegrasses seeded at a ratio of nine parts tall fescue to one bluegrass by weight.

Nitrogen fertilization is inversely related to the persistence of tall fescue in mixtures with Kentucky bluegrass. D. T. Duff (1964. Tall fescue and tall fescue-Kentucky bluegrass turf as influenced by cutting height, nitrogen application, and irrigation. M.S. Thesis. Ohio State Univ.) and A. J. Turgeon (personal communication) reported that applications of 195 kg/ha N resulted in a turf with more bluegrass than turf which received 0 or 97 kg/ha N. J. R. Hall (personal communication) observed that spring applications of N reduced tall fescue populations more than fall applications in a 90% tall fescue-10% Kentucky bluegrass mixture. Split N applications made in the spring and fall to provide 146 kg/ha/year of N reduced tall fescue populations 21% when compared with equivalent fall applications. The lowest rate of fertilization required to produce the quality of turf needed will aid in maintaining the balance among species. Clumpy growth and coarseness should not be a problem as long as good competition exists.

Mixtures of tall fescue with warm-season grasses have been tried to provide favorable winter color, but such mixtures are usually not successful. Generally, under medium-to-high management levels, the warm-season species compete excessively during the summer and tall fescue thins out. In the cooler portions of the transitional region, the warm-season species are lost by low temperature kill and tall fescue predominates. Tall fescue has been used with some success for winter overseeding into bermudagrass (*Cynodon dactylon* (L.) Pers.) turf on sport fields or other areas subjected to high traffic.

MAINTENANCE

A. Establishment Period

Seedling establishment of tall fescue is relatively slow (Cowan, 1956; Jones, 1958; Hughes and Nicholson, 1961) although it is a vigorous plant in later stages of growth. Several maintenance practices can be employed during the early establishment period to insure success of seedling development into a dense, mature turf. One of the most critical factors during the establishment phase is adequate soil moisture, especially at the soil surface. The surface soil layer should be maintained in a moist condition until germination is complete and young seedlings are well established. A light irrigation each day may be necessary, especially if the area is not mulched. If the area was irrigated to thoroughly wet the soil following seeding, the quantity of water required during the daily irrigations would be small. When the seedlings have grown to a height of 2.5 to 4.0 cm, irrigations can be reduced or stopped if an adequate moisture supply is available within 5.0 to 7.5 cm of the soil surface. If a light straw or hay mulch has been applied uniformly, it can be left in place to decompose. If too much mulch is applied or becomes piled up in areas, it should be removed when the grass is approximately 2.5 cm high.

Mowing should begin when seedlings reach a height one-third greater than the anticipated normal mowing height. For example, if the turf is to be maintained at a 5.0-cm cutting height, the new seedlings should be mowed when they reach a height of 7.5 cm. A severe setback of the plants may occur if seedlings are permitted to grow unusually tall and are then cut back to the normal mowing height, especially during warm weather. A light application of N fertilizer given at first mowing will enhance the establishment rate substantially. Application of about 25 kg/ha of water-soluble N is quite effective.

Weeds are not usually as much a problem in establishing tall fescue as in Kentucky bluegrass because tall fescue seed germinate much quicker and seedling growth is rapid. However, if weed populations are extensive enough to compete excessively with tall fescue seedlings, weed control measures will shorten the establishment period. Broadleaf weeds can be effectively controlled with the phenoxy-type herbicides. Herbicides in general are more toxic to young tall fescue seedlings than to more mature plants. Herbicides should be applied only if necessary and applications should be delayed as long as possible after germination. Grassy weeds are difficult to control with postemergence herbicides without severely damaging the turfgrass seedlings. Siduron [1-(2-methylcyclohexyl)-3-phenylurea] applied at seeding will effectively control grassy weeds and should be used when lawn seedings are at suboptimal times.

New swards of tall fescue are often severely weakened or destroyed by traffic before plants are sufficiently mature to tolerate the injurious effects imposed by wear. This frequently occurs on athletic fields and other heavy traffic areas. The excellent wear tolerance reported for tall fescue was based on tests of mature sod (Moorish and Harrison, 1948; Youngner, 1961; Shearman and Beard, 1975). Young tall fescue seedlings were much less wear-tolerant than plants that had developed tillers and crown tissue (J. J. Murray, unpublished data). A deliberate effort should be made to protect new swards from excessive traffic until a good sod has developed, especially during periods of high soil moisture. In the transitional zone tall fescue seeded in August or early September is generally not sufficiently developed for heavy use until the following year.

Table 14-1—Range in cutting height tolerance of 10 turfgrass species.†

Cutting height (cm)
0 — 1: Bentgrass
1 — 3: Bermudagrass
2 — 3: Zoysia
2 — 6: Red fescue
3 — 6: Carpetgrass
3 — 6: Kentucky bluegrass
3 — 6: Perennial ryegrass
4 — 8: Tall fescue
3 — 7: St. Augustinegrass
8 — 10: Canada bluegrass

† Compilation from Beard (1973), Musser (1962), Hanson and Juska (1969), and state experiment station publications.

B. Maintenance of Sod

Little documented research is available on maintenance of tall fescue for acceptable turfgrass appearance for a period of several years. In general, tall fescue has been considered a "low-maintenance" turfgrass. Perhaps this is due, to some extent, to its not being observed or tested as extensively as other species under high maintenance. However, the principles and concepts underlying good turfgrass cultural practices apply equally well to tall fescue and other cool-season turfgrasses. Certain aspects of physiology, anatomy, and morphology of tall fescue are unique and must be considered in a maintenance program to maximize turfgrass quality over an extended period of time.

Mowing

Turfgrasses vary considerably in their tolerance to height of mowing (Table 14-1). Tall fescue is an erect, upright-growing bunch grass with a rapid rate of vertical shoot elongation. It will not tolerate a mowing height as low as the more prostrate-growing turfgrasses such as Kentucky bluegrass or bentgrass (*Agrostis* spp.). Turfgrass quality of tall fescue has been best with mowing heights of 3.8 to 7.5 cm (Davis, 1958; D. T. Duff. 1964. Tall fescue and tall fescue-Kentucky bluegrass turf as influenced by cutting height, nitrogen application and irrigation. M.S. Thesis. Ohio State Univ.; Youngner, 1965; Juska et al., 1969; Burns, 1976; A. J. Turgeon, unpublished data). Generally, as the mowing height has been decreased from 7.5 cm, a corresponding increase in cultural practices was required to maintain turf quality. Mowing shorter than 3.8 cm will encourage germination and growth of weeds and other grasses in a short period of time.

The rapid shoot growth of tall fescue in combination with higher soil fertility levels will require mowing more frequently than would otherwise be the case. On home lawns, mowing once a week at 5.0 to 7.5 cm has provided satisfactory turf. Periods during the spring and fall may occur which will require mowing every 4 or 5 days. Less frequent mowing is necessary during periods of slow growth. Shorter-cut lawns require more frequent mowing to maintain a good appearance. Allowing tall fescue to become excessively tall and then cutting it back to the normal mowing height results in a stemmy-appearing turf that is more susceptible to weed invasion and has reduced recuperative potential. Gradually lowering the mowing height back to normal over several weeks results in less injury to the turf and better appearance. The frequency of mowing should be adjusted so that no more than 30 to 50% of the tissue is removed at any one mowing.

Mowability of tall fescue is not as good as that of Kentucky bluegrass. However, this is not a serious problem if the mower is kept sharp. A dull mower will shred the large vascular bundles of the leaf tissue, decreasing appearance of the turf. This is especially true on turf under low maintenance conditions. Mowing can be accomplished by either reel or rotary type mowers. The best type depends to some extent on the level of turf maintenance. In general, reel mowers give a cleaner cut that results in high quality tall fescue turf at cutting heights of 5.0 cm or less. Rotary mowers have performed best at higher mowing heights and on turf under low maintenance.

Clipping Management

Clippings of tall fescue mowed at proper intervals need not be removed from the turf. They decompose readily if water and fertility conditions are favorable and provide nutrients for subsequent growth. At Beltsville, Md. (J. J. Murray, unpublished data), where tall fescue clippings were removed, the N fertilization rate had to be increased by 122 kg/ha annually to produce turfgrass quality comparable to adjacent areas where clippings were returned. During the periods when the growth rate is quite rapid or when environmental conditions have interrupted the normal mowing frequency, an excessive amount of clippings may accumulate. When this occurs, the clippings must be removed to prevent smothering the turf. Thatch accumulation has not been a problem with tall fescue. Vertical mowing is generally not practiced. When it is practiced, the result is thinning of the stand and weed invasion, especially when practiced in the spring or early summer (J. J. Murray, unpublished data).

Fertilization

Turfgrass species vary considerably in their nutritional requirements, particularly N, to maintain optimal color, shoot density, and recuperative potential. Tall fescue is considered to have a lower N requirement than some other turfgrasses. However, it does respond well to N fertilization (Juska et al., 1969). Among turfgrass species, tall fescue falls in the medium category for relative annual N requirement (Table 14-2). However, tall

Table 14-2—Annual N requiremtns for several turfgrass species.†

Relative N level	N	Turfgrass species
	kg/ha	
Low	0–147	Bahiagrass
	49–98	Carpetgrass
	49–147	Red fescue
Medium	98–196	Zoysia
	98–293	Tall fescue
	196–293	Perennial ryegrass
High	196–293	Kentucky bluegrass
	245–342	Bentgrass
	245–489	Bermudagrass

† Summarized from Beard (1973), Hanson and Juska (1969), and state experiment station publications.

fescue is perhaps best characterized as being tolerant to a wide range of fertility levels.

Some contradiction exists as to the appropriate fertilization program for tall fescue. Perhaps, no one application rate or frequency is superior except under rather specific conditions. Fertilization programs are influenced by geographical locations, environmental conditions, appearance desired, intensity of use, soil physical and nutritional conditions, and cultural system utilized. Recommendations for tall fescue frequently range from 98 to 293 kg/ha of N annually.

A fertilization program that avoids the application of N in the early spring is recommended by most states in the transitional zone of the United States. Generally, these programs recommend the application of 50 to 75 kg/ha of readily available N in September, October, and either November-December or February-March. A 25-kg/ha application of N may be applied at any time between June and September if needed for additional color. If N sources are used having one-half or more of the N in slowly available form, apply 150 to 250 kg/ha in September and 50 to 100 kg/ha in late May or early June. Early spring applications are avoided because they encourage annual weedy grasses. Summer N fertilization should be practiced judiciously, especially during periods of heat stress or drought stress when growth is minimal. A fertilization program that has proved satisfactory under irrigation in northern portions of the transition zone consists of applying N at 50 to 75 kg/ha throughout the growing season. The frequency of applications is adjusted according to appearance and level of maintenance desired. High rates of N in late fall, especially in northern areas, may increase the chance for winter injury associated with low temperatures (Miller, 1969; Jung and Kocher, 1974).

Phosphate and potash must be maintained at adequate levels for satisfactory growth of tall fescue. A year's supply of these nutrients can be applied in a single application. However, utilization of complete fertilizers on the basis of the N program will generally supply adequate amounts of P and K. Periodic soil analysis will reveal if current levels of these nutrients are sufficient.

Irrigation

Tall fescue is one of the most drought-tolerant, cool-season turfgrasses (Youngner, 1965; Butler, 1975). It frequently sustains growth and remains green when Kentucky bluegrass goes dormant. In humid regions, it will survive without irrigation. However, under severe soil moisture stress, tall fescue becomes dormant and turns brown. Irrigation during periods of drought stress improves the appearance of tall fescue turf.

Good quality tall fescue turf at Beltsville, Md. (unpublished data) has been maintained by irrigating only after drought symptoms become severe enough to cause the grass to become semidormant. When this stress level is reached, water is applied to wet the soil to a depth of approximately 16 cm. Frequent, light watering encourages shallow roots and weed infestations. Disease development may also be greater under frequent irrigation, especially at higher fertility levels. Watering after drought dormancy causes poor appearance of the turf and encourages thinning of the stand and weed infestations.

Overseeding Practice

The practice of overseeding thin stands of tall fescue turf with more tall fescue helps maintain a thick healthy stand of grass. The effectiveness of overseeding with tall fescue is generally good. However, the basic procedures in overseeding turfgrasses must be followed. The best time to overseed tall fescue is late summer or early spring—late February or early March. Some cultivation by hoeing, slicing, vertical mowing, or raking is usually required to insure good contact between the soil and seed. The area should be fertilized following the cultivation operation and lime can also be applied if needed. After seeding, raking or dragging a mat and rolling will assist in obtaining good seed-soil contact. The area should be thoroughly irrigated immediately following seeding and kept moist until germination is complete.

PESTS

One of the best attributes of tall fescue as a turfgrass has been its relative resistance or tolerance to most common disease and insect pests that severely damage other cool-season turfgrasses. However, reports of injury from these pests have increased recently and perhaps they will become more important as the use of tall fescue increases. In pest control or prevention, emphasis should be on cultural practices that reduce the severity of turfgrass pests rather than reliance on chemicals. However, appropriate measures must be taken to achieve control should infestations become sufficiently severe to cause injury. Tall fescue has good-to-excellent tolerance to most commonly used turfgrass fungicides, insecticides, and herbicides when applied uniformly with properly calibrated equipment at the desired rate (Juska and Hanson, 1964; Berry and Buchanan, 1974; J. J. Murray, unpublished data).

The two most serious diseases of tall fescue turf are Helminthosporium blight (*Helminthosporium dictyoides* Drechsl.) and Brown patch (*Rhizoctonia solani* Kuhn). Both diseases can cause severe injury to new seedings (Couch, 1962). Helminthosporium blight is most severe during the spring of the year. The disease appears first as short longitudinal streaks of brown tissue. These net patterns eventually coalesce into dark brown blotches. Heavily diseased leaves ultimately turn yellow and die back from the tips. Symptoms of Brown patch on tall fescue may appear as circular brown patches or as a general foliar blight. Leaves generally turn light brown and then tan before dying. Symptoms usually begin at tips of leaves and progress downward with time. Several other turfgrass diseases may occur on tall fescue. For specific information on diseases see Chapter 15.

The incidence of insect damage to tall fescue is less frequent than that observed on many other cool-season turfgrasses. However, infestations of armyworms, cutworms, sod webworms, and white grubs occasionally occur and can cause substantial damage.

Weeds allowed to infest tall fescue turf disrupt the uniformity of color, leaf width, and growth habit and result in substantial degradation of turf appearance. In addition, weeds compete with the turfgrass for sunlight, soil moisture, and nutrients. Weed infestations are often the result of incorrect cultural practices such as mowing too closely or improper use of water or fertilizers. Thus, a key prerequisite in controlling weeds is the use of cultural practices that insure a vigorous, dense, healthy turf. Control of weeds with herbicides is only temporary unless the basic cause of their original presence is determined and corrected. Heavy weed infestations, if removed chemically, will leave large bare areas of soil which should be reseeded as soon as possible.

LITERATURE CITED

1. Anon. 1923. "Turfing Fescue" a new name misrepresenting an old grass. USDA Off. Sec., Press Release 1061-23.
2. Beard, James B. 1973. Turfgrass: Science and culture. Prentice-Hall, Inc., Englewood Cliffs, N.J. 658 p.
3. Berry, Charles D., and Gale A. Buchanan. 1974. Tolerance of *Phalaris tuberosa* L. and *Festuca arundinacea* Schreb. to preemergence herbicide treatment. Crop Sci. 14:96-99.
4. Burns, Robert E. 1976. Tall fescue turf as affected by mowing height. Agron. J. 68:274-276.
5. Butler, J. D. 1975. Drought tolerant grasses for turf use. p. 47-54. *In* Proc. 21st Annu. Rocky Mountain Reg. Turfgrass Conf. 23-24 Jan. 1975. Coop. Ext. Serv., Colorado State Univ., Ft. Collins.
6. Couch, H. B. 1962. Diseases of turfgrasses. Reinhold Publ. Corp., New York. 289 p.
7. Cowan, J. R. 1956. Tall fescue. Adv. Agron. 8:283-320.
8. Davis, R. R. 1958. The effect of other species and mowing height on persistence of lawn grasses. Agron. J. 50:671-673.

9. Grau, F. V. 1954. Tough grasses for hard wear. J. Housing p. 336-340.
10. Hanson, A. A., and F. V. Juska (eds.). 1969. Turfgrass science. Agronomy 14. Am. Soc. Agron., Madison, Wis.
11. Harper, J. C., II, and M. A. Hein. 1954. Better lawns. Crops Res. Div., ARS, USDA. Home & Garden Bull. 51.
12. Hughes, R., and I. A. Nicholson. 1961. Comparisons of grass varieties for surface seeding upland pasture types. II. Deep peat. J. Br. Grassl. Soc. 16:314-322.
13. Jones, Ll. I. 1958. Grassland agronomy. Welsh Plant Breed. Stn. Rep., 1950-56. Univ. Coll. of Wales, Aberystwyth. p. 94-128.
14. Juska, F. V. 1959. Evaluation of cool-season turfgrasses alone and in mixtures. Agron. J. 51:597-600.
15. ―――, and A. A. Hanson. 1959. Evaluation of cool-season turfgrasses. Park Maint. 12:18-20.
16. ―――, and ―――. 1964. Effect of preemergence crabgrass herbicides on seedling emergence of turfgrass species. Weeds 12:97-101.
17. ―――, ―――, and A. W. Hovin. 1969. Evaluation of tall fescue, *Festuca arundinacea* Schreb., for turf in the transition zone of the United States. Agron. J. 61:625-628.
18. Jung, G. A., and R. E. Kocher. 1974. Influence of applied nitrogen and clipping treatments on winter survival of perennial cool-season grasses. Agron. J. 66:62-65.
19. Lantz, H. L. 1955. Cool-season grasses—bluegrass, fescue, bent. Golf Course Rep. 23: 32-34, 36-37.
20. Miller, R. W. 1969. Tall fescue and tall fescue-bluegrass mixtures for athletic fields and other turfgrass areas. p. 9-16. *In* Proc. Ohio Turfgrass Conf. 1-3 Dec. 1969. Ohio Coop. Ext. Serv., Ohio State Univ., Columbus.
21. Moorish, R. H., and C. M. Harrison. 1948. The establishment and comparative wear resistance of various grasses and grass-legume mixtures to vehicular traffic. J. Am. Soc. Agron. 40:168-179.
22. ―――, and J. A. DeFrance. 1948. Greenswards in the cooler regions. p. 307-310. *In* Alfred Stefferud (ed.) Yearb. Agric., USDA.
23. Reed, F. J. 1950. Lawns and playing fields. Faber and Faber Ltd., 24 Russell Square, London. 212 p.
24. Shearman, R. C., and J. B. Beard. 1975. Turfgrass wear tolerance mechanisms. I. Wear tolerance of seven turfgrass species and quantitative methods for determining turfgrass wear injury. Agron. J. 67:208-211.
25. Underwood, J. K. 1964. Tennessee lawns. Univ. of Tennessee Ext. Bull. 326. 45 p.
26. Wood, G. M., and H. E. Buckland. 1966. Survival of turfgrass seedlings subjected to induced drouth stress. Agron. J. 58:19-23.
27. Youngner, V. B. 1961. Accelerated wear tests on turfgrasses. Agron. J. 53:217-218.
28. ―――. 1965. A report on tall fescue for turf. West. Landscape News 5:6, 20.
29. ―――. 1970. Turfgrass varieties and irrigation practices. Golf Super. 38:66, 68.

Chapter 15

Diseases and Nematodes

R. A. CHAPMAN
Department of Plant Pathology
University of Kentucky
Lexington, Kentucky

Despite intermittent enthusiasm for tall fescue (*Festuca arundinacea* Schreb.) as forage during the past century, it was only about 40 years ago that serious development of it as a forage and turfgrass began (Tabor, 1952; Cowan, 1956). Among the characteristics that attracted attention to the possibility of exploiting tall fescue was its relative freedom from severe damage by certain diseases that plague meadow fescue (*F. elatior* L.), which has been intensively cultivated for a much longer period of time. In order to avoid gathering false information about diseases and nematodes of tall fescue, I have attempted to use and cite references that clearly apply to *F. elatior* var. *arundinacea, F. arundinacea,* or one of the named cultivars of the species.

DISEASES

The vast majority of recognized diseases of tall fescue are caused by fungi. I did not find a report of a bacterial disease and I found little information about viral diseases. The principal concern about viral diseases has been whether tall fescue serves as a reservoir or overwintering host of two viruses that are significant pathogens of cereals. Oswald and Houston (1953) found that 'Kentucky 31' is a host for barley yellow dwarf virus (BYDV) and designated it a symptomless carrier. Bruehl and Toko (1957) showed that it was susceptible to two strains of BYDV and Weerapat et al. (1972) presented evidence that the virus caused leaf yellowing in Kentucky 31 in spring and fall when temperatures were relatively cool. The other virus-tall fescue interrelationship that has received some attention involved maize dwarf mosaic virus (MDMV). No infection followed mechanical inoculation of tall fescue with MDMV (Shepherd, 1965; Mackenzie et al., 1966).

Crown rust, netblotch, cercospora leaf spot, and rhizoctonia leaf scald (brown patch in turf) are the most conspicuous fungal diseases of tall fescue and have attracted most of the attention of plant pathologists.

Copyright © 1979 ASA-CSSA-SSSA, 677 South Segoe Road, Madison, WI 53711 USA. *Tall Fescue.*

Puccinia coronata Cda., the cause of crown rust, is coextensive with meadow and tall fescue. The disease is a typical leaf rust, appearing as light yellow flecks that eventually enlarge and develop into reddish-brown pustules containing urediospores of the fungus. Later, the characteristic crown-like teliospores that give the rust its common name are formed in dark colored telia surrounding the pustules. Heavily infected leaves turn yellow at the tips and the chlorosis progresses toward the sheaths. Tall fescue is generally resistant to this disease whereas meadow fescue is generally susceptible. Its resistance was one of the characteristics of tall fescue that attracted attention to the possibility of exploiting it as a forage grass (Piper, 1924; Hardison, 1945c).

Development of distinct cultivars of tall fescue that have been distributed into widespread climatic environments combined with inherent variability of pathenogenicity of the fungus have resulted in a wide range of susceptibility to crown rust within tall fescue. Kreitlow and Myers (1946, 1947) reported 'Alta' and Kentucky 31 highly resistant-immune when inoculated under greenhouse conditions. When crown rust was first observed in the field in Kentucky in 1968, 'Fawn' and 'Goar' were highly susceptible, Alta was intermediate, and Kentucky 31, 'Kenwell', and several ryegrass (*Lolium* sp.)-fescue hybrids were resistant (Buckner and Burrus, 1970). Berry and Gudauskas (1972) reported severe outbreaks of crown rust on Goar in Alabama. They screened 6,000 seedlings and obtained a few plants that were resistant.

Crown rust is the only significant rust of tall fescue. Two other species of *Puccinia*, *P. graminis* Pers. in Great Britain and *P. striiformis* Westend. in Washington, have been recorded on tall fescue, but their effectiveness as pathogens has not been evaluated (Sampson and Western, 1954; Dietz and Hendrix, 1962).

Helminthosporium dictyoides Drechs., the cause of netblotch, is also coextensive with tall and meadow fescues. The characteristic net-like patterns of streaks of dark brown tissue in young lesions on leaves give the disease its common name. Later the streaks coalesce into dark brown spots and heavily infected leaves turn yellow and die back from the tips. Loss of foliage can be severe when conditions are optimal for development of the disease. The fungus is more active and the distinctive symptoms are more evident in the cooler spring, fall, and winter months than during midsummer (Drechsler, 1923; Luttrell, 1951). Optimal temperatures for pathogenicity are 15 to 20 C (Flores et al., 1969).

Tall fescue is generally more resistant than meadow fescue to *H. dictyoides,* and this is another characteristic that attracted attention to the development of tall fescue for forage (Hardison, 1945c; Kreitlow et al., 1950). However, the distinction between the reactions of the two grasses to *H. dictyoides* is not as great as that between their reactions to *P. coronata,* and there is considerable variation among cultivars of tall fescue. This is exacerbated by variability in pathogenicity of the fungus.

Braverman and Graham (1960) distinguished three physiologic forms of the fungus: 1) *H. dictyoides* f. sp. *dictyoides* from tall and meadow fescues; 2) *H. dictyoides* f. sp. *perenne* from annual (*Lolium multiflorum* Lam.) and perennial (*L. perenne* L.) ryegrasses; 3) *H. dictyoides* f. sp. *phlei* from timothy (*Phleum pratense* L.). The first two infected tall fescue, the last did not. Henson and Buckner (1957, 1958) rated reactions of three inbred lines of tall fescue to *H. dictyoides* during an epiphytotic of netblotch in Kentucky in 1957. Fourteen clones of a second generation family were moderately susceptible, 15 clones of a third generation family were highly susceptible, and 15 clones of another third generation family were highly resistant. Kenwell has been rated moderately resistant when compared with a highly resistant inbred line of tall fescue (Buckner and Burrus, 1968) and highly resistant when compared with Alta and Kentucky 31 (Flores et al., 1969).

H. dictyoides is primarily a pathogen of fescues and ryegrasses. The genus *Helminthosporium* contains numerous other species that are important pathogens of many graminaceous hosts, and the reactions of tall fescue to some of them have been determined. Moore and Couch (1961) reported that *H. vagans* infected tall fescue under greenhouse conditions. Gentry (1968) and Gentry et al. (1968) found a range of reactions to *H. vagans* among Kenwell, least susceptible; Kentucky 31 and Alta, intermediate; Goar, most susceptible. There was a highly significant negative correlation between peroline content of the cultivars and susceptibility. Flores et al. (1969) confirmed the relative susceptibilities of Kenwell, Kentucky 31, and Alta. Optimal temperatures for pathogenicity were 20 to 25 C, and *H. vagans* was more active in the field in midsummer than in spring or fall. This situation is the reverse of that with *H. dictyoides*.

Tall fescue has been inoculated with several other species of *Helminthosporium*. *H. siccans* Drechs., *H. rostratum* Drechs., *H. leersiae* Atk., *H. oryzae* B. deHaan, and three unidentified isolates, two from barley (*Hordeum vulgare* L.) and one from oats (*Avena sativa* L.) were infective (Whitehead and Calvert, 1959; Braverman and Graham, 1960; Nelson and Kline, 1961). Luttrell (1953) reported very slight infection by *H. sativum* P.K.B., but Nelson and Kline (1961) reported that *H. sorokinianum* Sacc. ex Sorokin (syn. *H. sativum*) was not infective. *H. carbonum* Ullstrip, *H. cynodontis* Marig., *H. homorphus* Luttrell and Rogerson, *H. maydis* Nisik. and Miyake, *H. sacchari* (B. deHaan) Butl., *H. sorghicola* Lefebvre and Sherwin, *H. turcicum* Pass., *H. victoriae* Meehan and Murphy, and several unidentified isolates were not infective (Braverman and Graham, 1960; Nelson and Kline, 1961).

Rhizoctonia solani Kuhn causes rhizoctonia leaf scald (brown patch in turf) of many grasses and cereals including tall fescue (Sprague, 1950; Sampson and Western, 1954; Couch, 1973). Foliage is destroyed rapidly during periods of high temperature accompanied by high humidity. Severe epiphytotics have been reported on Alta in Maryland (Allison et al., 1949), on tall fescue in Kentucky (Anonymous, 1950, 1951), and on Alta and Kentucky 31 in Louisiana (Atkins, 1952). Johnson (1953) regarded leaf scald the most serious disease of tall fescue in the southern U.S.

R. solani is a ubiquitous fungus that has a wide host range and isolates from different plants differ greatly in their pathogenicity to particular grasses (Sprague, 1950). Gentry (1968) and Gentry et al. (1968) reported a range of susceptibility to *R. solani* among Kenwell, least susceptible; Kentucky 31 and Alta, intermediate; Goar, most susceptible, when they were inoculated with several isolates of the fungus. These relative reactions are in the same order as those of the same cultivars inoculated with *H. vagans,* and the same highly significant negative correlation between perloline content of the cultivars and susceptibility occurred.

R. solani is a significant pathogen of other species of turfgrasses, and Alta tall fescue ranked in the middle of a group of 18 turfgrasses that were moderately to highly susceptible to 29 isolates from various sources (Shurtleff, 1953).

Hardison (1945c) observed a mild leaf spot on tall fescue in Kentucky in 1943 and 1944. He described the pathogen and named it *Cercospora festucae* (Hardison, 1945d). Subsequently, Johnson and Valleau (1949) synonomyzed *C. festucae* Hardison with *C. apii* Fres. on the basis of cross infectivity tests. Whitehead (1950) and Whitehead and Holt (1950) reported a severe outbreak of cercospora leaf spot on Alta in Texas in 1949. In this epiphytotic the typical gray-centered, purple-brown margined eyespot lesions coalesced, resulting in considerable necrosis of leaf tissue. Approximately 10 and 30% of 1-year-old seedlings of Kentucky 31 and Alta, respectively, were killed. All plants of mature clones of Alta in a spaced planting were infected and 36% were severely damaged or killed. They provided an amended description of the fungus and retained the name *C. festucae* Hardison. Luttrell (1951) estimated as much as 20% damage from this disease in tall fescue nurseries in Georgia. On the other hand, Sprague (1950) observed it as a mild leafspot in Oregon.

Tall fescue is damaged by several other fungi that attack numerous grasses, especially when they are managed as turf (Sprague, 1950; Howard et al., 1951; Britton, 1969; Couch, 1973). These include the snow molds *Typhula incarnata* Lasch ex Fr. (typhula blight, snow scald, winter scald); *Fusarium nivale* (Fr.) Snyder and Hansen (pink snow mold); *Sclerotinia borealis* Bubak and Vieugel (sclerotinia patch). *F. nivale* also causes fusarium patch during periods of cool, wet weather without snow cover. Fairy rings, caused by various basidiomycetes and slime molds, principally *Physarum cinereum* (Batsch.) Pers. and *Mucilago spongiosa* (Leyss.) Morg. occur with tall fescue as well as other grasses. Tall fescue ranked in the middle of a group of 12 species of turfgrasses in susceptibility to *Ophiobolus graminis* Sacc., the cause of ophiobolus patch (Couch, 1973). Tall fescue was susceptible to *Pythium aphanidermatum* (Edson) Fitzpatrick (cottony blight) when inoculated with the fungus in a greenhouse (Freeman and Horn, 1963), and has been reported susceptible to *P. debaryanum* Hesse, *P. graminicola* Subrm., and *P. irregulare* Buism., causes of seedling blights and/or root necrosis (Sprague, 1950). *P. ultimum* Trow. did not infect tall fescue when inoculated in a greenhouse (Moore and Couch, 1961).

Two fungal diseases are particularily important when tall fescue is managed for seed production. *Gloeotinia temulenta* (Prill. and Del.) Wilson, Noble, and Gray, the cause of blind seed which results in reduced germination of seed of various grasses, especially ryegrasses, has been reported on tall fescue in New Zealand (Neill and Hyde, 1942) and the United States (Sprague, 1950). Hardison (1978) considers it a significant factor in seed production fields in Oregon. *Fusarium tricinctum* f. sp. *poae* (Pk.) Snyder and Hansen has long been considered the cause of silver top (white heads, white top) which reduces seed production of various grasses. It has been reported on tall fescue in the United States (Sprague, 1950; Couch, 1973). The mite, *Pediculopsis (Siteroptes) graminum* Reuter is a vector of the fungus. Hardison (1959, 1976) presented considerable evidence that these two organisms are secondary associates of silver top and concluded that insects are the primary cause of the disease.

Three clavicepitaceous fungi, *Claviceps purpurea* (Fr.) Tul., *Balansia epichloe* (Weese) Diehl, and *Epichloe typhina* (Pers.) Tul., that occur on tall fescue as well as on many other graminaceous plants, have attracted considerable attention because of their possible relationship to the tall fescue toxicity problem (see Chapter 13). *C. purpurea* causes ergot in which fungal sclerotia replace seed of infected flowers. The sclerotia contain alkaloids and, depending on the quantity ingested, various manifestations of ergotism can occur in animals grazing infected plants. The symptoms of ergotism closely resemble those of fescue foot which has occurred sporadically, especially in cattle grazing soil-bank or other previously ungrazed fields as emergency winter pasture. It is in such fields that infection by *C. purpurea* might be sufficient to produce toxic plant tissue and/or abundant sclerotia. *C. purpurea* is confined to the flowers of infected plants and the likelihood of ergot alkaloids occurring outside the sclerotia seems remote. Maag and Tobiska (1956) reported the occurrence of ergot alkaloids in hay from plants that showed no visual evidence of infection by *C. purpurea,* but Cunningham (1949) and Riggs (1966) found no evidence of ergot alkaloids in tissue of infected plants. Riggs et al. (1965, 1968) reported that the alkaloid content of tall fescue was not altered as the result of infection by *C. purpurea* and the alkaloid content of sclerotia of the fungus from tall fescue was similar to that of sclerotia from rye (*Secale cereale* L.) and ryegrass-tall fescue hybrids. Although ergot is widespread in tall fescue, it is not especially abundant and true ergotism probably occurs infrequently in animals grazing it. It is not considered to be part of the fescue foot problem (Yates, 1971).

B. epichloe has been observed much less frequently than *C. purpurea* on tall fescue. It was abundant on smutgrass (*Sporobolus poiretti* (Roem. and Schult.) Hitchc.), which was a common weed in a tall fescue pasture where cattle were affected by the summer syndrome of tall fescue toxicity in Georgia (Bacon et al., 1975; Porter et al., 1975). It was found occasionally on the tall fescue in this field. This fungus produces spore-bearing stomata on leaves of infected plants, and cattle would have an opportunity to ingest considerable fungal material while grazing. Alkaloids similar to those in sclerotia of *C. purpurea* have been detected in *B. epichloe* (Porter et al., 1977), and it seems reasonable to assume that this fungus could be toxic.

However, in this field tall fescue was a minor source of the fungus and if the fungus was the cause of the toxicity it may be misleading to label the problem a fescue toxicity.

E. typhina, the cause of "choke" of many grasses including several species of *Festuca,* was first reported in tall fescue by Neill (1941). The fungus is a systemic endophyte and infected plants are symptomless. Because the characteristic ascospore-producing stage of the fungus has not been observed on tall fescue, its identification is based on the morphology of the intercellular mycelium in plant tissue. Pith of culms is especially useful and the mycelium in infected plants looks like that of *E. typhina* (Sampson, 1933; Neill, 1940). Conidia produced in cultures of the fungus isolated from infected plants are characteristic of *Sphacelia typhina* (Pers.) Sacc., the asexual stage of *E. typhina.* The fungus has been detected in Kentucky 31, 'Kenhy', and two ryegrass-fescue hybrids. Evidence that *E. typhina* may be involved in the summer syndrome of fescue toxicity of cattle was presented by Bacon et al. (1977).

The sporadic occurrence of fescue foot has many attributes of a mycotoxin problem (Yates, 1971). Several nonspecific, frequently nonphytopathogenic fungi isolated from toxic tall fescue produced metabolites that were toxic to mice and rabbits in laboratory assays used for screening mycotoxin production. Isolates of the genus *Fusarium* seem to be especially suspect (Keyl et al., 1967; Ellis and Yates, 1971; Yates, 1971).

Records of the occurrence of several other fungi on tall fescue have been made. They include *Fusarium acuminata* Ell. and Ev. and *Gloeosporium bolleyi* Sprague in the northwest U.S. (Sprague, 1950). These fungi are associated with root rots of a large number of graminaceous plants. Sarasola and Campi (1947) were successful in infecting *F. arundinacea* with *Rhynchosporium secalis* (Oud.) Davis, the cause of leaf scald of a number of small grains and grasses. *Phleospora idahoensis* Sprague, the cause of stem eyespot on red fescue (*F. rubra,* L.), was found on introduced tall fescue in western Canada (Smith and Elliott, 1970). 'Manade', Fawn, and Alta were affected similarily and 3-year-old plants of Manade were more severely diseased than 2-year old plants. *Urocystis agropyri* (Preuss) Schroet., the cause of flag smut of a number of small grains and grasses, was found on tall fescue in Britain (Webster and Iqbal, 1971). This may be the only record of a smut on tall fescue. *Sclerotium rolfsii* Sacc., the cause of white mold of many plants, and *Colletotrichum graminicola* (Ces.) G. W. Wills., the cause of anthracnose of many graminaceous hosts, were reported on tall fescue in Georgia by Luttrell (1951). He also found fungi in the genera *Aspergillus, Curvularia, Fusarium, Marasmius,* and *Thielavia* associated with crown rot of tall fescue.

Reports of the absence of infection of tall fescue by certain fungi should be noted. Hardison (1944, 1945a, b) inoculated tall fescue with conidia of *Erysiphe graminis* DC, the cause of powdery mildew of a large number of graminaceous plants. No infection occurred. Graham (1952) reported similar results when he inoculated tall fescue with *Stagnospora maculata* (G.) Sprague, the cause of purple leafspot of orchardgrass (*Dactylis glomerata* L.).

NEMATODES

Tall fescue in fields and lawns is associated with a variety of genera and species of plant parasitic nematodes that feed, grow, and reproduce on or in roots. Poor growth is the readily observable symptom of root damage caused by nematodes. It is a useful but nonspecific indicator of a nematode problem because poor growth can be caused by a myriad of biotic and abiotic factors acting individually or in concert. Identifying a nematode as a possible causal agent of damage usually depends upon determining whether tall fescue is a host of the nematode because plant parasitic nematodes are obligate parasites and amount of root injury is usually positively correlated with population development of the nematode. However, development of high populations of nematodes on roots of tall fescue does not always indicate significant pathogenicity. Many plant parasitic nematodes are not very pathogenic and tall fescue can tolerate large numbers of some without apparent ill effect.

Suitability of tall fescue as a host for three root-lesion nematodes was determined in greenhouse experiments. *Pratylenchus penetrans* (Cobb) reproduced an Alta, but its effect on plant growth was not determined (Jensen, 1953). *P. scribneri* Steiner reproduced on Ky 45-50 without significantly affecting growth of the grass (Minton, 1965) and *P. zeae* Graham did not reproduce on Kentucky 31 (Endo, 1959).

Townshend et al. (1973) rated four cultivars of tall fescue for suitability as hosts of *Pratylenchus neglectus* (Rensch), a pin nematode, *Paratylenchus projectus* Jenkins, and a spiral nematode, *Helicotylenchus dihystera* (Cobb) on the basis of numbers of nematodes recovered from soil in a 3-year-old turfgrass trial in Ontario. Reproduction by *P. neglectus* was nonexistant on 'Backafall', poor on 'S-170', and moderately good on Kentucky 31 and Manade. *P. projectus* did not reproduce on Backafall and S-170, reproduced poorly on Kentucky 31, and moderately good on Manade. Reproduction of *H. dihystera* was poor on Backafall and Manade and moderately good on Kentucky 31 and S-170. None was rated a good host for any of the nematodes and an evaluation of effects on plant growth was not made. Tall fescue was included in a list of hosts of the cereal cyst nematode, *Heterodera avenae* Wollenweber (Holdeman and Watson, 1977).

Greenhouse experiments designed to study population development and effects of nematodes on tall fescue in more detail have been conducted. *P. projectus* reproduced abundantly on Kentucky 31 and reduced top growth after 3 months in a sandy loam soil in pots in a greenhouse (Coursen and Jenkins, 1958). McGlohon et al. (1961) obtained similar results from a 9-month greenhouse experiment during which the grass was clipped to 10 cm every 3 to 6 weeks. Reduction of growth of Kentucky 31 was greater in two sandy loam soils than in a clay loam soil even though population development of *P. projectus* was much greater on plants in the clay loam soil. A lance nematode (*Hoplolaimus galeatus* Cobb) and the stubby-root nematode (*Trichodorus christiei* Allen) were more injurious than *P. projectus* in all three soils, and there was no difference in the amount of injury between plants growing in the two types of soil. Both nematodes reproduced abundantly. The stunt nematode (*Tylenchorhynchus claytoni* Steiner) and a spiral nematode (*Helicotylenchus dihystera* Cobb) reproduced abundantly on Kentucky 31 without affecting top growth of the grass (McGlohon et al., 1961).

A different approach to determining whether nematodes affect growth of tall fescue is to compare yields of forage from nematicide-treated field plots with those from untreated plots. Hoveland et al. (1975) compared yields of Goar and Kentucky 31 from untreated plots and plots treated with methyl bromide (MB) and carbofuran (CN) in a field of fine sandy loam in Alabama. *H. galeatus, T. christiei,* and *T. claytoni* were the predominant plant parasitic nematodes in the field. Second-year forage yields from the MB-treated plots were 107% and from the CN-treated plots 39% greater than the yield from the untreated plots. Numbers of all three nematodes were usually, but not always, significantly smaller in the treated plots, especially in those treated with MB. Stands survived drought and persisted better in treated (especially in MB-treated) than in untreated plots. This information should be interpreted cautiously. MB is a general biocide and CN is an insecticide as well as a nematicide. Both, especially MB which evoked the greater response, could have reduced populations of many organisms other than nematodes that adversely affect the growth of tall fescue.

One of the noteworthy aspects of the host-parasite interrelationships between tall fescue and plant parasitic nematodes is its resistance to certain root-knot nematodes, species of *Meloidogyne.* Martin (1958) found *M. javanica* (Treub.) on tall fescue in Africa. McGlohon et al. (1961) reported that Kentucky 31 was highly resistant to *M. javanica, M. arenaria* (Neal), *M. hapla* Chitwood, and *M. incognita* (Kofoid and White). Chapman (1973) found that Alta, Fawn, Goar, Kenhy, Kenwell, and S-170 were as resistant as Kentucky 31 to *M. incognita* and *M. hapla.* These species of *Meloidogyne* are important pathogens of many vegetable, ornamental, fruit, and field crops, and tall fescue is a useful rotation crop where these nematodes are an economic problem and crop rotation is a feasible aid for controlling them.

Tall fescue is a host for two species of *Meloidogyne* that are primarily pathogens of graminaceous plants. Dickerson (1966) and Williams and Laughlin (1968) reported that *M. graminis* Sledge and Golden) reproduced on Kentucky 31. Radewald et al. (1970) reported reproduction by *M. naasi* Franklin on Alta. Chapman (1973, unpublished data) found Fawn, Goar, Kenhy, and Kenwell as susceptible as Alta and S-170 resistant to *M. naasi.* The effects of these nematodes on growth of tall fescue have not been evaluated.

DISEASE AND NEMATODE CONTROL

During its brief history of intensive cultivation, tall fescue has had relatively few serious problems with diseases and nematodes in its areas of adaptation. Part of the process of adaptation has been accommodation to effects of potential pathogens and these have probably played significant roles in delineating the areas of adaptation for this cool-season perennial grass. As the length of time and intensity of cultivation of tall fescue increase within these areas, problems with crown rust, netblotch, rhizoctonia leaf scald, and cercospora leaf spot will probably increase and certain now minor or unknown parasites may emerge as significant threats.

Efforts are being made to use tall fescue as a forage grass outside its areas of adaption, especially as winter pasture in the southern U.S. Crown rust, netblotch, and rhizoctonia leaf scald have, on occasion, been severe and there is evidence that nematodes may be significant factors for forage production in this area. The grass must survive summers to be useful as winter pasture, and the summer climate is not one to which tall fescue is well adapted. It is not surprising that, under these conditions, diseases and nematodes become more significant than they are where tall fescue is better adapted to its environment.

Control of diseases and nematodes of tall fescue used for forage has been accomplished primarily by using productive cultural and management practices, and by selecting and developing resistant cultivars. These methods will continue to be of paramount importance. Pesticides, especially fungicides, are useful on tall fescue in turf.

Changes in specific disease and nematode control methodology are unpredictable. For example, open burning of seed production fields has long been a very successful method for controlling blind seed, silver top, and ergot of tall fescue and other grasses, but such open burning seems likely to be prohibited and alternative methods are being sought (Hardison, 1976, 1978). Specific management practices and cultivars will have to be developed for each new climatic area if the range of successful establishment of tall fescue is to be extended significantly. No recommendations for specific chemicals are included here because the use of pesticides is a rapidly changing set of procedures and current information about their use must be consulted.

Tall fescue is a relatively new agricultural species. The obvious diseases are being confronted. The task now is to be alert and identify the significance of potential pathogens.

LITERATURE CITED

1. Allison, J. L., H. S. Sherwin, I. Forbes, Jr., and R. E. Wagner. 1949. *Rhizoctonia solani*, a destructive pathogen of Alta fescue, smooth bromegrass and birdsfoot trefoil. Phytopathology 39:1.
2. Anonymous. 1950. 63rd Annu. Rep. of the Kentucky Agric. Exp. Stn. p. 9.
3. ————. 1951. 64th Annu. Rep. of the Kentucky Agric. Exp. Stn. p. 8.
4. Atkins, J. G., Jr. 1952. Forage crop *Rhizoctonia* cross inoculation tests. Phytopathology 42:282.
5. Bacon, C. W., J. K. Porter, and J. D. Robbins. 1975. Toxicity and occurrence of *Balansia* on grasses from toxic fescue pastures. Appl. Microbiol. 29:553–556.
6. ————, J. K. Porter, J. D. Robbins, and E. S. Luttrell. 1977. *Epichloe typhina* from toxic tall fescue grasses. Appl. Environ. Microbiol. 34:576–581.
7. Berry, C. D., and R. T. Gudauskas. 1972. Susceptibility of tall fescue, *Festuca arundinacea* Schreb. to crown rust. Crop Sci. 12:101–102.

8. Braverman, S. W. 1967. Disease resistance in cool-season forage, range, and turfgrasses. Bot. Rev. 33:329-378.
9. ————, and J. H. Graham. 1960. *Helminthosporium dictyoides* and related species on forage grasses. Phytopathology 50:691-695.
10. Britton, M. P. 1969. Turfgrass diseases. *In* A. A. Hanson and F. V. Juska (eds.) Turfgrass science. Agronomy 14:288-335. Am. Soc. of Agron., Madison, Wis.
11. Bruehl, G. W., and H. V. Toko. 1957. Host range of two strains of the cereal yellow-dwarf virus. Plant Dis. Rep. 41:730-734.
12. Buckner, R. C., and P. B. Burrus, II. 1968. 80th Annu. Rep. of the Kentucky Agric. Exp. Stn. p. 25.
13. ————, and ————. 1970. 82nd Annu. Rep. of the Kentucky Agric. Exp. Stn. p. 26.
14. Chapman, R. A. 1973. Southern and northern root-knot nematodes did not reproduce on tall fescue. 85th Annu. Rep. of the Kentucky Agric. Exp. Stn. p. 94.
15. Couch, H. B. 1973. Diseases of turfgrasses. 2nd Ed. Robert E. Krieger Publishing Co., New York.
16. Coursen, B. W., and W. R. Jenkins. 1958. Host-parasite relationships of the pin nematode, *Paratylenchus projectus*, on tobacco and tall fescue. Plant Dis. Rep. 42:865-872.
17. Cowan, J. R. 1956. Tall fescue. Adv. Agron. 8:283-320.
18. Cunningham, I. J. 1949. A note on the cause of tall fescue lameness in cattle. Aust. Vet. J. 25:27-28.
19. Dickerson, O. J. 1966. Some observations on *Hypsoperine graminis* in Kansas. Plant Dis. Rep. 50:396-398.
20. Dietz, S. M., and J. W. Hendrix. 1962. Reaction of grass to stripe rust at Pullman, Wash. Phytopathology 52:730.
21. Drechsler, C. 1923. Some graminicolous species of *Helminthosporium*. I. J. Agric. Res. 24:641-740.
22. Ellis, J. J., and S. G. Yates. 1971. Mycotoxins of fungi from fescue. Econ. Bot. 25:1-5.
23. Endo, B. Y. 1959. Responses of root-lesion nematodes, *Pratylenchus brachyurus* and *P. zeae* to various plants and soil types. Phytopathology 49:417-421.
24. Flores, J. M., R. A. Chapman, and L. Henson. 1969. Susceptibility of detached and attached leaf blades of tall fescue to infection by two species of *Helminthosporium*. Phytopathology 59:1010-1011.
25. Freeman, T. E., and G. C. Horn. 1963. Reaction of turfgrasses to attack by *Pythium aphanidermatum*. Plant Dis. Rep. 47:425-427.
26. Gentry, C. E. 1968. Interrelationship of *Rhizoctonia solani* Kuhn, environment, and genotype on the alkaloid content of tall fescue, *Festuca arundinacea* Schreb. Ph.D. Disser., Univ. of Kentucky (Disser. Abstr. Int. 30:1446B. 1969-70).
27. ————, L. Henson, and R. A. Chapman. 1968. The interrelationships of *Rhizoctonia solani* and *Helminthosporium vagans*, nitrogen, and variety on alkaloids in *Festuca arundinacea*. Phytopathology 58:1051.
28. Graham, J. G. 1952. Purple leafspot of orchardgrass. Phytopathology 42:653-656.
29. Hardison, J. R. 1944. Specialization of pathogenicity in *Erysiphe graminis* on wild and cultivated grasses. Phytopathology 34:1-20.
30. ————. 1945a. Specialization of pathogenicity in *Erysiphe graminis* on *Poa* and its relation to bluegrass improvement. Phytopathology 35:62-71.
31. ————. 1945b. Specialization in *Erysiphe graminis* for pathogenicity on wild and cultivated grasses outside the tribe Hordeae. Phytopathology 35:394-405.
32. ————. 1945c. Observations on grass diseases in Kentucky, Septemer, 1942 to September, 1944 and a preliminary check list. Plant Dis. Rep. 29:76-85.
33. ————. 1945d. A leafspot of tall fescue caused by a new species of *Cercospora*. Mycologia 37:492-494.
34. ————. 1959. Evidence against *Fusarium poae* and *Siteroptes graminum* as causal agents of silver top of grasses. Mycologia 51:712-728.
35. ————. 1976. Fire and flame for plant disease control. Annu. Rev. Phytopathol. 14:355-379.
36. ————. 1978. Chemical suppression of *Gloeotinia temulenta* apothecia in field plots of *Lolium perenne*. Phytopathology 68:513-516.
37. Henson, L., and R. C. Buckner. 1957. Resistance to *Helminthosporium dictyoides* in inbred lines of *Festuca arundinacea*. Phytopathology 47:523.

38. ――――, and R. C. Buckner. 1958. Field resistance to *Helminthosporium dictyoides* Drechs. in inbred lines of *Festuca arundinacea* Schreb. 70th Annu. Rep. of the Kentucky Agric. Exp. Stn. p. 23.
39. Holdeman, Q. L., and T. R. Watson. 1977. The oat cyst nematode, *Heterodera avenae* Wollenweber 1924, a root parasite of cereal crops and other grasses. California Dep. Food and Agric. 82 p.
40. Hoveland, C. S., R. Rodriguez-Kabana, and C. D. Berry. 1975. Phalaris and tall fescue production as affected by nematodes in the field. Agron. J. 67:714–717.
41. Howard, F. L., J. B. Rowell, and H. L. Keil. 1951. Fungus diseases of turfgrasses. Univ. Rhode Island Agric. Exp. Stn. Bull. 308. 56 p.
42. Jensen, H. L. 1953. Experimental greenhouse host range studies of two root-lesion nematodes, *Pratylenchus vulnus* and *Pratylenchus penetrans*. Plant Dis. Rep. 37:384–387.
43. Johnson, E. M., and W. D. Valleau. 1949. Synonymy in some common species of *Cercospora*. Phytopathology 39:763–770.
44. Johnson, H. W. 1953. Leaf diseases of grasses in the South. p. 259–262. *In* Plant diseases, the Yearbook of Agriculture, 1953. USDA, Washington, D.C.
45. Keyl, A. C., J. C. Lewis. J. J. Ellis, S. G. Yates, and H. L. Tookey. 1967. Toxic fungi isolated from tall fescue. Mycopathol. Mycol. Appl. 31:327–331.
46. Kreitlow, K. W., and W. M. Myers. 1946. Reactions to crown rust in *Festuca elatior* and *F. elatior* var. *arundinacea*. Phytopathology 36:404.
47. ――――, and ――――. 1947. Resistance to crown rust in *Festuca elatior* and *F. elatior* var. *arundinacea*. Phytopathology 37:59–63.
48. ――――, H. Sherwin, and C. L. Lefebvre. 1950. Susceptibility of tall and meadow fescues to *Helminthosporium* infection. Plant Dis. Rep. 34:189–190.
49. Luttrell, E. S. 1951. Diseases of tall fescue grass in Georgia. Plant Dis. Rep. 35:83–85.
50. ――――. 1953. Spot blotch of brome grass, tall fescue grass, and barley. Plant Dis. Rep. 37:150–151.
51. Maag, D. D., and J. W. Tobiska. 1956. Fescue lameness in cattle. II. Ergot alkaloids in tall fescue grass. Am. J. Vet. Res. 17:202–204.
52. MacKenzie, D. R., C. C. Wernham, and R. E. Ford. 1966. Differences in maize dwarf mosaic virus isolates of the northeastern United States. Plant Dis. Rep. 50:814–818.
53. Martin, G. C. 1958. Root-knot nematodes (*Meloidogyne* spp.) in the Federation of Rhodesia and Nyasaland. Nematologica 3:332–349.
54. McGlohon, N. E., J. N. Sasser, and R. T. Sherwood. 1961. Investigations of plant parasitic nematodes associated with forage crops in North Carolina. North Carolina Agric. Exp. Stn. Tech. Bull. 148.
55. Minton, N. A. 1965. Reaction of white clover and five other crops to *Pratylenchus scribneri*. Plant Dis. Rep. 49:856–859.
56. Moore, L. D., and H. B. Couch. 1961. *Pythium ultimum* and *Helminthosporium vagans* as foliar pathogens of Graminae. Plant Dis. Rep. 45:616–619.
57. Neill, J. C. 1940. The endophyte of rye-grass (*Lolium perenne*). N.Z. J. Sci. Tech. 21:280A–291A.
58. ――――. 1941. The endophyte of *Lolium* and *Festuca*. N.Z. J. Sci. Tech. 23:185A–195A.
59. ――――, and E. O. C. Hyde. 1942. Blind-seed disease of ryegrass. II. N.Z. J. Sci. Technol. 24:65A–71A.
60. Nelson, R. R., and D. M. Kline. 1961. The pathogenicity of certain species of *Helminthosporium* to species of the Graminae. Plant Dis. Rep. 45:644–648.
61. Oswald, J. W., and B. R. Houston. 1953. Host range and epiphytology of the cereal yellow dwarf disease. Phytopathology 43:309–313.
62. Piper, Charles V. 1924. Forage plants and their culture. The Macmillan Company, New York.
63. Porter, J. K., C. W. Bacon, J. D. Robbins, and H. C. Higman. 1975. A field indicator in plants associated with ergot-type toxicities in cattle. J. Agric. Food Chem. 25:88–93.
64. ――――, C. W. Bacon, J. D. Robbins, D. S. Himmelsbach, and H. C. Higman. 1977. Indole alkaloids from *Balansia epichloe* (Weese). J. Agric. Food Chem. 25:88–93.
65. Radewald, J. D., L. Pyeatt, F. Shibuya, and W. Humphrey. 1970. *Meloidogyne naasi*, a parasite of turfgrass in southern California. Plant Dis. Rep. 54:940–942.
66. Riggs, R. K. 1966. The interrelationship of tall fescue and *Claviceps purpurea* upon the development of alkaloids in each. Ph.D. Diss., Univ. of Kentucky (Diss. Abstr. Int. 31:1643B. 1970–71).

67. ─────, L. Henson, and R. A. Chapman. 1965. Cross infectivity and alkaloid production of isolates of *Claviceps purpurea*. Phytopathology 55:1073.
68. ─────, ─────, and ─────. 1968. Infectivity of and alkaloid production by some isolates of *Claviceps purpurea*. Phytopathology 58:54–55.
69. Sampson, K. 1933. The systemic infection of grasses by *Epichloe typhina* (Pers.) Tul. Trans. Br. Mycol. Soc. 18:30–47.
70. ─────, and J. H. Western. 1954. Diseases of British grasses and herbage legumes. Cambridge Univ. Press, London.
71. Sarasola, J. A., and M. D. Campi. 1947. Reaccion de algunas Cebadas con respecto a *Rhynchosporium secalis* en Argentina. Rev. Invest. Agric. Buenos Aires 1:243–260.
72. Shepherd, R. J. 1965. Properties of a mosaic virus of corn and johnson grass and its relation to the sugarcane mosaic virus. Phytopathology 55:1250–1256.
73. Shurtleff, M. C. 1953. Susceptibility of lawn grasses to brown patch. Phytopathology 43:110.
74. Smith, J. D., and C. R. Elliott. 1970. Stem eyespot on introduced *Festuca* spp. in Alberta and British Columbia. Can. Plant Dis. Surv. 50:84–87.
75. Sprague, R. 1950. Diseases of cereals and grasses in North America. Ronald Press, New York.
76. Tabor, P. 1952. Tall fescue grows up. Crops Soils 4(8):9–11.
77. Townshend, J. L., J. L. Eggens, and N. K. McCollum. 1973. Turfgrass hosts of three species of nematodes associated with forage crops. Can. Plant Dis. Surv. 53:137–141.
78. Webster, J., and S. H. Iqbal. 1971. *Urocystis agropyri* (Preuss) Schroet. on *Festuca arundinacea*. Trans. Br. Mycol. Soc. 56:159.
79. Weerapat, P., D. T. Sechler, and J. M. Poehlman. 1972. Host study and symptom expression of barley yellow dwarf virus in tall fescue (*Festuca arundinacea*). Plant Dis. Rep. 56:167–168.
80. Whitehead, M. D. 1950. Cercospora leaf spot, a severe disease of tall fescue and smooth brome. Phytopathology 40:791–792.
81. ─────, and O. H. Calvert. 1959. *Helminthosporium rostratum* inciting ear rot of corn and leaf-spot of thirteen grass hosts. Phytopthology 49:817–820.
82. ─────, and E. C. Holt. 1950. *Cercospora* leaf spot of tall fescue and smooth brome. Phytopathology 40:1023–1026.
83. Williams, A. S., and C. W. Laughlin. 1968. Occurrence of *Hypsoperine graminis* in Virginia and additions to the host range. Plant Dis. Rep. 52:162–163.
84. Yates, S. G. 1971. Toxin-producing fungi from fescue pasture. p. 191–206. *In* S. Kadis, A. Ciegler, and S. J. Ajl (eds.) Microbial toxins. Academic Press, New York.

Chapter **Conservation**

16 ORUS L. BENNETT
SEA-USDA
West Virginia University
Morgantown, West Virginia

The role of grasses in conserving our soil and water resources has been recognized for many years. Grasses are among the first species to invade areas disturbed by man and serve to quickly protect the soil surface against both water and wind erosion. The more than 100 *Festuca* species have played an important role in filling the conservation needs throughout the world. Tall fescue (*Festuca arundinacea* Schreb.) is well-known as a soil conservation aid because of its deep, penetrating root system, tolerance to adverse soil conditions, and wide adaptation to various climates (Buckner and Cowan, 1973). According to the Soil Conservation Service (SCS), (USDA, Soil Conservation Service, 1969), tall fescue is used for pastures, hay, crop rotations, soil building, recreational areas, stabilization of waterways, slopes, banks, cuts, fills, and strip mine spoils.

SOIL STABILIZATION

Tall fescue has been widely accepted for soil stabilization against water and wind erosion. It will grow on poor droughty soils and has a wide tolerance to soil acidity (Fleming et al., 1974). Significantly, tall fescue grows in spite of these conditions; not because of them. A good soil fertility maintenance program will aid materially in producing satisfactory growth under conditions where soil stabilization is urgently needed. Musser (1962) indicated that tall fescue will grow well in shade and is widely used under these conditions.

Two features make tall fescue useful in conservation work. First, it produces a large amount of coarse, tough roots and, secondly, it forms a dense ground cover rapidly. Hafenrichter et al. (1968) noted that in the upper 20 cm of Palouse silt loam, a deep prairie soil, a 6-year-old stand of tall fescue produced more than 7,800 kg/ha of roots/acre. Even when clipped at 3-week intervals, it produced 5,600 kg/ha of roots. Bennett and Doss (1960) found that tall fescue produced from 6,160 kg/ha of roots under high soil moisture conditions to over 10,000 kg/ha of roots under relatively dry soil moisture conditions. Benefits to the soil from this extensive root system include improved soil structure, decreased soil density, and increased resistance to erosion. According to Hafenrichter et al. (1968),

Copyright © 1979 ASA-CSSA-SSSA, 677 South Segoe Road, Madison, WI 53711 USA. *Tall Fescue.*

Fig. 16-1—Contour strip cropping with corn in between strips of tall fescue and waterways of tall fescue sod.

tall fescue has been successfully used as a cover crop in establishing irrigated orchards where the shade is not too dense.

A. Contour Strip Cropping and Farm Waterways

If contour strip cropping is to be effective in reducing soil erosion, a good plant cover is essential on the protective strips and waterways. Tall fescue has gained wide acceptance for these purposes in the United States and to a lesser extent in other parts of the world. Tall fescue is especially adapted to these uses because it makes an excellent ground cover, even under rather adverse soil conditions. Nevertheless, it responds to good fertilizer and management practices (Reid et al., 1967). Tall fescue establishes rapidly, thus providing excellent waterway protection, and is ideal for cover in low, wet lands. In hill country, like the Appalachia region, fescues are widely recommended for strip cropping and waterways protection by the SCS and other action agencies dealing directly with the farmer. Because of its growth habits, it will withstand frequent traffic by farm machinery and clipping for hay or as a pasture species. It may be used alone or in combination with other grasses and legumes and grows in environments from Maine to northern Florida (VanArsdall and Wiggins, 1961). Throughout the United States, much land is cropped on the contour with strips of meadows and protected waterways. Many of these areas contain mixtures of grasses, such as tall fescue, in combination with legumes that provide hay and pasturage as well as control erosion (Fig. 16-1).

B. Disturbed Land Areas

Surface Mining Spoils

Conservative estimates indicated that more than a third of the total U.S. land surface has the potential to be seriously disturbed by man in the future. This estimate includes mining for sand, gravel, stone, clay, metals, minerals, and fossil fuels and disturbance by large construction projects like the interstate highway system. Presently the total U.S. land surface disturbed by mining alone currently exceeds 1,620,000 ha. These mined areas are subject to accelerated erosion which introduces sediments and acid pollutants to surface waters, thus affecting the usefulness of the surrounding areas. To avoid erosion, seriously disturbed land surfaces should be vegetated as quickly as possible with a desirable plant cover.

Recent research has identified several plant species adapted to the environmental conditions generally found on disturbed land areas. Tall fescue, either alone or in combination with other grasses and legumes, is widely used for stabilization of disturbed land areas throughout the United States and other parts of the world. In the eastern U.S. seriously disturbed land areas are usually characterized by poor physical and chemical conditions, representing a rather hostile plant growth environment. The addition of all major elements including N, P, K, Ca, and Mg is usually necessary for good growth. Strip-mined land areas in the eastern U.S. are generally characterized by low pH and toxic concentrations of Fe, Al, and Mn. Under western conditions, Na and Ca are usually excessive. Tall fescue has been grown successfully under conditions existing in both the eastern and western U.S. Bennett (1973) stated that if the fertility and management needs of tall fescue are met, it can be grown on strip-mined lands and other disturbed land areas under a wide range of environmental conditions.

Excellent results have been obtained with tall fescue using relatively small lime and fertilizer rates on strip mine spoils with a pH of 4.0 and above (Fig. 16-2). Palazzo and Duell (1974) indicated that tall fescue production increased as pH increased from 4.2 to near neutrality and then decreased at pH 7 and above. To prevent severe erosion and pollution from sedimentation, tall fescue should be seeded alone or with other grasses and legumes as quickly as possible after mine spoils or disturbed areas have been graded to final form. Jones et al. (1975a) conducted a 3-year study using tall fescue combined with various clovers on a surface mine spoil in West Virginia. They found that average first year dry-matter yields ranged from 6.4 metric tons/ha for a mixture of sweet clover (*Melilotus officinalis* L.) and 'Kentucky 31' tall fescue to a low of 1.4 metric tons/ha for a mixture of chemung crownvetch (*Coronilla varia* L.) and Kentucky 31 tall fescue. When red clover (*Trifolium pratense* L.) and white clover (*T. repens* L.) were used alone, yields averaged 3.7 metric tons/ha for the red clovers and 2.4 metric tons/ha for the white clovers. When tall fescue was added as a companion grass, yields increased 32% for the red clover mixture and 63%

Fig. 16-2—Harvesting tall fescue plots on acid strip mine spoil in West Virginia.

Table 16-1—Ground cover and forage yield from selected legumes and legume-grass mixtures on a surface mine spoil.†

Forage	Ground cover 1972 Legume fraction	Ground cover 1972 Total	Ground cover 1973 Legume fraction	Ground cover 1973 Total	Yield‡ 1971	Yield‡ 1972	Yield‡ 1973	Average
	%				metric tons/ha			
Chesapeake red clover alone	76*	86a-c	70a-c	83a-c	2.9b-d	1.6ef	2.8b-d	2.4
-with Ky 31 tall fescue	37b-e	98a	40c-f	100a	3.9a-d	4.9ab	5.3a-c	4.7
Kenland red clover alone	72a	81a-c	75a-c	86a-c	4.6a-d	2.7b-f	4.2a-d	3.8
-with Ky 31 tall fescue	26c-e	99a	31d-f	98ab	3.7a-d	4.9ab	5.6ab	4.7
Pennscott red clover alone	79a	91a	85ab	92a-c	4.2a-d	2.1d-f	3.7a-d	3.3
-with ky tall fescue	52a-d	99a	46c-e	100a	5.7ab	4.0b-e	5.8ab	5.2
Mammoth red clover alone	59a-c	71bc	45c-e	55d	3.0b-d	2.2d-f	2.5cd	2.6
-with Ky 31 tall fescue	20de	99a	39c-f	100a	6.1ab	4.5a-d	4.1a-d	4.9
Alsike clover alone	84a	87a-c	40c-f	55d	2.5cd	2.8b-f	3.5a-d	2.9
-with Ky 31 tall fescue	20de	100a	47c-e	98ab	3.2a-d	4.8ab	4.0a-d	4.0
White dutch clover alone	79a	88ab	67a-d	82a-c	2.3cd	0.8f	2.0d	1.7
-with Merion bluegrass plus creeping red fescue	36b-e	100a	52b-e	100a	3.1a-d	1.9ef	3.5a-d	2.8
Ladino clover alone	67ab	86a-c	53b-e	71cd	2.1cd	1.3f	3.1a-d	2.2
-with Merion bluegrass plus creeping red fescue	34b-e	96a	53b-e	100a	4.9a-d	0.8f	5.1a-c	3.6
Wild white clover alone	76a	86a-c	60a-d	76b-d	2.7cd	0.8f	1.8d	1.8
-with Merion bluegrass plus creeping red fescue	17e	99a	41c-f	97ab	3.6a-d	1.6ef	2.9b-d	2.7
Crownvetch alone	80a	94a	90a	96ab	1.4d	2.0ef	6.0a	3.1
-with Ky 31 tall fescue	21de	99a	65a-d	99ab	2.9cd	4.0b-e	4.7a-d	3.9
Sweet clover (yellow) alone	19e	67c	11ef	58d	3.3a-d	2.4c-f	1.8d	2.5
-with Ky 31 tall fescue	15e	97a	7f	97ab	6.4a	6.7a	3.3a-d	5.5

* Any two values in the same column having a letter in common are not significantly different.
† Jones et al., 1975.   ‡ Average of three replicates.

for the white clover mixture (Table 16-1). Although yields were highest from a mixture of sweet clover (*M. indica* L.) and tall fescue during the first and second years, yield was mainly tall fescue, since sweet clover constituted only 15% of the total ground cover by the end of the second year. In the third year the sweet clover was only a minor constituent of the mixture. In contrast, in the third year the slow starting crownvetch produced the top yield and constituted 65% of the ground cover when seeded with a Kentucky 31 tall fescue.

Bennett et al. (1972) indicated that Kentucky 31 tall fescue produced excellent growth on low pH strip mine areas both on the bench and outer slopes. They found that a combination of plant species, including tall fescue with a legume, would be desirable. Studies showed that surface mulching was necessary on mine spoils to insure germination and to protect seedling growth. Webb (1975) also indicated that revegetation and growth of tall fescue and other species were aided by waste from a paper mill, when used as a surface mulch.

To establish tall fescue on strip-mined areas, especially steep outer slopes and eroded areas, Mezapowskyj and Brider (1970) suggested hydroseeding as an easy and convenient way of seeding and obtaining successful stands. Fertilizer and lime, if necessary, could be added to the mixture at seeding.

Bennett et al. (1972) found that under most conditions it is extremely important to obtain a quick cover of grasses and legumes to protect disturbed land areas from erosion. If tree species are desired, these can be interseeded or planted later (Thorn, 1973). Plaas (1968) conducted a study using Kentucky 31 tall fescue for interplanting of four tree species and found that a fescue ground cover did not affect tree survival, since survival for all species was not appreciably better on plots where the ground cover had been removed.

Tall fescue is an acceptable plant species for revegetation of disturbed land areas throughout the United States, since it competes aggressively with other species. However, N fertilization is essential if continued survival is expected. According to most reclamation handbooks (Reclamation Handbook. Dep. of Natural Resources, Div. of Reclamation, Charleston, W. Va. 1971) tall fescue is now recommended for seeding on most strip mine areas throughout Appalachia and other regions (Cook et al., 1974; Davis et al., 1965). In Ohio, Struthers (1960) found that tall fescue rated high on both stand and vigor in spring seedlings along with orchardgrass (*Dactylis glomerata*), bromegrass (*Bromus inermis*), and redtop (*Agrostis tenus*). In other studies in Ohio, Sutton (1970) used tall fescue for establishing vegetative cover on barren areas after toxic spoil had been covered with soil materials to depths of 5, 10, 15, 20, and 25 cm. After 1 year the areas were seeded to sweet clover, red clover, lespedeza (*Lespedeza cuneatea*), orchardgrass, tall fescue, and birdsfoot trefoil (*Lotus corniculatus* L.). No plant growth was obtained on the 5-cm soil depth; however, lespedeza and tall fescue were the predominant plant species that remained on all other treatments. Few plant roots of any species were found growing into the acid spoil below the top soil.

Sutton (1973) found that tall fescue could be established on extremely acid spoil material using 46 cm of top soil cover; however, sweet clover and lespedeza made more growth than tall fescue under these conditions. In West Virginia, Bennett (1971) listed tall fescue as one of the grasses that showed promise on low pH strip mine areas. He found that low pH is detrimental to plant growth, but low pH per se may not be the main problem associated with revegetation. Elements, such as Fe, Al, and Mn, under very acid conditions, were toxic to plants at low pH levels. Fleming et al. (1974) and Foy et al. (1967) also pointed out the possibility of reduced growth on low pH mine spoils where Al and Mn were problems.

Deep-Mine Tailings

The refuse pile—the mountains of slate, shale, and sandstone separated from the deep mine coal—has been an unsightly trademark of the coal industry for years. The waste material has usually been transported to nearby dumping sites, and over a period of years the refuse pile became larger and larger until a high mountain of material developed. These refuse piles have been problem areas for many years. Subject to severe water and wind erosion, they cause untold pollution to surrounding streams. Spontaneous combustion often sets the refuse on fire deep inside the pile, sending foul smelling smoke across the landscape. During recent years added emphasis has been given to ways of leveling and revegetating these areas, but only a few reported attempts of these efforts have been made.

Smith (1972) reported on work by the Idamay Coal Co. No. 44 in Marion Co., West Virginia, in which they have leveled and revegetated 24 ha of land formed by a coal refuse pile. The refuse was dumped and spread in 4.6-m layers and then covered with 46 cm of soil. A cover of switchgrass (*Panicum virgatum* L.) and 'Alta' tall fescue provided excellent protection for the refuse pile. The sides of the pile were sloped to fit the natural contour of the surrounding landscape and covered with soil. As the pile grew upward, the slopes were protected with seedings of switchgrass, Alta fescue, perennial ryegrass (*Lolium perenne* L.), and yucca (*Yucca filamentosa* L.). These layers were later covered with trees. Revegetation work is now being done on old refuse piles by the state departments of natural resources throughout most of the states of Appalachia and other regions.

Capp and Adams (1971) used a mixture of Kentucky 31 tall fescue, redtop (*Agrostis alba* L.), orchardgrass, annual ryegrass, and birdsfoot trefoil to revegetate a 0.4-ha coal mine refuse area treated with fly ash near Morgantown, W. Va. Kentucky 31 tall fescue was apparently the predominant species. The spoil pH was 4.6 before treatment and 7.5 after treatment with 124 to 1,344 metric tons/ha of fly ash. Hay yields averaged 2.44 metric tons/ha for the period studied.

Highway Roadbank Stabilization

The construction and improvement of interstate highway systems has added materially to the existing problems associated with land disturbance (Rovine, 1975). Estimates show that completion of the national system of interstate and scenic highways will add more than 400,000 ha to the existing 1 million ha of roadside areas. Between 60 and 75% of the right-of-ways consist of unpaved areas of exposed soil. The need to control erosion is immediate and should be carried out with as little delay as possible to prevent serious erosion. Today, erosion control is almost as important as design and location (Hottenstein, 1970). Adapted grasses and legumes play an extremely important role in achieving an effective and harmonious transition between the highway right-of-way and the adjoining countryside. Tall fescue has been used extensively throughout the United States for maintaining highway rights-of-way and to enhance overall appearance (Shears, 1971) (Fig. 16-3). Blazer et al. (1962) found that tall fescue is used extensively for both berm and the roadside bank areas. It is adapted to both northern and

Fig. 16-3—Roadbank stabilization with tall fescue.

southern exposures. However, growth is usually better during the summer months on northern exposures in southern latitudes. On berm areas, tall fescue provides a tough, skid-resistant sod, which can be economically maintained, and its appearance is pleasing to both highway users and to adjacent residents (Hottenstein, 1970; Johnson, 1957; Owens, 1960; and Johnson and Gurley, 1961). The modern high speed transportation arteries associated with the interstate highway system must have clear unobstructed recover areas adjacent to the travel lanes to provide safe zones for drivers who swerve or are forced from the pavement to roadside areas. Tall fescue provides a highly desirable turf for this use.

Construction Sites

Land disturbance, made necessary by various construction projects, is a continuous process and has become a part of our everyday, complex, way of living. Protection of these sites presents one of the greatest challenges of our time. If left unprotected, sediment produced by erosion from these sites represents a serious source of pollution for our surface waters. A large and increasing source of sediment are the non-farm lands that are disturbed by man. Each year thousands of hectares of land in transition from agricultural and forestry use to urban, transportation, industrial, and related developmental uses are subject to erosion. Grasses must be used to control this erosion (Partain, 1974). Tall fescue is used extensively for such purposes throughout the United States and other parts of the world where this species is adapted. The local unit of the SCS is one of the best sources of information on the establishment of grass cover for disturbed areas for construction sites. Many state offices are recommending tall fescue as one of the more desirable grass species for use on construction sites to prevent serious water and wind erosion. Heavier seeding rates than those for agricultural lands are required and recommended to quickly and effectively establish the dense sod or turf needed to resist wear and tolerate the pressures on these sites. Generally, continued success with tall fescue on construction sites will depend on a heavy fertilization program, good maintenance, and sometimes, supplemental irrigation. Musser (1962) reported that tall fescue is now being widely used under conditions where heavy duty turf is needed. According to the *West Virginia Erosion and Sediment Control Handbook for Urban Areas,* one important point is that seeding or planting of an area should be started as soon as possible after the area has been disturbed. A time limit in construction contracts for the length of time a site can remain unprotected would be highly desirable. Many SCS guides recommend seeding perennial species, like redtop, weeping lovegrass (*Eragrostis curvula* (Schrad.) Nees), tall fescue, or perennial ryegrass for temporary cover at high seeding rates and good fertilization for quick stabilization and cover of the areas. Tall fescue is usually seeded at the rate of 45 to 60 kg/ha. On steep slopes it is recommended that tall fescue be seeded with a mulch using a hydroseeder or other techniques for establishing a desirable mulch.

Park and Recreational Areas

Juska and Kreitlow (1972) recognized that Kentucky 31 tall fescue formed a tough, durable turf throughout much of the transitional zone where neither cool- nor warm-season grasses are especially well adapted. They also mentioned that tall fescue is seldom seriously injured by insects or diseases, and that tall fescue is widely used as a turf grass on home lawns, play areas, athletic fields, airfields, parks, roadsides, and other areas where turf is desired. It is considered wear-resistant and compares favorably in shade tolerance with fine leaf fescues. Juska (1969) found that tall fescue makes an acceptable lawn where Kentucky bluegrass (*Poa pratensis* L.) and other fine-leafed lawn grasses do poorly because of its greater tolerance to adverse soil reactions, drainage, sun and shade, disease, insects, and droughts. In parks and recreation areas where shade is a problem, chewing fescue (*Festuca rubra* var. *commutata*) and red fescue (*F. rubra* L.) may be more desirable (Buckner and Cowan, 1974).

Establishing and maintaining adequate vegetative cover to provide site protection and improve site aesthetics is a basic management problem on developed forest recreation areas. Concentrated recreation use, coupled with limited operating and maintenance budgets, often results in the death of understory vegetation through trampling, soil compaction, and erosion. The end result is a recreation site that is dusty in dry weather and muddy and eroding in wet weather. Cordell et al. (1974) studied the survival of sown and volunteer grasses under various stand conditions on the Indian Boundary Campground, Cherokee National Forest in southeastern Tennessee. Three species of grass (creeping red fescue, tall fescue, and Kentucky bluegrass) were studied along with staggered campground opening days (1, 2, and 4 years after sowing), at three levels (10, 40, and 70%) of overstory canopy reduction. Studies indicated that heavy overstory shade severely limited the establishment and survival of grass and that volunteer grasses made a better showing than sown species without annual fertilizer applications. Kentucky bluegrass and tall fescue survived better after 5 years of recreational use than did shade fescue. The initial survival of creeping red fescue was much higher than either tall fescue or Kentucky bluegrass, but survival after 5 years of recreational use was very poor. Creeping red fescue also had an adverse effect on the development of volunteer grasses.

SOIL IMPROVEMENT

A. Rotational Systems and Green Manure Crops

Runoff from land in row crops means a loss of water usually needed for plant growth. Runoff also causes loss of soil and plant nutrients. Pasture and meadow mixtures of grasses and legumes protect the soils the entire year, whereas row crops offer little protection during the critical months. Where grasses and legumes are grown, the soil has a more porous structure, soil organic matter is increased, and water can be absorbed more rapidly by the soil. Furthermore, a greater percentage of the soil surface is covered by leaf and stems and the splashing action of raindrops is minimized. Adams and Barnett (1965) showed that both runoff and soil losses are negligible from tall fescue sod on moderate slopes. They pointed out that grasses, like tall fescue, play an important role in conservation rotation systems on red and yellow podzolic soils characteristic of the southern Piedmont. According to Wolfolk et al. (1948) tall fescue has become a common grass species for mulching, cover crops, pasture improvement, or as a green manure crop.

In Indiana, Kohnke (1950) recommended a mixture of grasses and legumes which could be used for both erosion control and for future pastures on mine spoils. Tall fescue was recommended in the seeding mixtures as one of the desirable grass species. Kohnke (1950) also suggested that a thick growth of grasses and legumes for 5 to 10 years is a good way to create conditions for crop rotation involving orchards and at the same time create good soil tilth. According to Stevens (1964), North Carolina tobacco (*Nicotiana tabacum* L.) farmers like tall fescue in their rotation systems. Tall fescue roots aid in maintaining soil porosity and the surface mulch of leaves protects the surface soil from the dispersing action of raindrops and plugging the soil pores.

B. Changes in Chemical and Physical Properties of Soil

The protective action of grasses for reducing runoff and erosion and maintaining good infiltration is well known throughout the world. Equally as important is the effect of grasses and grasslands on soil properties. The effects of grasses and grasslands on the chemical properties of the soil was done at Rothamsted, England (Costin, 1964). In these studies, soil N increased under permanent pasture for about 200 years before reaching equilibrium. The most rapid increase for soil N occurred the first few years and about half the equilibrium value was obtained within 25 years (Richardson, 1938). Several workers found that grasses will mobilize soil nutrients and concentrate them near the soil surface in available forms. Russell (1960a, b) in studies to determine soil fertility changes in long term experimental plots at Kybyvolite, South Australia, found beneficial changes in pH, total N, P, organic C, and bulk density. William and Lipsett (1960) found a significant build-up of available K under pastures in Australia.

From a soil conservation viewpoint, some of the physical changes associated with effects of grass on the ability of the soils to absorb water and resist erosion are more important than their chemical effects. Soil structure is extremely important in this regard. It is well known that grasses produce a stable aggregate-structured soil mainly because of their dense fibrous root systems. In the long-term grass plots of Rothamsted Park, 76 to 83% of the soils have aggregates greater than 1 mm diameter as compared with the Broad bulk wheat (*Triticum aestivum* L.) plots with only 6%. Greacen (1958) compared soil aggregation under conditions of continuous cultivation with improved pastures of various ages and virgin pasture in Australia. Continuously cultivated areas had very low stable aggregate values as compared with virgin pastures which had very high water stable aggregates at all depths. He found that under Australian conditions, up to 10 years under pasture may be needed for effective structural improvement of the top soil, however, improvement was considerable in the surface 15 cm within 3 years. The reverse change, i.e., the deterioration of structure with cultivation of grasslands, can be rapid. This decline in structure is usually associated with a sharp decrease in organic matter and N content and an increase in the bulk density.

The deep, penetrating fibrous root system of tall fescue makes it one of the better grasses for aid in soil building. Tall fescue can and does play an important role in changing the chemical and physical properties of soils. Jones et al. (1975a, b) conducted studies on newly graded mine spoil to determine effect of various forage grasses on the transition from spoil to a soil. The study was a regraded spoil bank of the surface-mined Pocahontas No. 3 coal seam near Beckley, W. Va. The initial pH of this spoil ranged

Table 16-2—Determinations of soil pH 4.5 years after application of amendments, White Oak Mountain, W. Va. Initial spoil pH averaged 3.8 to 4.0.†

Soil amendment	Sampling depth	Species				
		Brome-grass	Orchard-grass	Ky 31 Fescue	Timothy	Average
	cm	pH				
1. Rock phosphate	0–7.5	5.46	5.29	5.46	5.44	5.41
	7.5–15	5.12	4.89	5.14	5.04	5.04
	15.0–22.5	4.86	4.78	5.06	5.03	4.93
2. Rock phosphate plus pot. chloride	0–7.5	5.46	5.38	5.59	5.47	5.48
	7.5–15	5.14	5.02	5.08	5.10	5.09
	15.0–22.5	5.02	4.91	4.99	4.98	4.98
3. Lime plus super-phosphate	0–7.5	6.74	6.10	6.62	6.39	6.46
	7.5–15	5.66	5.07	5.50	5.38	5.40
	15.0–22.5	5.24	4.92	5.14	5.01	5.08
4. Super-phosphate	0–7.5	5.29	5.03	5.11	5.16	5.15
	7.5–15	5.26	4.92	4.99	5.06	5.06
	15.0–22.5	5.17	4.83	4.92	5.03	4.99

† From: Jones et al., 1975.

Table 16-3—Organic matter† accumulations from forage sods of bromegrass, orchardgrass, tall fescue, and timothy during 4.5 years, White Oak Mountain, W. Va.‡

Species	Soil depth (cm)		
	0–7.5	7.5–15	15–22.5
	% organic matter		
Bromegrass	1.81§	1.40	1.20
Orchardgrass	2.01	1.44	1.39
Tall fescue	1.79	1.41	1.34
Timothy	1.81	1.55	1.41

† Initial organic matter content of the regraded spoil prior to seeding ranged from 0.2 to 0.3%.
‡ From Jones et al. (1975).
§ Each value in the table is the average of eight determinations, composited from 60 field samples.

from 3.8 to 4.0, with an organic matter content of from 0.2 to 0.3%. The available $CaCO_3$ averaged 216 kg/ha; P, 5 kg/ha; K, 61 kg/ha. Treatments included various rates of limestone and raw rock phosphate or super phosphate. They reported results from bromegrass, Kentucky 31 fescue, timothy (*Phleum pratense* L.), and orchardgrass. Measurements made 4.5 years after the test area was seeded indicated that the spoil pH had been increased from about 4.0 to between 5.1 and 6.5 in the surface 7.6 cm of spoil material (Table 16-2). This increase in pH was irrespective of lime or rock phosphate applications. It also appeared that the spoil material where Kentucky 31 tall fescue had been grown increased slightly more in pH than did the pH for bromegrass, orchardgrass, or timothy. After 4.5 years, profile samples were taken at 7.6-cm increments to a depth of 23 cm to determine organic matter which increased from an insignificant amount (0.2%) to more than 2% for orchardgrass (Table 16-3). Kentucky 31 tall fescue increased soil organic matter content up to 1.79% in the surface 7.6 cm and to 1.34% at the 15- to 23-cm depth. After 4.5 years the organic matter content in this

Table 16-4—Infiltration measurements on four forage sods—White Oak Mt., Summer 1975.†

Species	Site	Initial rate	Minimum rate
		cm/hour	
Bromegrass	A‡	0.53	0.15
	B	0.13	0.15
Orchardgrass	A	0.19	0.10
	B	0.74	0.30
Tall fescue	A	0.20	0.10
	B	0.51	0.23
Timothy	A	0.30	0.33
	B	0.25	0.08
Adjacent	A	2.34	0.56
undisturbed	B	1.88	1.12
land area	C	2.34	1.68

† The A and B represents two different locations within the same area.   ‡ From Jones et al. (1975).

Table 16-5—Bulk density determinations (coarse fragments excluded) from random sites within each grass species and estimates of available water holding capacity.†

Species	Method	Bulk density	Available soil water
		g/ml	cm/cm
Bromegrass	clod	1.55‡	0.10
	7.6 cm cores	1.17§	0.06
Orchardgrass	clod	1.63	0.08
	7.6 cm cores	1.34	0.07
Tall fescue	clod	1.65	0.08
	7.6 cm cores	1.33	0.06
Timothy	clod	1.68	0.09
	7.6 cm cores	1.36	0.09

† From Jones et al. (1975).   ‡ Average from four random sampling sites (0 to 150 cm).
§ Average from six random sampling sites (0 to 75 cm).

"young mine spoil" would compare favorably with many productive agricultural soils.

Jones et al. (1975a, b) also made infiltration measurements on the mine spoil areas where the four forage sods had been grown (Table 16-4). They found infiltration rates ranging from a high of 0.74 cm/hour to a low of 0.07 cm/hour. While these rates are considered very low, considerable improvements had been made in the overall infiltration rate during the 4-year period. Another measure of the chemical and physical properties of the spoil is indicated by the bulk density and the available soil water (Table 16-5). The initial available water holding capacity of regraded mine spoil in this area is usually about 5 to 8% immediately after mining, and bulk density can vary considerably and be extremely high in the surface 5.0 to 7.5 cm. These bulk density measurements after 4 years of sod are only slightly higher than many agricultural soils of the area. The water holding capacity of these mine spoils was obviously increased significantly from the increased organic matter and soil particle weathering (Table 16-6) with average available soil moisture in the 11 to 18% range. Jones et al. (1975a, b) concluded that acid mine spoils can be beneficially altered in a relatively short time by using appropriate forage grasses, like Kentucky 31 tall fescue and orchardgrass.

Table 16-6—Soil water retention by species and depths—White Oak Mt., W. Va., 1975.[†]

Species	Sampling depth	Atm tension					Available moisture
		1/3	1	3	5	15	
	cm	mm of Hg					%
Bromegrass	0– 7.5	21.92	17.98	13.45	10.74	8.92	13.00
	7.5–15.0	26.94	17.33	12.64	10.11	8.24	18.70
	15.0–22.5	22.74	17.08	12.49	9.67	8.63	14.11
Orchardgrass	0– 7.5	21.69	18.26	13.43	11.04	9.08	12.61
	7.5–15.0	20.07	17.67	12.77	10.45	8.83	11.24
	15.0–22.5	19.86	17.59	12.78	10.79	8.12	11.74
Tall fescue	0– 7.5	21.31	18.28	13.83	11.33	9.41	11.90
	7.5–15.0	20.63	17.87	13.57	11.24	9.19	11.44
	15.0–22.5	20.50	17.86	13.49	11.30	9.26	11.24
Timothy	0– 7.5	23.94	17.82	14.06	11.36	9.23	14.71
	7.5–15.0	22.26	17.74	14.06	11.17	10.30	11.96
	15.0–22.5	23.01	17.91	13.77	11.09	10.07	12.94

[†] From Jones et al. (1975).

SPECIAL CONTRIBUTIONS

A. Use on Marginal Land Areas for Pasture and Forage Production

Tall fescue is used extensively for pasture improvement of marginal land areas, since it can be successfully established under rather adverse conditions. Marginal lands are usually shallow, low in fertility, and may be too steep or rough for production of row crops or for intensive production. Bailey and Scott (1949) reported that the extensive deep root systems of fescues were ideal for holding steep Class VI land and for pastures on Class IV land in the southeastern states. Rampton (1950) found that the combination of tall fescue and subterranean clover (*Trifolium subterraneum* L.) has more than doubled the carrying capacity of pastures on western Oregon hill lands. Brooks (1950) reported that ladino-tall fescue pastures gave excellent results in mountainous areas of northern Georgia when fertilized with P, K, and lime. With the release of Kentucky 31 tall fescue, Cowan (1956) indicated that much of the hill country of Kentucky, Tennessee, Virginia, and West Virginia was brought into productive use.

Tall fescue can be established on marginal land areas covered with considerable amounts of logs, stumps, trash, brush, or weeds by burning and then seeding in the ashes. According to Rampton (1945) this method of seeding is usually more successful when done in the fall. On low fertility soils and droughty conditions, Tabor (1952) obtained good results by seeding tall fescue in rows 76 to 91 cm apart and interseeding between rows with white clover. Success has been obtained on marginal land areas by broadcasting the seed evenly and following with a light harrowing. Interseeding with the zip-seeder into existing grass stands on marginal land areas shows considerable promise. The no-till method can be used successfully either with or without the aid of herbicides depending on kind of vegetation present. Several no-till seeders are now on the market for seeding improved grasses and legume pasture mixture on rough marginal land areas.

According to Burton and DeVane (1953) tall fescue heritability studies have shown wide variability existing for certain characteristics. The opportunity for selecting superior genotypes with greater forage production for marginal land areas than those now available seems to be very good. Tall fescue is gaining wide usage and acceptance in many U.S. areas (Bryan et al., 1970; Buckner and Cowan, 1973; Smiley, 1972; Wellhauser, 1962; Wheaton, 1967). This is especially true for extending the grazing season in the hill country of Appalachia (USDA-SCS, 1969). According to Rommann et al. (1974) and Crawford et al. (1967) a pasture establishment system, developed in the Missouri Ozarks by Crawford and Bjugstad (1967), involved aerial herbicide application in late spring for brush control, followed by burning in September, and then aerial seeding with tall fescue. This same system is presently used in eastern Oklahoma (Rommann et al., 1974).

The practice of frost seeding of tall fescue on rough hill pasture areas in northern Appalachia has been practiced with considerable success. Seed and concentrated fertilizers can be applied on steep mountain pasture with either fixed wing aircraft or by helicopter. Seeding usually follows a prior herbicide application during summer to control brush and weeds.

B. No-Till Crop Production

During recent years, several research workers (Adams et al., 1970; Bennett et al., 1976; Carreker et al., 1972; Harrold et al., 1970; Phillips, 1971; Schwab and Fouss, 1967; Jones et al., 1968) investigated the possibility of producing corn (*Zea mays* L.) and other row crops by seeding directly into a previously killed sod or in corn stubble. This system for corn production significantly decreases the amount of runoff and erosion until the mulch produced by the grass or sod species is lost (Onstad, 1972). The system offers the advantage of decreasing runoff and erosion (Harrold et al., 1970; Harrold and Edwards, 1972), increasing the amount of available soil moisture for plant growth (Bennett et al., 1973; Blevins et al., 1971), and enabling row crop production on hilly or mountainous terrain previously thought to be useable for row crop production (Fig. 16-4). Two approaches have been utilized for no-till corn production. In the first system the sod was killed to eliminate any competition from regrowth of the sod species. The second approach was to place the sod in a state of dormancy by carefully controlled applications of herbicides (Bennett, 1970). In the second system the choice of sod species is a predominant factor because many grass species are killed with low rates of herbicides, while others are able to tolerate considerably higher rates. According to Bennett et al. (1976) a desirable species would have a wide herbicidal tolerance range, remain in a state of semidormancy for extended periods, and then rapidly recover. Tall fescue has been successfully used under both systems. Studies were conducted on a Cecil fine sandy loam soil in Georgia by Carreker et al. (1972) in which corn was no-till planted with and without irrigation in live, stunted, and killed fescue. They utilized N rates to 140 kg/ha in 1969, and 112, 224,

Fig. 16-4—Corn planted in tall fescue sod using the no-till technique in West Virginia.

and 448 kg/ha in 1970. They obtained no corn yields without irrigation when planted in a live fescue sod and about 1,000 kg/ha on killed sod. There was no response to N levels of 112, 224, and 448 kg/ha. However, with irrigation yields were directly proportional to N levels and to the amount of live sod. The excellent yields from corn in tall fescue grass sods showed that corn can be grown on the sloping fields of the Piedmont and related areas without causing serious soil erosion. In other studies, Carreker et al. (1973) indicated that no-till corn production in tall fescue sod seems to offer an excellent system for increasing production of grain and forage while providing for erosion control and a safe place for utilization of poultry litter. They conducted studies where corn was no-till planted in both live fescue sod and sod that was killed with 2.2 and 0.28 kg/ha of atrazine, [2-chloro-4-(ethyl-amino)-6-(isopropyl-amino)-*s*-triazine] and paraquat (1,1′-dimethyl), respectively. Subplot treatments received four different rates of poultry litter along with 335 kg/ha of N. Over 2 years they found that the corn population was higher and yields were better in the killed sod than in live sod. After the corn was harvested, live fescue sod regrowth was estimated at between 2,000 kg/ha in 1970 and greater than 6,000 kg/ha in 1971. They concluded that no-till corn in tall fescue with irrigation and poultry litter will produce higher yields of corn, conserve soil and water, and turn a waste product into a useable resource. Adams and Barnett (1965), using perennial grasses and legumes to control runoff and erosion in corn production, found that fescue decreased annual runoff and erosion to less than 5 cm and 0.34 metric tons/ha, respectively. Adequate environmental protection was not obtained with clean, tilled corn.

Table 16-7—Hay yields of five grass species before and after corn was planted and atrazine was applied at Point Pleasant, W. Va.†

Species	1966†	1967 Atrazine-kg/ha		1968 Atrazine-kg/ha	
		1.7	3.4	1.7	3.4
		kg/ha			
Bromegrass	3,964 a*	4,051 a	3,170 b	4,910 a	2,426 b
Timothy	3,928 a	0	0	1,778 bc	0
Ky. Bluegrass	2,412 b	0	0	131 e	0
Orchardgrass	3,561 a	2,474 bc	669 d	4,439 a	768 d
Ky 31 Tall fescue	3,398 ab	3,075 b	2,190 c	4,378 a	1,261 cd

* Values followed by same letter for treatments within same year are not significantly different according to Duncan's multiple range test.　† Yields of grasses at beginning of study before herbicides were applied and before corn was planted.　‡ From Bennett et al. (1976).

Spain et al. (1965) reported a system of planting corn for silage and grain in dormant tall fescue sod. Yield results were erratic and mainly depended upon degree of sod dormancy. Studies by Bennett et al. (1976) indicated that tall fescue sod could be successfully used for no-till corn production, either with or without killing the sod. In a study comparing no-till corn and hay production from five sod species, they found that higher herbicide rates were needed to control growth of a vigorous fescue sod than that for bluegrass or timothy. Stands of tall fescue and hay yields were maintained over 3 years in a double-cropping system with interseeded no-till corn (Table 16-7). In these studies, sod-planted corn had significantly more yield (Table 16-8) than conventionally planted corn with bromegrass, orchardgrass, and tall fescue producing excellent regrowth and hay yields when fields were treated with 1.7 kg/ha of atrazine after a corn crop.

According to Britt (1974) drought decreases grain yields because rainfall is not evenly distributed in most parts of the country. A cropping system is needed for increasing the production of grain and forage, for controlling erosion, and providing better utilization of soil moisture and plant nutrients (Barnes, 1969; Moschler et al., 1975). Early work in minimum tillage by Jones et al. (1968) and in corn sod systems by Adams et al. (1970) indicated that corn could be grown in tall fescue sod while conserving the soil and water resources. Tall fescue is easily established, maintained, and managed. According to Carreker et al. (1972), when tall fescue is killed, fertilized, and irrigated, the sod is an excellent place for corn production. Carreker indicated that an intensive conservation program in the rolling Piedmont area of the United States has resulted in the seeding of much of this area into forage species, like fescue, or to trees. However, presently more grasslands are being viewed as a potential and highly desirable avenue for expansion of corn grain production using the sod-planting technique.

Presently, grass sods like tall fescue are being utilized more and more for interseeding of annual row crops for hay and silage production. With the production of new seeding equipment, like the zip-seeder and sod-seeder, annual crops like sorghum (*Sorghum bicolor (L.) Moench) hybrids* (Hart et al., 1971), millets (*Panicum milliaceum* L.), sweet sorghums, grain

Table 16-8—Yields of silage from sod-planted corn as affected by grass species and rates of atrazine and N at Point Pleasant, W. Va.‡

Sod species	1966				1967			
	1.7 kg/ha atrazine		3.4 kg/ha atrazine		1.7 kg/ha atrazine		3.4 kg/ha atrazine	
	135 N	225 N	135 N	225 N	135 N	225 N	135 N	225 N
	metric tons/ha							
Bromegrass	19.5*	33.2	30.0	35.2	52.7	65.6	62.6	72.7
Timothy	34.1	41.7	38.1	39.9	58.1	59.2	58.1	63.2
Ky. bluegrass	30.5	31.8	38.6	40.4	59.6	54.7	51.6	46.2
Orchardgrass	29.6	32.3	36.3	38.8	0	0	32.5	46.2
Ky 31 tall fescue	43.1	42.6	44.4	46.4	26.0	18.4	43.3	66.4
Plowed	28.4	28.2	26.3	27.1	47.75	46.0	34.5	30.7
Average for N × atrazine	30.8	34.9	35.6	37.9	40.7	40.6	47.1	54.2
Average for rates of atrazine	32.9		36.7		40.5		50.6	
Average for rate of N × year	33.2	36.4			43.9	47.4		

* Probability levels for 1966 and 1967 yields, respectively, were as follows: between species, P >0.001 and 0.0001; rates of atrazine, P >0.0028 and 0.0021; species × atrazine, P >0.23 and 0.001; rates of N, P >0.0002 and 0.04; species × N, P = 0.001 and 0.25.   † L.S.D. 0.05, 4.95 and 7.11 metric tons/ha for 1966 and 1967, respectively.   ‡ From Bennett et al. (1976).

sorghums, and other grasses and legumes are being interseeded directly into existing sods. This interseeding can often be done without destroying the existing sod which may produce significant regrowth after the annual crop is removed. According to Blakely (1974), corn is planted in the southern states in stands of grass like tall fescue or ryegrass that have been suppressed with herbicides. The corn is harvested for silage and the grass makes a regrowth for soil protection and pasture. Under such a production management system, year-round soil protection can be maintained.

C. Aesthetic Values

The aesthetic value of grasses has been recognized by man since the beginning of time. Partain (1974) says that of all plants, grass is the most important to man, and that its importance extends beyond mere economics or as a direct or indirect source of food. Grasses have been recognized for their soil protection, beauty, inspiration, leisure, and as a symbol of well being for man. Grasses are the protector of nature, since it is the first species to invade areas disturbed and left by man and prevent untold loss of soil and natural resources by water and wind erosion. Tall fescue has certainly taken its place among the more important grass species of the world because of its versatility and usefulness. It is used on land areas where beauty and a pleasing landscape is required. It is used extensively for lawns (Scherry, 1975), play areas, golf courses, along public highways, and in forested areas to give a break in the landscape. Highway planners and construction engineers give a high priority in using grasses such as tall fescue in locations where the aesthetic value is needed. For example, highway designers recognize that a highway is more than just an artery for transporting people. It must be pleasing to the eye for the motorist, as well as protective of the entire right-of-way. The right-of-way must blend in with the surrounding countryside to give the motorist a feeling of safety and contentment. Partain (1974) indicated that there is increasing evidence that monotony of the landscape along highways, especially modernized speed expressways, tends to increase drowsiness for the driver, thus creating a hazard not only to himself but to others. He suggested that combinations of grass, trees, and shrubs have a real value in combating this driver monotony. Grasses like tall fescue form the basis for interesting, yet safe and attractive views for the motorist. Today, the soil surface protection, safety, and beauty are all a part of the well-designed and executed travelways.

When using tall fescue to increase aesthetic value of property, we generally think of the home lawns. However, its use in contour farming, natural waterways, and for open areas in surrounding farm lands is a creation of untold beauty. One of rural America's most beautiful scenes can be the curvaceous and patchwork patterns of farms in the gently rolling countryside. Grass species usually form an important part of these contours. The aesthetic value of tall fescue can be seen and appreciated at home, work, or play. Tall fescue is now being used extensively for certain areas on the golf course, like fairways and rough areas, especially where shade is prevalent. Musser (1962) indicated that tall fescue is also suitable under fairway conditions where watering is not practical.

It is certainly anticipated that tall fescue will continue to be one of the most versatile grass species used for soil stabilization, reclamation, pastures, no-till crop production, recreation, and aesthetic values in areas where it is adapted.

LITERATURE CITED

1. Adams, W. E., and A. P. Barnett. 1965. The role of grass in conservation rotations on red and yellow podzolic soils. p. 560-566. *In* Int. Grassl. Congr. Proc. 9th, Sao Paulo, Brazil.
2. ———, James E. Pallas, Jr., and R. N. Dawson. 1970. Tillage methods for corn sod systems in the southern Piedmont. Agron. J. 62:646-649.
3. Bailey, R. Y., and L. B. Scott. 1949. Using tall fescue in soil conservation. USDA Leaflet No. 254.
4. Barnes, P. L. 1969. Increase corn yields in drought areas. Hoards Dairyman 114:359+.
5. Bennett, O. L. 1970. 1 field + 1 year = 2 crops. Crops Soils 23(1):15-17.
6. ———. 1971. Grasses and legumes for revegetation of strip mined areas. p. 23-26. *In* D. M. Bondurant (ed.) Proc. Revegetation and economic use of surface mined land and mine refuse. Pipestem, W. Va. 2-4 Dec. 1970. West Virginia Univ., Morgantown.
7. ———. 1973. Vegetation to heal scars. p. 249-253. *In* Soil Cons. Soc. Am. Proc. 27th Annual Meeting, Hot Springs, Ark. 30 Sept.-3 Oct. 1973. Soil Cons. Soc. Am., Ankeney, Iowa.
8. ———, and Basil Doss. 1960. Effect of soil moisture level on root distribution of cool-season forage species. Agron. J. 52:204-207.
9. ———, J. N. Jones, Jr., W. H. Armiger, and P. E. Lundberg. 1972. New techniques for revegetation of strip mine areas. p. 50-55. *In* Soil Cons. Soc. Am. Proc. 27th Annual Meeting. Hot Springs, Ark. 30 Sept.-3 Oct. 1973.

10. ―――, E. L. Mathias, and P. E. Lundberg. 1973. Crop responses to no-till management practices on hilly terrain. Agron. J. 65:488-491.
11. ―――, ―――, and C. B. Sperow. 1976. Effect of rates of atrazine, nitrogen, and sod species on double cropping for hay and no-tillage corn production. Agron. J.
12. Blakely, B. D. 1974. Pollution control. Contribution of grasslands. p. 196-203. *In* H. B. Sprague (ed.) Grasslands of the United States. Iowa State Univ. Press.
13. Blazer, R. E., G. W. Thomas, C. R. Brooks, G. J. Shoop, and J. B. Martin, Jr. 1962. Turf establishment and maintenance along highway cuts. School of Engineering and Applied Sciences, Univ. of Virginia. Reprint No. 6.
14. Blevins, R. L., Doyle Cook, S. H. Phillips, and R. E. Phillips. 1971. Influence of no-tillage on soil moisture. Agron. J. 63:593-596.
15. Britt, P. R. 1974. No-till stands tall. Soil Conserv. 39:21.
16. Brooks, O. L. 1950. Ladino-tall fescue pastures for North Georgia Proc. Assoc. South. Agric. Workers 47:184.
17. Bryan, W. B., W. F. Wedin, and R. L. Vetten. 1970. Evaluation of reed canarygrass and tall fescue as spring, summer, and fall saved pasture. Agron. J. 62:75-80.
18. Buckner, Robert C., and J. Ritchie Cowan. 1973. The fescues. *In* M. Heath, P. S. Metcalf, and R. L. Barnes (ed.) Forages. Iowa Univ. Press, Ames, Iowa.
19. Burton, G. W., and E. H. Devane. 1953. Estimating heritability in tall fescue (*Festuca arundinacea*) from replicated clonal material. Agron. J. 45:478-481.
20. Capp, John P., and L. M. Adams. 1971. Reclamation of coal mine wastes and strip spoil with fly ash. p. 48-53. *In* D. M. Bondurant (ed.) Proc. Revegetation and Economic Use of Surface Mined Land and Mine Refuse Symp. Pipestem, W. Va. 2-4 Dec. 1970. West Virginia Univ., Morgantown.
21. Carreker, J. R., J. E. Box, R. N. Dawson, E. R. Beaty, and H. D. Morris. 1972. No-till corn in fescue grass. Agron. J. 64:500-503.
22. ―――, S. R. Wilkinson, J. E. Box, Jr., R. N. Dawson, E. R. Beaty, H. D. Morris, and J. B. Jones, Jr. 1973. Using poultry litter, irrigation, and tall fescue for no-till corn production. J. Environ. Qual. 2:497-500.
23. Cook, C. W., R. M. Hyde, and P. L. Sims. 1974. Revegetation guides for surface mined areas. Colorado State Univ. Range Sci. Dep. Science Series No. 16.
24. Cordell, H. K., G. A. James, and G. L. Tyre. 1974. Grass (*Festuca rubra* var. Heterophyllea, *Festuca arundinacea, Poa pratense*) established on developed recreation sites. J. Soil Water Conserv. 29(6):268-271.
25. Costin, A. B. 1964. Grasses and grasslands in relation to soil conservation. p. 236-258. *In* C. Barnard (ed.) Grasses and grasslands. MacMillan & Co. Ltd., London.
26. Cowan, J. Ritchie. 1956. Tall fescue. Adv. Agron. 8:283-320.
27. Crawford, H. S., and A. J. Bjugstad. 1967. Establishing grass range in the southwest Missouri Ozarks. U.S. North Central For. Exp. Stn. U.S. For. Serv. Res. Note N.C. 22.
28. Davis, G., R. W. Ruble, J. F. Ibberson, W. H. Wheeler, E. G. Musser, R. D. Shipman, and W. G. Jones. 1965. A guide for revegetating bituminous strip mine spoils in Pennsylvania. Res. Comm. on Coal Mine Spoil Revegetation in Pennsylvania.
29. Fleming, A. L., J. W. Schwartz, and C. D. Foy. 1974. Chemical factors controlling the adaptation of weeping lovegrass and tall fescue to acid mine spoils. Agron. J. 66:715-719.
30. Foy, C. D., A. L. Fleming, G. R. Burns, and W. H. Armiger. 1967. Characterization of differential aluminum tolerance among varieties of wheat and barley. Soil Sci. Soc. Am. Proc. 31:513-521.
31. Greacen, E. L. 1958. The soil structure profile under pastures. Aust. Agric. Res. 9:129.
32. Hafenrichter, A. L., J. N. Schwendiman, H. L. Harris, R. S. MacLauchlan, and H. W. Miller. 1968. Grasses and legumes for soil conservation in the Pacific Northwest and Great Basin States. Agric. Handb. No. 339, SCS, USDA.
33. Harrold, L. L., G. B. Triplett, and W. M. Edwards. 1970. No-tillage corn, characteristics of the system. Agric. Eng. 51:128-131.
34. ―――, and W. M. Edwards. 1972. Severe rainstorm test of no-till corn. J. Soil Water Conserv. 27:36.
35. Hart, R. H., Henry J. Retzer, Richard F. Dudley, and G. E. Carlson. 1971. Seeding sorghum × sudangrass hybrids into tall fescue sod. Agron. J. 63:378-380.
36. Hottenstein, W. L. 1970. Erosion control, safety, and esthetics on the roadsides—summary of current practices. Public Roads 36(2).
37. Johnson, A. W. 1957. Erosion control along American highways. Am. Soc. Agric. Eng.

38. Johnson, J. R., and G. W. Gurley. 1961. Tall fescue. Georgia Agric. Coll. Ext. Circ. 478:1-4.
39. Jones, J. N., Jr., J. E. Moody, G. M. Shear, W. W. Moschler, and J. H. Lillard. 1968. The no-tillage system for corn (*Zea mays* L.). Agron. J. 60:17-20.
40. ―――, W. H. Armiger, and O. L. Bennett. 1975a. A two-step system for revegetation of surface mine spoils. J. Environ. Qual. 4:232-236.
41. ―――, ―――, and ―――. 1975b. Forage grasses aid the transition from spoil to soil. p. 185-194. *In* Third Symp. on Surface Mining and Reclamation, Vol. II, NCA/BCR Coal Conf. and Expo II, Louisville, Ky. 21-23 Oct. 1975.
42. Juska, F. V. 1969. Evaluation of tall fescue (*Festuca arundinacea* Schreb.) for turf in the transition zone of the U.S. Agron. J. 61:625-628.
43. ―――, and K. W. Kreitlow. 1972. Selecting lawn grasses from bahia to zoyoia. Yearb. Agric., USDA. p. 111-118.
44. Kohnke, Helmut. 1950. The reclamation of coal mine spoils. Adv. Agron. 2:318-349.
45. Mezapowskyj, M., and Ross Brider. 1970. Hydroseeding on anthracite coal-mine spoils. USDA For. Serv. Res. Note NE-124.
46. Moschler, W. W., D. C. Martens, and G. M. Shear. 1975. Residual fertility in soil continuously field cropped to corn by conventional tillage and no-till methods. Agron. J. 67: 45-48.
47. Musser, H. Burton. 1962. Turf management. McGraw Hill Book Co., Inc., New York. p. 114-115.
48. Onstad, C. A. 1972. Soil and water losses as affected by tillage practices. Am. Soc. Agric. Eng. Trans. 15:287-289.
49. Owens, H. B. 1960. Landscape design and the modern highway. Natl. Res. Counc., Highway Res. Board, Comm. Roadside Development Rep. Bibliography 26. p. 11-17.
50. Palazzo, H. A., and R. W. Duell. 1974. Responses of grasses and legumes to soil pH. Agron. J. 66:678-682.
51. Partain, Lloyd, E. 1974. Grass for protection, safety, beauty, and recreation. p. 204-215. *In* H. B. Sprague (ed.) Grasslands. Iowa State Univ. Press, Ames, Iowa.
52. Phillips, S. H. 1971. Corn moves up Kentucky slopes. Hoards Dairyman 116:127 F10.
53. Plaas, W. T. 1968. Tree survival and growth of fescue covered spoil bank. U.S. For. Serv. Res. Note NE-90.
54. Rampton, H. H. 1945. Alta fescue production in Oregon. Oregon Agric. Exp. Stn. Bull. 427.
55. ―――. 1950. Alta fescue, for high yielding pastures. Crops Soils 2(7):18-19, 25.
56. Reid, R. L., E. K. Odhuba, and G. A. Jung. 1967. Evaluation of tall fescue pasture under different fertilization treatments. Agron. J. 59:265-271.
57. Richardson, H. L. 1938. The nitrogen cycle in grassland soils: With special reference to the Rothamsted Park grass experiment. J. Agric. Sci. 28:73.
58. Romine, Russel. 1975. The future of roadside management general views and specific moves. Rural Urban Roads 13(2):11-12.
59. Rommann, L. M. 1974. Converting brush to tall fescue. Oklahoma State Univ. Exp. Stn. Facts No. 2563.
60. Russell, J. S. 1960a. Soil fertility changes in the long term experimental plots at Kybyvolite, South Australia. I. Changes in pH, total nitrogen, organic carbon, and bulk density. Aust. J. Agric. Res. 2:902.
61. ―――. 1960b. Soil fertility changes in the long term experimental plots of Kybyvolite, South Australia. II. Changes in phosphorus. Aust. J. Agric. Res. 2:925.
62. Scherry, R. W. 1975. Fescues, the hard working grasses for home lawns. Home Garden. 58:24-25.
63. Schwab, G. O., and J. L. Fouss. 1967. Tile flow and surface runoff from drainage systems with corn and grass cover. Am. Soc. Agric. Eng. Trans. 10 No. 4:492-493.
64. Shears, George A. 1971. Experience in revegetation of highway cuts and fills. p. 33-36. *In* D. M. Bondurant (ed.) Proc. Revegetation and Economic Use of Surface Mined Land and Mine Refuse Symp. Pipestem, W. Va. 2-4 Dec. 1970. West Virginia Univ., Morgantown.
65. Smiley, C. 1971. Cattle and grass; basis for stability in a time of change. Farm Q. 26:14-17.
66. Smith, Gordon S. 1972. Checking the impact of mining. Soil Conserv. 37:230-231.
67. Spain, J. M., G. C. Klingman, and C. R. Martin. 1965. Dormant sod planting. Prog. Rep. N.C. State Univ., Raleigh, N.C.

68. Stevens, W. W. 1964. North Carolina tobacco growers like tall fescue in rotations. Soil Conserv. 29:139-140.
69. Struthers, Paul H. 1960. Forage seedings help reclaim areas on spoil banks. Ohio Farm Home Res. 45(1):12-13.
70. Sutton, Paul. 1970. Restoring productivity of coal mine spoils. Ohio Rep. No. 62.
71. ———. 1973. Reclamation of toxic strip mine spoil banks. Ohio Rep. p. 18-20.
72. Tabor, P. 1952. Fescue in rows, clover in middle. Soil Conserv. 18:39-40.
73. Thorn, James. 1973. On the strip mining front: Achievement in the trenches. Soil Conserv. 38(6):126-127.
74. USDA-Soil Conservation Service. 1969. Using fescue for year-round grazing. Inf. Sheet No. 11.
75. USDA-Soil Conservation Service. 1969. Plants for conservation in the Northeast—Tall fescue. Conservation Plant Sheet NE-3.
76. VanArsdall, H. E., and L. Wiggins. 1961. Tall fescue, a winter grass for Florida. Soil Conserv. 26:235-236.
77. Webb, D. M. 1975. Revegetation possible for abandoned strip mined lands. Maryland Conserv. p. 7-9.
78. Wellhausen, H. W. 1962. Tall fescue. Arkansas Agric. Ext. Leaflet 340:18.
79. Wheaton, H. N. 1967. Tall fescue. Crops Soils 20(3):14-15.
80. William, C. H., and J. Lipsett. 1960. The buildups of available potassium under pastures. Aust. J. Agric. Res. 2:473.
81. Wolfolk, E. J., E. F. Costello, and B. W. Allred. 1948. The major range type grasses. Yearb. Agric. USDA. p. 205-211.

Chapter 17 The Future of Tall Fescue

A. A. HANSON
SEA-USDA
Beltsville Agricultural Research Center
Beltsville, Maryland

During the past 3 decades tall fescue (*Festuca arundinacea* Schreb.) has moved from relative obscurity to a prominent position among the top 10 grasses grown in the United States for forage, pasture, general purpose turf, and soil conservation. The recent acceptance of tall fescue can be traced through a cursory examination of various Extension Service publications. The Virginia Agricultural Extension Service Bulletin 194, issued January 1952, does not mention tall fescue as a component in any recommended pasture mixture; a revised version, issued in February 1962, includes tall fescue as a component in one of the pasture mixtures recommended for each geographical region of the state. The story of its rise to prominence differs considerably from that associated with most introduced grasses. The unique aspects of the acceptance of tall fescue in the United States tell us much about probable future trends and aid in assessing the merits of identifiable research needs.

The unending search for better grasses started shortly after the settlement of the Eastern Seaboard. This search, necessitated by the low quality and unreliability of indigenous native species, might be compared with the quest for the Holy Grail; there were those, from the gentleman planter of the Colonial Period to latter-day Crusaders, who have continued to search for the "perfect grass." Many valuable grasses have been introduced from foreign countries, and almost without exception, have been promoted vigorously by those who first recognized their potential worth. The virtues of a particular grass often appeared endless in the eyes of early converts. With time, each new grass became identified with a specific zone or area of adaptation. Experience led to an understanding of the uses for which the grass was well suited.

In the northeastern and north central regions of the United States, timothy (*Phleum pratense* L.) reigned supreme from the early Colonial Period until the horse was replaced by the farm tractor. The 1930's represented a period when attention was focused on the use of a series of grasses for hay, pasture, and soil conservation. Some of these grasses were newly introduced and others could be traced to the early Colonial Period. These included orchardgrass (*Dactylis glomerata* L.) and smooth brome (*Bromus inermis* Leyss.) in many of the northeastern and north central states, crested wheatgrass (*Agropyron cristatum* L.) in the northern Great Plains and dallisgrass (*Paspalum dilatatum* Poir.) in the southeastern region. Subse-

---

Copyright © 1979 ASA-CSSA-SSSA, 677 South Segoe Road, Madison, WI 53711 USA. *Tall Fescue.*

quently, local and regional attention has been directed to improved bermudagrass (*Cynodon dactylon* (L.) Pers.) cultivars, rust-resistant annual ryegrasses (*Lolium multiflorum* Lam.), bahiagrasses (*Paspalum* spp.), Russian wildrye *(Elymus junceus* Fisch.), buffelgrass (*Cenchrus ciliaris* L.), lovegrasses (*Eragrostis* spp.), digitgrasses (*Digitaria* spp.), blue panicgrass (*Panicum antidotole* Retz.), and many others. It is questionable, however, that any of these or other grasses came into use with the same degree of acrimony that was associated with the advent of tall fescue. Tall fescue had enthusiastic supporters who saw in this grass a near-perfect answer to the need for cover and pasture on eroded cropland, for restoring the productivity of depleted grazing land, and for extending the grazing season for herds that were being maintained with inadequate supplies of forage and other conserved feeds. Conversely, others scoffed at the potential role of tall fescue in animal production systems. They claimed, with considerable justification, that tall fescue was much inferior to other cool-season grasses with respect to both palatability and quality.

The spread of tall fescue in Oregon following the first commercial seed harvest of 'Alta' tall fescue in 1936 did not attract immediate national attention. The situation changed dramatically with the rapid increase of 'Kentucky 31' tall fescue in the upper South during the mid-1940's, a growth in use that was associated with the promotional activities of seed growers, extension specialists, and soil conservationists. It was at this time that the voices of reason were often lost in a plethora of rash claims and counterclaims. Nevertheless, in a few years, tall fescue had become firmly established as an integral part of forage production systems in a large portion of the southeastern U.S. The fact that tall fescue continued to increase in importance in comparison with other perennial cool-season grasses can be attributed to a long list of desirable agronomic characteristics. These desirable attributes, described in detail in this text, include excellent seed yields, good yields of herbage, excellent persistence, adaptation to a wide range of soil conditions, compatibility with various management practices, a comparatively long grazing season, and the absence or near absence of susceptibility to serious disease and insect damage. Research workers found that forage intake and animal performance was improved by close, intensive grazing. In addition, the versatility of the grass was demonstrated in its adaptation to the practice of stockpiling herbage for late fall and winter grazing.

The increasing frequency of reports of fescue-foot in the 1950's gave rise to another round of breast-beating and diatribes over the real and imagined deficiencies of tall fescue. It was soon realized, however, that the incidence of fescue-foot was comparatively low and that the disorder did not pose a major threat to livestock production. Thus, the hectarage and use of tall fescue did not decline because of concern over the fescue-foot problem. Similarly, the suggestion that animals on tall fescue pastures might be suffering from varying degrees of fescue toxicity, which could affect animal performance and in extreme cases lead to fescue-foot, had little impact on the frequency of new seedings. In practice, farmers had learned to count on tall fescue as a reliable pasture species. Today there is every indication that hectarages within the species' range of adaptation will continue to expand as livestock producers search for least-cost approaches to improving the carrying capacity and production of depleted pastures.

Starting in the mid-1940's, the Soil Conservation Service (SCS) encouraged the use of tall fescue in a wide array of conservation plantings. The persistence of tall fescue in these and other seedings prompted research workers to evaluate tall fescue for turf purposes and for soil stabilization and cover along highway rights-of-way. On the basis of extensive tests, tall fescue was shown to be especially valuable for general purpose turf throughout the broad transition zone that marks the border between the zones of best adaptation for cool-season and warm-season grasses. In the transition zone, the use of tall fescue has increased at a phenomenal rate as a grass cover on highway rights-of-way and around various industrial sites and public installations. The major expansion into these uses started in the early 1950's and continued thereafter; today, thousands of hectares are occupied by pure stands of tall fescue planted solely as a low maintenance grass cover.

The use of tall fescue in home lawns has received relatively little attention. However, research has demonstrated the value of tall fescue for lawn use in selected areas throughout the transition zone and upper South. There is every reason to expect that the use of tall fescue as a lawn grass will continue to increase. This trend will be supported by improved understanding of suitable management practices, and especially by the development of cultivars selected specifically for turf purposes.

Judging from the recent past, there appears to be no basis for predicting anything other than a bright future for tall fescue. The situation reminds me of a statement attributed to the late H. A. Schoth, who made the original selection of Alta tall fescue in 1923. When he was asked in 1946 what needed to be done to improve tall fescue, he is reported to have stated that Alta was a damn fine grass that didn't need any (expletive deleted) improvement. It is immaterial if this assertion is correct or not, because the fact remains that the two ecotypes that were released as Alta and Kentucky 31, before 1946, occupy practically the entire hectarage seeded to tall fescue. New cultivars selected over the past 30 years have made little or no inroads in the use of these two cultivars.

Can we conclude that observed trends will continue? The answer to this question must be a guarded "yes," based on the changing pattern of pest damage that could reduce persistence, and on the mounting need to improve the quality of herbage available for ruminant livestock.

Damage to tall fescue stands and herbage from disease and insect attacks appears to be increasing. Hard evidence is available in the frequency of reports of rust damage in a species long considered to be immune to the disease. This situation is not uncommon. There are many examples of newly introduced species that were considered to have few, if any, pest problems associated with their culture. In time, pest damage increased in frequency and severity as these species were accepted for use and occupied increasing hectarages. Thus, in the not-too-distant future we will need new cultivars with appreciably better levels of pest resistance if we are to capitalize on the valuable attributes of tall fescue.

Very little research has been conducted in the United States to improve the palatability and quality of tall fescue. The one exception is provided by the long-term studies conducted at the Kentucky Agricultural Experiment Station in cooperation with the USDA. This cooperative program has contributed to notable advances in selecting for improved palatability and in investigating the prospects of transferring palatability and quality from ryegrasses into a background of tall fescue germplasm. The failure to make concomitant advances in our understanding of the factor(s) that contribute to intake and nutritional disorders, such as fescue-foot, has hampered the progress of all breeding efforts. As reported in previous chapters, significant progress has been made in the past few years in identifying some of the factors that may influence nutritional disorders and the intake of herbage. Many questions must be answered, however, before plant breeders can move forward with complete confidence in breeding for improved herbage quality.

Research workers in the United States and Europe have been guilty of underestimating the complexity of the problems involved in the improvement of tall fescue, and the author of this chapter is no exception. It will take a concerted effort, even with the development of reliable screening techniques for the traits, correlated with quality, to incorporate these features into cultivars that retain the desirable agronomic characteristics of tall fescue. This statement especially applies to intergeneric hybrids, in which the end-products of selection may range from those that are indistinguishable from Kentucky 31 to those that should be managed as distinct species.

Information assembled in this text provides evidence of the progress that can be achieved in developing improved cultivars and management practices. These concepts and leads must be pursued if we are to meet the forage needs that have been projected for our livestock industry, and if we are to realize the full promise that tall fescue holds for turf and conservation purposes. The goals are attainable, but only if research is supported at a level commensurate with current and projected contributions of tall fescue to U.S. agriculture and to our environment. The status quo is not good enough!

SUBJECT INDEX

A

Acetylloline, 45
Acid soils, 158, 295, 321, 323
Adaptability, 5-6, 41
Adaptation, 9, 10, 12, 14, 16, 26
   Africa, 27, 115
   Asia, 23
   Australia, 25
   Canada, 25
   Europe, 23, 115
   Japan, 25
   Mexico, 26
   New Zealand, 25
   north central U.S., 17
   northeast U.S., 17
   South America, 27
   southern U.S., 18
   United Kingdom, 24
   western U.S., 21
Adaptation-turf, 295
Aesthetic value, 336
Africa, 27
Alkaloids, 45-46, 117-119, 235, 256-258, 273-278
Alta, 3, 111, 125, 342
Aluminum, 69
Aluminum toxicity, 16, 68, 158, 321
Ammonium toxicity, 278-279
Anatomy, 33-35
Aneuploids, 97, 104-105
Animal abortion, 254
Animal grades, 239, 243
Animal health, 61
Animal performance, 115-117, 120
Animal response, 165
Anthesis, 77
Apical dominance, 80
Asheville, 128
Asia, 23
Australia, 25
Auxin, 78

B

B-chromosomes, 98-99, 102
Backafall, 129
Bacterial diseases, 307
Balansia epichloe, 311
Barley yellow dwarf virus, 307
Beef cattle backgrounding, 129-130, 233, 235, 239-240

Beef cattle finishing, 233, 242
Beef cows, 201
   feeding large bales, 212
   summer pasture, 202, 206
   winter pasture, 207-208, 210
   year-round, 214
Beef cows and calves, 201-216
   creep feeding, 203, 206, 214-216
   summer pasture, 202, 206
Bermudagrass, 159, 190-191, 213, 215, 240
Bioassays, 120
Biological fixed N, 53
Bivalents, 93, 96
Blind seed, 311
Bovinae, 1, 9
Breeding, 1, 111, 296, 342
Breeding procedures, 111-112
Bridge crosses, 122
Broadcast plantings, 156
Bromus arundinaceus, 32
Brown patch, 309
Brudzynska, 130
Burning, 148

C

Cadmium, 68
Calcium, 65-66, 119
Calcium requirement, 64-65
Calving, 208, 211
Canada, 25
Canopy, 80, 86-87
Carbohydrates, 179
Carbon-nitrogen ratio, 52
Carrying capacity, 206, 208, 216
Cellulolytic enzymes, 118
Cellulose digestion, 235
CER, 84, 87
Cercospora festucae, 310
Cercospora leaf spot, 310
Chromosome number, 93
Chromosome pairing, 93-98
Claviceps purpurea, 311
Climatic adaptation, 10, 12
Clipping management, 302
Clonal evaluation, 112
Colchicine treatment, 131
Colletotrichum gramincola, 312
Conception rate, 207-208
Conservation, 5, 319, 343

Construction sites, 326
Contours, 320
Copper, 67
Cricket feeding trials, 118
Crown rust, 123, 307–308
Cultivars, 172, 174, 180, 190–192, 235, 238, 296
  for turf, 296
  with other grasses, 171–174, 181–183, 191, 195
Cutting (see defoliation)
Cytogenics, 36–38
Cytology, 36–38
Cytotaxonomy, 5

D

Dairy cattle, 222
  digestibility, 226
  dry cows and heifers, 227
  intake, 226
  milk persistancy, 222–223, 225–227
  summer pasture, 222
  TDN, 226
  winter pasture, 224
Deep-mine tailings, 158, 324
Deferred growth, 237
Defoliation, 77, 81, 83, 89
  harvest interval, 171–175, 180
  number of harvests, 171–174
  rest period, 173, 181
  stage of growth, 171–172, 174–175, 177, 180, 191
  stubble cutting height, 173, 175, 191
  time of harvest, 171, 180–181
Digestibility, 117, 210, 226
Digestible dry matter, 238
Digestible energy, 236
Disease control, 314–315
Disease resistance, 122, 123, 307
Diseases in turf, 304, 307, 343
Distribution of growth, 181, 189, 190, 196
Disturbed land areas, 321, 324
Dry lot finishing, 245

E

Economic return, 152
Ecotypes, 93, 96, 98–99
Ecotypic variation, 115
Edaphic adaptation, 10, 14, 16, 26
Elevation, 22, 27
English bluegrass, 2
Enzyme, 84
Epichloe typhina, 312

Ergot, 311
Erysiphe graminis, 312
Establishment, 143
Europe, 123

F

F_1 hybrids, 13
Fairy rings, 310
Fall and winter growth, 115, 122
Fall calving, 211
Family selection, 118
Fat necrosis, 247, 282–285
Fawn, 125–126
Feeding on pasture, 241, 243–244
Fergus, E. N., 3
Fertilization, 144, 146, 241–243, 299, 302–303
  broiler litter, 284
  factors influence, 302–303
  programs, 302–303
Fescue foot, 116, 247–248, 255, 342
  nitrogen causes, 255
Fescue toxicity, 207, 247, 342
Festal, 130
Festuca
  arundinacea subsp. *arundinacea,* 24, 26
  arundinacea var. *atlantigena,* 97, 99, 102
    forma *pseudomairei,* 37
  arundinacea var. *cirtensis,* 37, 97, 99, 103
  arundinacea subsp. *fenas,* 36
  arundinacea var. *glaucescens,* 36–37, 95, 97, 99, 102
  arundinacea subsp. *interrupta,* 24
  arundinacea var. *letourneuxiana,* 37, 97, 99
  arundinacea subsp. *orientalis,* 24
  arundinacea subsp. *uechtritziana,* 36
  elatior, 1, 2, 31, 35, 36
  elatior subsp. *arundinacea,* 32, 35
  elatior var. *arundinacea,* 32, 36
  elatior subsp. *pratensis,* 2, 36
  fenas, 36
  genus, 31
  gigantea, 31, 132
  gigantea × *F. arundinacea,* 101, 132
  hybrids, 36–38
  mairei, 37
  ovina, 31
  pratensis, 1, 23, 32, 35, 36, 37, 93
  pratensis × *F. arundinacea,* 100
  pratensis subsp. *apennina,* 36

SUBJECT INDEX

pratensis var. *apennina*, 35–37
pratensis subsp. *pratensis*, 36
rubra, 31, 93, 95
scariosa, 37
section *bovinae*, 31, 37
  festuca, 31, 37
  ovinae, 31, 37
  scariosae, 37
Festuceae, 1
Flexibility of leaves, 119
Flooding, 15, 26, 41
Forage quality, 115–121, 124, 132, 214, 227
  acid detergent fiber, 192
  animal performance, 194
  crude fiber, 177
  crude protein, 171, 180, 195
    seasonal, 177, 194
    stockpiled forage, 179, 183, 184, 192
  digestibility, 171, 179, 183, 184, 192, 195
  potassium, 184
Forage yield (see yield)
Formylloline, 45
Fortune, 124
Frequency, 301–302
Fructosan, 44
Fungal diseases, 307–312
Fusarium
  acuminata, 312
  nivale, 310
  tricinctum, 311

G

Garton's own leafy, 130
Gene banks, 115
General combining ability, 112
Genetic diversification, 115
Genetics, 36–38
Genomic constitution, 95–96, 111
Geographic adaptation, 10, 12
Geographic distribution, 32–33
Germination, 75–76, 156
Giant fescue, 31, 101, 132
Gloeosporium bolley, 312
Gloeotinia temulenta, 310
Goar, 127
Grass tetany, 62, 264
Grassland renovation, 160
Grasslands 4710, 128
Grazing days, 235
Grazing trials, 116–117, 120
Green manure, 328
Growth analysis, 89

H

Harvest interval, 171–175, 180
Harvesting practices, 171–191
Hay yield, 163, 165
Height, 299, 301–302
Heinricks, Max, 4
Helicotylenchus dinystera, 313
Helminthsporium
  carbonum, 309
  cynodontis, 309
  dictyoides, 308–809
  homorphus, 309
  leersiae, 309
  maydis, 309
  oryzae, 309
  rostratum 309
  sacchari, 309
  sativum, 309
  siccans, 309
  sorghicola, 309
  sorokinianum, 309
  turcicum, 309
  vagans, 309
Herbicides, 145, 147, 163, 334
Heritability, 105–106, 113, 118, 123
Heterodera avenae, 313
Heterosis, 313
Hexaploid, 5, 115, 122
Highway roadbank stabilization, 325
Hitchcock, 1
Hokuryo, 129
Homoeologous pairing, 96–98, 102, 104–105
Hopolaimus galeatus, 313
Hypomagenesemia, 264, 271

I

Identification, 35–36
In vitro digestibility, 118–119
Inbreeding, 114
Inbreeding depression, 114
Infrared reflectance, 119
Inhibitory substances, 120
Insects in turf, 305, 343
Intergeneric and interspecific hybridization, 130–134
Intergeneric hybrids, 103–105
Interseeding into bermudagrass, 159, 190–191
Interspecific hybrids, 100–103
Intraspecific hybrids, 90–100
Introduction of tall fescue, 32
Iron, 67
Irradiance, 85
Irrigation, 20–21, 78, 217–218, 241–243, 298, 304, 334

SUBJECT INDEX

J
Japan, 25
Johnstone, William C., 3-4

K
Karyotype, 94-95
Kenhy, 105, 131-132
Kenmont, 126-127
Kentucky, 3-4, 31, 111, 125, 158, 342
Kenwell, 126

L
Laboratory methods, 118-119
LAI, 88-90
Leaves, 78, 81, 87
Legumes, 165, 175-176, 188-190, 239, 241
Lime requirement, 64-65
Loline, 45, 256, 273, 275-276
Lolium, 31, 35-37
　multiflorum, 35, 37, 96
　multiflorum × *F. arundinacea,* 96, 104, 106
　perenne, 24, 31, 37
　perenne × *F. arundinacea,* 98, 103, 131
　(perenne × *F. pratensis)* × *F. arundinacea,* 105
Ludine, 130
Ludion, 130

M
Magnesium, 266-268
　animal metabolism, 268
Magnesium requirements, 65-66, 119
Maize dwarf mosaic virus, 307
Manade, 130
Management, 86, 90
Manganese, 67
Marginal land, 332
Maris Jebel, 122
Maris Kasba, 122
Maturity effects, 121
Meadow fescue, 1, 2, 31, 35, 36
Meiosis, 93-95, 98-105
Meiotic behavior, 5
Meiotic irregularities, 114-115, 122, 131
Melik, 129
Meloidogyne
　arenaria, 314
　graminis, 314
　hapla, 314
　incognita, 314
　javanica, 314
　naasi, 314
Methemoglobin, 280
Mexico, 26
Micronutrient requirements, 67-68
Mineral deficiencies, 119
Missouri, 96, 127-128
Monoculture vs. mixtures in turf, 298-299
Monosomics, 96-97
Morphology, 33-35
Motall, 130
Mowability, 302
Mowing, 299, 301-302
Mucilago spongiosa, 310
Mulching, 298
Multivalents, 93
Mycotoxins, 258-260, 312

N
Nematode control, 314-315
Nematodes, 123, 313-314
Netblotch, 16, 308
New plantings, 158
New Zealand, 25
Nitrate, 43, 45, 47, 280-281
　accumulation, 42-44
　toxicity, 279-282
Nitrogen, 77, 118, 267
　accumulation in fescue sod, 53-54
　accumulation in plants, 42
　cycling, 52
　dry matter yield response to rate, 48-51
　effect of N on alkaloid, 45-46
　　feed value, 47-48
　　other mineral nutrients, 46-47, 65
　　root growth, 51
　　soluble carbohydrates, 44-45
　function in plants, 42
　recovery, 50-51
Nitrogen fertilization, 146, 159, 188, 191, 233
　vs. legume nitrogen, 163, 164, 185, 188, 189, 190
　yield response, 174, 176, 181, 182, 195
Nitrogen/sulfur ratios, 54-55
Nomenclature, 32
Non-protein nitrogen, 42, 272, 278, 285
No-till crop production, 333
Number of harvests, 171, 174

SUBJECT INDEX

Nutrient cycling
  calcium, 67
  magnesium, 67
  nitrogen, 52-54
  phosphorus, 60-61
  potassium, 64
  sulfur, 56
Nutritive value index, 119, 134

O

Ophiobolus graminis, 310
Organic acid accumulation, 42
Ottawa Syn A, 130
Overseeding, 304

P

Palatability, 116-117, 119, 126-127, 344
Paratylenchus projectus, 313
Park and recreational areas, 327
Pasture systems, 235, 236, 237
Peppard, J. C., Seed Company, 4
Perlolidine, 275-276
Perloline, 45, 117-118, 235, 256, 273, 276, 285
Persistence of stock, 172-175, 191
Pests, 304-305, 343
pH, 64
Phleospora idahoensis, 312
Phosphorus, 119
  animal requirement, 57
  cycling, 60
  dry matter yield response, 57-59
  effect on N recovery, 60-61
  function in plants, 57
Photoperiod, 77
Photorespiration, 84-85
Photosynthesis, 61, 77, 84-85
Physarum cinareum, 310
Plant introduction, 115
Ploidy, 5, 93, 97, 101, 115, 122
Polyhaploids, 97
Popularity, 6
Post-harvest management, 147
Potassium, 61-64, 119, 266-268
  animal health, 61-62
  antagonism to magnesium uptake, 65
  cycling, 64
  dry matter yield response, 57-59
  function in plant, 57
Potassium/calcium and magnesium, 119
Pratylenchus
  neglectus, 313

  penetrans, 313
  scribreri, 313
  zeae, 313
Progeny testing, 112
Puccinia
  coronata, 122, 308
  graminis, 308
  striformis, 308
Pulawska, 130
Pyrrolizidine alkaloids, 45, 256, 273, 275-277
Pythium
  aphanidermatum, 310
  debaryanum, 310
  graminicola, 310
  irregulare, 310
  ultimum, 310

Q

Quality
  assay methods, 118-120
  factors, 115, 116, 132, 177, 192
  palatability, 116, 117, 119, 126, 127, 344
  TNC, 81-84, 87-91
  turf, 124

R

Rainfall, 13, 16, 21, 25, 27
Reed fescue, 2
Regions of culture, 23
Regulation theory, 96
Renovating tall fescue, 162-163
Renovation of old sods, 160
Reproductive isolation, 100
Respiration, 83, 85
Rest period, 173, 181
Rhizoctonia leaf scald, 309
Rhizoctonia solani, 309-310
Rhizome, 78
Rhynchosporium secalis, 312
Root structure, 15
Roots, 51, 76, 77, 319, 329
Rotational systems, 328
Row plantings, 143-144, 156
Rozelle, 130
Runoff, 328

S

S-170, 5, 128
Schoth, H. A., 4
Schreber, 1
Sclerotinia borealis, 310

Sclerotium rolfsii, 312
Seasonal growth, 77–78
Seasonal yield, 189, 236
Seed
  certification, 148
  dormancy, 75, 78
  harvest, 148
  management, 142–143
  marketing, 152
  moisture, 150
  production, 6
  production areas, 141–142
  size and seeding rate, 156
  storage, 75, 151
  viability, 75, 151
  yield, 123–124, 141
Seedbed, 296
Seeding, 143, 297, 298
  depth, 297
  rate, 297
  time, 297
Seedling and stand development, 157
Selium, 68
Sewage sludge, 67
Sheep, 216
  summer pasture, 201, 217
  winter pasture, 201, 219
Silica accumulation, 68–69
Silver top, 311
Sink, 87
Skim plowing, 145
Slaughter grades, 233, 240, 242–244
Slime molds, 310
Sludge, 67
Snow molds, 310
Sod bound, 144
Sod production, 159, 298
Soils, 158, 321, 323, 329
  improvement, 328
  moisture, 14, 20
  oxygen, 41
  stabilization, 319
Solubility methods, 118–119
Source, 87
South America, 27
Special contributions, 332
Species relationship, 114
Sphecelia typhia, 312
Spoils, 158, 321, 324
Spring calving, 208
Stage of growth, 171–172, 174–175, 177, 180, 191
Stagnospora maculata, 312
Stand establishment, 156

Stem eyespot, 312
Stockpiled tall fescue, 177
  dry matter loss, 177, 182–183
  period of growth accumulation, 177–179, 181, 183
  precipitation effects, 183
  seed crop residue, 179–180
  stockpiling definition, 177
  summer defoliation, 180
  temperature effects, 183–184
Stockpiling or deferring, 237
Stomata, 86
Stubble cutting height, 159, 173, 175, 191
Suiter farm, 155
Summer syndrome, 247, 277
Sulfur
  cycling, 56
  dry matter response, 55–56
  function in plant, 54–55
Supplemental feeding of animals, 243
Surface mining spoils, 321

T

Tall English bluegrass, 2
Tall fescue legume mixtures, 201, 203, 207, 212–213, 216–217, 239, 244, 323, 328
Tall fescue toxicity, 311
Tall fescue vs. other grasses, 171–174, 181–183, 191, 195
Taxonomy, 31–38
Temperature, 13, 78, 86, 88
Temperature effects, 155
Tetraploids, 101
Tillering, 78–81, 147, 172, 174, 177
Time of harvest, 171, 180, 181
TNC, 81–84, 87–90
Toxicity, 116, 311
Toxicosis, 247, 285
Transition zone, 5
Trichodorus christlieli, 314
Trisomics, 97
Trispecific hybrids, 105
Turf, 124–125, 294, 343
  establishment, 296
Turfgrass, 293–306
  attributes for turf, 294
    growth habit, 293–294
    pH tolerance, 295
    seedling vigor, 295–296, 299
    wear tolerance, 295, 301
  history of use, 294
  utilization, 294

SUBJECT INDEX

Tylenchorhynchus claytoni, 314
Typhula incarnata, 310

U
United Kingdom, 24
Univalents, 93
Urocystis agropyre, 312

V
Vegetative propagation, 113
Virus diseases, 307
Vulpia, 31, 37

W
Waterways, 320
Weed control, 145, 147
Weeds in turf, 300, 305

Winter injury, 18, 22, 174
Worldwide distribution, 7, 23–27

Y
Yamanami, 129
Yield
  distribution, 77–78, 189, 236
  limitations, 87–91
  plant types, 121, 122, 172–174
  renovated swards, 163, 165, 190, 191
  response to defoliation, 173, 174, 176
    fertilization, 48, 51, 55, 57, 62, 188
  stockpiled, 177, 178, 180–184

Z
Zapadnaya (western), 130
Zinc, 67